Null Curves and Hypersurfaces of Semi-Riemannian Manifolds

Null Curves and Hypersurfaces of Semi-Riemannian Manifolds

Krishan L Duggal
University of Windsor, Canada

Dae Ho Jin
Dongguk University, Korea

NEW JERSEY · LONDON · SINGAPORE · BEIJING · SHANGHAI · HONG KONG · TAIPEI · CHENNAI

Published by

World Scientific Publishing Co. Pte. Ltd.

5 Toh Tuck Link, Singapore 596224

USA office: 27 Warren Street, Suite 401-402, Hackensack, NJ 07601

UK office: 57 Shelton Street, Covent Garden, London WC2H 9HE

British Library Cataloguing-in-Publication Data
A catalogue record for this book is available from the British Library.

NULL CURVES AND HYPERSURFACES OF SEMI-RIEMANNIAN MANIFOLDS

ISBN-13 978-981-270-647-8
ISBN-10 981-270-647-X

Printed in Singapore.

Preface

Since the second half of the twentieth century, the Riemannian and semi-Riemannian geometries have been active areas of research in differential geometry and its applications to a variety of subjects in mathematics and physics. Recent survey in Marcel Berger's book [15] includes the major developments of Riemannian geometry since 1950, citing the works of differential geometers of that time.

During mid 70's, the interest shifted towards Lorentzian geometry, the mathematical theory used in general relativity. Since then there has been an amazing leap in the depth of the connection between modern differential geometry and mathematical relativity, both from the local and the global point of view. Most of the work on the global Lorentzian geometry has been described in a standard book by Beem and Ehrlich [10] and in their second addition in 1996, with Easley. They concentrated on geodesic and metric completeness, the Lorentzian distance function, and the Morse index theory for Lorentzian manifolds. Ehrlich and his collaborators [33] are still actively working on the volume comparison theorems for Lorentzian manifolds, using warped product technique which was introduced by Bishop and O'Neill [17] in 1969.

In 1996, Duggal-Bejancu published a book [28] on the lightlike (degenerate) geometry of submanifolds needed to fill an important missing part in the general theory of submanifolds. That book included a series of papers on a specific technique of introducing a non-degenerate screen distribution to define the induced geometric objects such as linear connection, second fundamental form, needed to obtain the Gauss-Codazzi type equations for a lightlike submanifold. Since then several researchers have done further work on lightlike geometry by direct use of Duggal-Bejancu's technique and also, in general, there has been an increase in papers on the geometry and physics of null curves and hypersurfaces using several ways corresponding to a given problem.

The objective of this book is to present a comprehensive up to date information on the differential geometry of null curves and lightlike hypersurfaces.

The motivation comes from considerable new information on the geometry of these two interrelated topics and their use in mathematical physics. Indeed, see Ferrández *et al.* [35, 36, 37, 38, 39, 40, 41] on null curves, soliton solutions and relativistic particles involving the curvature of $3D$ null curves; Gutiérrez *et al.* [49, 50, 51, 52] on null conjugate points along null geodesics and Perlick [90, 91, 92, 93, 94] on a variational principle for light rays and many referred therein.

The works of all above cited researchers, along with the works of present authors (see Duggal [23, 24, 25, 26, 27] on globally null manifolds and null geodesics; Duggal-Jin [30] and Jin [67, 68, 69, 70, 71] on geometry of null curves), is the main source of inspiration in writing this book.

Moreover, these topics are suitable for those graduate students who know the theory of non-null curves and surfaces and are interested in their null counterparts. To the best of our knowledge, there does not exist any other text book covering the material included in this book which is just within the understanding (neither too high nor too elementary) of a graduate student. A senior level undergraduate course in differential geometry is the sole prerequisite. A fresh and improved version of the material appeared in [28, Chapters 3 and 4] is included to make this volume a self contained book. Our approach, in this book, has the following special features:

- The preliminaries are introduced as needed at the appropriate places. We expect that this approach will help the readers to understand each chapter independently without knowing all the prerequisites in the beginning.

- This is the first-ever graduate text book on null curves and hypersurfaces.

- The book contains a large variety of solved examples and exercises which range from elementary to higher levels.

- The sequence of chapters is arranged so that the understanding of a chapter stimulates interest in reading the next one and so on.

There are eight chapters whose subjects are clear from the contents. Equations are numbered within each chapter and its section. To illustrate this, we have introduced a triplet (a, b, c) for each equation such that a, b and c stand for the chapter, the section and the number of equation in that section accordingly. Likewise, say in chapter a, theorem $b.c$ means theorem number c in section b. Each chapter is accompanied by a set of notes of background material (may not be familiar to some readers) followed by a set of typical exercises.

Overall this book is aimed at graduate students, research scholars and faculty interested in differential geometry. As a text book, it is suitable in sequence for the following two consecutive graduate semester courses meeting 3 hours a week:

First semester. Chapters 1, 2, 3 and 4. Prerequisite. Any senior undergraduate course in Differential Geometry.
Second semester. Chapters 5, 6, 7 and 8 and reading of related research papers.

Both the authors are grateful to all authors of books and articles whose works have been used in this volume. The first author (Krishan L. Duggal) is thankful to the Natural Sciences and Engineering Research Council of Canada for financial support. The second author (Dae Ho Jin) was supported by Dongguk University Research Fund. He also wishes to thank the University of Windsor for appointing him visiting professor (Jul. 98 – Aug. 99) and its Department of Mathematics and Statistics for providing him hospitality and kind support.

Finally, it is a great pleasure to thank the World Scientific Publishers for their effective cooperation and care in preparing this book. Constructive suggestions (towards the improvement of the book) by unknown reviewers is appreciated with thanks. Any comments and suggestions by the readers will be gratefully received.

Krishan L. Duggal
Dae Ho Jin

Contents

Chapter 1

The concept of null curves

In this chapter we review on semi-Riemannian manifolds, covariant and exterior derivatives, followed by the concept of a null curve C in a semi-Riemannian manifold (M, g), where g is its non-degenerate metric tensor of a constant index. We show that there exist three types of quasi-orthonormal basis along C.

1.1 Smooth manifolds

Let V be a real n-dimensional vector space with a symmetric bilinear mapping $g : V \times V \to \mathbf{R}$. We say that g is positive (negative) definite on V if $g(v, v) > 0$ $(g(v, v) < 0)$ for any non-zero $v \in V$. On the other hand, if $g(v, v) \geq 0$ $(g(v, v) \leq 0)$ for any $v \in V$ and there exists a non-zero $u \in V$ with $g(u, u) = 0$, we say that g is positive (negative) semi-definite on V.

Let $B = \{u_1, \ldots, u_n\}$ be an arbitrary basis of V. Then, g can be expressed by an $n \times n$ symmetric matrix $G = (g_{ij})$, where

$$g_{ij} = g(u_i, u_j), \qquad (1 \leq i, j \leq n).$$

G is called the *associated matrix* of g with respect to the basis B. We assume that rank $G = n \iff g$ is non-degenerate on V. The non-degenerate g on V is called a *semi-Euclidean metric (scalar product)*. Then (V, g) is called a *semi-Euclidean vector space*. For a semi-Euclidean $V \neq \{0\}$, there exists an orthonormal basis $E = \{e_1, \ldots, e_n\}$ such that

$$g(v, v) = -\sum_{i=1}^{q}(v^i)^2 + \sum_{a=q+1}^{q+p} (v^a)^2, \qquad (1.1.1)$$

where $q + p = n$ and (v^i) are the coordinate components of v with respect to E. Thus, with respect to (1.1.1), G is a diagonal matrix of canonical form:

$$diag(-\ldots-+\ldots+). \qquad (1.1.2)$$

1

The sum of these diagonal elements (also called the *trace* of the canonical form) is called the *signature* of g and the number of negative signs in (1.1.2) is called the *index* of V. Throughout this book, we set the form of the signature of g as given by (1.1.2), unless otherwise stated. Also, we denote by $\mathbf{R}_q^n$ a semi-Euclidean space of constant index $q > 0$ and by $\mathbf{R}^n$ a Euclidean space.

Smooth manifolds. Given a set M, a *topology* on M is a family $\mathcal{T}$ of subsets of M such that

1. the empty set $\emptyset$ and M are in $\mathcal{T}$,

2. the intersection of any two members of $\mathcal{T}$ is in $\mathcal{T}$,

3. the union of an arbitrary collection of members of $\mathcal{T}$ is in $\mathcal{T}$.

In the above case, $(M, \mathcal{T})$ is called a *topological space* whose elements are the open sets of $\mathcal{T}$. As M depends on the choice of $\mathcal{T}$, M can have many topologies. In the sequel, we assume that M is a topological space with a given $\mathcal{T}$. M is a *Hausdorff topological space* if for every $p, q \in M$, $p \neq q$, there exist non intersecting neighborhoods $\mathcal{U}_1$ and $\mathcal{U}_2$ respectively. A neighborhood of p in M is an open set that contains p. A system of open sets of $\mathcal{T}$ is called a *basis* if its every open set is a union of the set of the system.

Definition 1.1. *An n-dimensional manifold M is a topological Hausdorff space whose each point has a neighborhood homeomorphic to an open set in $\mathbf{R}^n$.*

A manifold is simply understood to be a set M with the property that each point of M can serve as the origin of local coordinates valid in an open neighborhood which is homeomorphic to an open set in $\mathbf{R}^n$.

The Hausdorff condition is not necessary, although is assumed most often. The open neighborhood of each point admits a *coordinate system* which determines the position of points and the topology of that neighborhood. For a smooth transformation of two such coordinate systems and also taking care of the intersecting neighborhoods, we need the concept of *differentiable manifolds* as follows:

A homeomorphism $\varphi : M \to \mathbf{R}^n$, mapping an open set $\mathcal{U}$ of M onto an open set $\varphi(\mathcal{U})$ of $\mathbf{R}^n$, is called a *chart*. By assigning to each point x in $\mathcal{U}$ the n local coordinates $x^1, \ldots, x^n$, we call $\mathcal{U}$ a *local coordinate neighborhood*. Let x be the point of the intersection $\mathcal{U}_1 \cap \mathcal{U}_2$ of two local coordinate neighborhoods $\mathcal{U}_1$ and $\mathcal{U}_2$ with respect to charts φ_1 and φ_2. We say that φ_1 and φ_2 are C^k- compatible if $\mathcal{U}_1 \cap \mathcal{U}_2$ is non-empty and $\varphi_2 \circ \varphi_1^{-1} : \varphi_1(\mathcal{U}_1 \cap \mathcal{U}_2) \to \varphi_2(\mathcal{U}_1 \cap \mathcal{U}_2)$ and its inverse are C^k.

Definition 1.2. *An n-dimensional differentiable manifold M is a set of points together with a family $\mathcal{A} = \{\mathcal{U}_\alpha, \varphi_\alpha\}$ of local coordinate neighborhoods such that*

1. *The union of $\mathcal{U}_\alpha$'s is M,*

2. *Any two charts of M are C^k-compatible,*

3. *Any two members of $\mathcal{A}$ are C^k-compatible.*

The family $\mathcal{A}$ is called a maximal atlas on M. M is called a *smooth manifold* if M is C^∞. An atlas $\mathcal{A} = \{\mathcal{U}_\alpha, \phi_\alpha\}$ of M is said to be locally finite if for each p in M, there is a local coordinate neighborhood $\mathcal{U}$ which intersects with only finitely many out of $\mathcal{U}_\alpha$'s. Another atlas $\mathcal{B} = \{\mathcal{V}_\beta, \psi_\beta\}$ of M is called a refinement of the atlas $\mathcal{A}$, if each $\mathcal{V}_\beta$ is contained in some $\mathcal{U}_\alpha$. M is *paracompact* if for every atlas $\mathcal{A}$ there is a locally finite refined atlas $\mathcal{B}$ of $\mathcal{A}$. In this book all manifolds will be assumed smooth and paracompact.

A trivial example of a manifold is $\mathbf{R}^n$. Other examples are 2-sphere $\mathbf{S}^2$, cylinder, tori, and Minkowski spacetime. To illustrate this, consider $\mathbf{S}^2$ in $\mathbf{R}^3$, with coordinates (y^i), centered at $(0, 0, 0)$ having radius a. We need only two charts, with respect to the rectangular coordinates (x, y), as follows:

$$y^1 = \frac{2a^2 x}{x^2 + y^2 + a^2}, \qquad y^2 = \frac{2a^2 y}{x^2 + y^2 + a^2},$$

$$y^3 = \epsilon a \frac{x^2 + y^2 - a^2}{x^2 + y^2 + a^2}, \qquad (\epsilon = \pm 1).$$

Note that the usual spherical coordinates (ϕ, θ) fail to give a one-one mapping at the poles, where θ is undetermined. It is left as an exercise to show that $\mathbf{S}^2$ is a smooth manifold in $\mathbf{R}^3$. Similarly, one can show that $\mathbf{S}^n$ in $\mathbf{R}^{n+1}$ is a smooth manifold. See at the end of this chapter some exercises on manifolds.

Let $F(x)$ denote the set of all real-valued smooth functions defined on some neighborhood of a point x of M. A *tangent vector* of M at x is a linear mapping $X_x : F(x) \to \mathbf{R}$. The set of all tangent vectors, at x, forms a vector space, called the *tangent space* denoted by $T_x M$, where $\dim(T_x M) = \dim(M) = n$, such that n vectors $\{\partial_{x^1}, \ldots, \partial_{x^n}\}$ form the basis (called a coordinate basis) of $T_x(M)$ with respect to a local coordinate system (x^i), $i = 1, \ldots, n$, on a neighborhood $\mathcal{U}$ of x. Let $f \in F(x)$. Then, the derivative $X_x f$ is given by

$$X_x f = \sum_{i=1}^{n} \frac{\partial f}{\partial x^i}(x) X_x(x^i) = X_x(x^i)(\frac{\partial}{\partial x^i})_x f.$$

Therefore, it follows that $X_x = X^i(x)(\partial_{x^i})_x$, where $X^i(x) = X_x(x^i)$ are the components of X_x. The *tangent bundle* of M is defined by

$$TM = \cup_{x \in M} T_x M, \tag{1.1.3}$$

which is a real $2n$-dimensional smooth manifold. A *vector bundle* is a smooth manifold M together with a vector space attached at each point of M. A *vector field* on M is called a *smooth section* of its respective vector bundle. For example a 2-sphere $\mathbf{S}^2$ in $\mathbf{R}^3$ together with the collection of its normals is a vector bundle. Another example is the tangent bundle, defined by (1.1.3) since it has a collection of all the tangent vector spaces $T_x M$ at each x of M. The local coordinates on any vector bundle are given by a pair $(x^i; y^a)$, where (x^i) and (y^a) are the local coordinates of M and its attached vector space respectively. We denote by $F(M)$ the algebra of smooth functions on M and by $\Gamma(E)$ the $F(M)$ module of smooth sections of a vector bundle E (same notation for any other vector bundle) over M. The *Lie bracket* $[X, Y]$ of any two vector fields X and Y is given by

$$[X, Y]f = X(Yf) - Y(Xf), \qquad \forall f \in F(M). \qquad (1.1.4)$$

Using the local expressions $X = X^i \partial_i$ and $Y = Y^j \partial_j$, it is easy to show that

$$[X, Y] = (X^i \partial_i Y^j - Y^i \partial_i X^j) \partial_j$$

is also a vector field on M, with the following properties:

1. $[aX + bY, Z] = a[X, Z] + b[Y, Z];\ a, b \in R,$ (bilinear)

2. $[X, Y] = -[Y, X],$ (skew-symmetry)

3. $[X, [Y, Z]] + [Y, [Z, X]] + [Z, [X, Y]] = 0,$ (Jacobi identity)

4. $[fX, hY] = fh[X, Y] + f(Xh)Y - h(Yf)X,$ $\forall f, h \in F(M).$

It follows that the set of all tangent vector fields on M, denoted by $\Gamma(M)$, has a *Lie-algebra* structure with respect to the Lie-bracket operation.

Every $T_p(M)$ has a *dual vector space* or *cotangent space* $T_p^*(M)$ of the same dimension and its elements, called *differential 1-forms*, are linear maps

$$\omega : T_x(M) \to \mathbf{R}.$$

Let $f \in F(M)$ and $X \in \Gamma(M)$. We know from above that $X(f)$ is a derivative on $F(M)$ and conversely, every derivative on $F(M)$ comes from some vector field on M. In other words, $X(f) = df(X)$ where d is the symbol for ordinary derivative. Relating this with ω, we conclude that

$$\omega = df : T_x(M) \to \mathbf{R} , \qquad \omega = \omega_i dx^i,$$

where the differentials dx^i form a *dual basis* of $T_x^*(M)$, satisfying $(dx^i)(\frac{\partial}{\partial x^j}) = \delta_j^i$ with respect to a coordinate system (x^i).

A *linear connection* on M is a map $\nabla : \Gamma(M) \times \Gamma(M) \to \Gamma(M)$ such that

$$\nabla_{fX+hY} Z = f(\nabla_X Z) + h(\nabla_Y Z), \qquad \nabla_X f = Xf,$$
$$\nabla_X(fY + hZ) = f\nabla_X Y + h\nabla_X Z + (Xf)Y + (Xh)Z,$$

for arbitrary vector fields X, Y, Z and smooth functions f, h on M. ∇_X is called *covariant derivative operator* and $\nabla_X Y$ is called *covariant derivative* of Y with respect to X. Define a tensor field ∇Y, of type $(1,1)$, and given by $(\nabla Y)(X) = \nabla_X Y$, for any X. Also, $\nabla_X f = Xf$ is the covariant derivative of f long X. The covariant derivative of a 1-form ω is given by

$$(\nabla_X \omega)(Y) = X(\omega(Y)) - \omega(\nabla_X Y). \qquad (1.1.5)$$

For local expressions, we consider the natural basis $\{\partial_i\}$, $i \in \{1, \ldots, n\}$, on a coordinate neighborhood $\mathcal{U}$ and set $\nabla_{\partial_j} \partial_i = \Gamma_{ji}^k \partial_k$, where Γ_{ji}^k are n^3 local components of ∇ on M. For $X = X^i \partial_i$, $Y = Y^j \partial_j$ and $\omega = \omega_i dx^i$ we have

$$\nabla_X f = X^i \partial_i f, \qquad \nabla_X Y = Y_{;k}^i X^k \partial_i, \qquad (1.1.6)$$

$$Y_{;k}^i = \partial_k Y^i + \Gamma_{kj}^i Y^j, \qquad \omega_{i;j} = \partial_j \omega_i - \Gamma_{ji}^k \omega_k, \qquad (1.1.7)$$

where ; is a symbol for the covariant derivative. A vector field Y on M is said to be parallel with respect to a linear connection ∇ if for any vector field X on M it is covariant constant, i.e., $\nabla_X Y = 0$. With respect to a natural basis $\{\partial_i\}$, we say that Y is parallel on M if and only if its local components Y^i, satisfy the following differential equation:

$$\partial_j Y^i + \Gamma_{kj}^i Y^k = 0, \qquad (1.1.8)$$

where Γ_{kj}^i are the connection coefficients of ∇ on M. More details on the above material may be seen, for example, in Kobayashi and Nomizu [75].

1.2   Semi-Riemannian manifolds

Let M be a real n-dimensional smooth manifold with a type $(0, 2)$ non-degenerate symmetric tensor field g. Thus g assigns smoothly, to each point x of M, a symmetric bilinear form g_x on the tangent space $T_x M$. Suppose g_x is non-degenerate on $T_x M$ and of constant index q at all points $x \in M$ so that $T_x M$ becomes an n-dimensional semi-Euclidean space. Then, (M, g) is called a *semi-Riemannian manifold* endowed with a *semi-Riemannian metric* g. In general, there are three categories of vectors fields, namely,

Spacelike   if $g(X, X) > 0$ or $X = 0$,
Timelike   if $g(X, X) < 0$,
Lightlike   if $g(X, X) = 0$ and $X \neq 0$.

In particular, M is *Riemannian* or *Lorentzian* according as $q = 0$ or 1 respectively. In case, $0 < q < n$, then, M is called a *proper semi-Riemannian manifold*. In general, the type into which a given vector field X falls is called the *causal character* of X. A simple way of understanding causal character of curves is as follows:

A null curve C in M has a causal structure if all its tangent vectors are null (lightlike); similarly for non-null curves. In case of Lorentz manifolds, we refer [10] for an important roll of the causal character of spacetime manifolds in relativity.

It is well-known that, although Riemannian metrics always exist on a paracompact manifold, but, in general, the existence of non-degenerate metrics can not be assured. This is why we assume that each semi-Riemannian manifold is paracompact, with a non-degenerate metric tensor field. Indeed, let (M, g) be a real n-dimensional smooth manifold with a symmetric tensor field g of type $(0, 2)$. Thus g assigns smoothly, to each point x of M, a symmetric bilinear form g_x on the tangent space $T_x M$. Suppose g_x is non-degenerate on $T_x M$ and the index of g_x (see equation (1.1.2)) is constant for all $x \in M$ so that $T_x M$ becomes an n-dimensional semi-Euclidean space. The tensor field g satisfying the above conditions is called a *semi-Riemannian metric* or a *metric tensor field* and (M, g) is called a *semi-Riemannian manifold*.

Observe that Riemannian metrics always exist on a paracompact manifold M. In fact, suppose $\{\mathcal{U}_\alpha, \phi_\alpha\}_{\alpha \in I}$ is a smooth atlas of M such that $\{\mathcal{U}_\alpha\}_{\alpha \in I}$ is a local finite open cover of M. Consider g_α as a Riemannian metric on $\mathcal{U}_\alpha$ given by

$$g_\alpha(x)(u, v) = \sum_{i=1}^{m} u^i v^i,$$

with respect to the natural frames field $\{\partial_{x^i}\}$. Then $g = \sum_{\alpha \in I} f_\alpha \, g_\alpha$ is the desired Riemannian metric where $\{f_\alpha\}_{\alpha \in I}$ is the partition of unity subordinated to the covering $\{\mathcal{U}_\alpha, \phi_\alpha\}_{\alpha \in I}$. The proof of this result is based on the positive definiteness of g_α and thus does not hold, in general, for a non-degenerate metric. However, the existence of a Riemannian metric g on M and a unit vector field E_o on M enables one to construct a Lorentz metric on M (see O'Neill [82, page 148]). In fact, consider the associate 1-form ω_o to E_o with respect to g, that is,

$$\omega_o(X) = g(X, E_o), \qquad \forall X \in \Gamma(TM),$$

and define $\bar{g}$ by

$$\bar{g}(X, Y) = g(X, Y) - 2\omega_o(X)\,\omega_o(Y), \qquad \forall X, Y \in \Gamma(TM).$$

Then, it is easy to check that $\bar{g}$ is a Lorentz metric on M.

The following result is very important for the study of null subspaces.

Proposition ([82]). *Let g be a proper semi-Euclidean metric on an n-dimensional vector space V of index q. Then, there exists a subspace $\bar{W}$ of V of dimension $min\{q, n - q\}$ and no larger, such that $g|\bar{W} = 0$.*

Proof. Let $E = \{e_1, \cdots, e_n\}$ be an orthonormal basis of V. Define g by

$$g(x, y) = -\sum_{i=1}^{q} x^i y^i + \sum_{a=p+1}^{n} x^a y^a, \qquad \forall \, x, y \in \mathbf{R}_q^n,$$

where (x^i) and (y^i) are the coordinates of x and y. Suppose $2q < n$. Now define a q-dimensional subspace

$$\bar{W} = Span\{u_1 = e_1 + e_{q+1}, \ldots, u_q = e_q + e_2\}.$$

It follows that $g|\bar{W} = 0$. Choose a null vector $N = \sum_{i=1}^{n} N^i e_i$ such that $g(N, u_a) = 0, \forall a \in \{1, \ldots, q\}$. Thus, $N^1 = N^{q+1}, \ldots, N^q = N^{2q}$. Since $\|N\| = 0$ and $\{e_1, \ldots, e_{2q}\}$ and $\{e_{q+1}, \ldots, e_n\}$ are timelike and spacelike respectively, we conclude that $N^{2q+1} = \ldots = N^n = 0$. Hence, $N = \sum_{a=1}^{q} N^a u_a$.

Thus, there is no subspace larger than $\bar{W}$ on which g vanishes. Similarly, for $2q \geq n$.

See chapter 7 (section 1) for some more information on curvature tensors of a semi-Riemannian manifold which we need for the study of lightlike hypersurfaces. The reader may find more information on semi-Riemannian manifolds in [28, chapter 2] or any other standard book on differential geometry.

1.3 Introduction to null curves

Let C be a smooth curve immersed in an $(m+2)$-dimensional proper semi-Riemannian manifold $(M = M_q^{m+2}, g)$ of a constant index $q \geq 1$. Then, with respect to a local coordinate neighborhood $\mathcal{U}$ on C and a local parameter t, C is given by

$$x^i = x^i(t), \quad i \in \{0, \ldots, m+1\}, \quad \text{rank}\,(dx_t^0 \cdots dx_t^{m+1}) = 1, \quad \forall t \in I,$$

where I is an open interval of a real line and we denote each $\frac{dx^i}{dt}$ by dx_t^i. The non-zero tangent vector field on $\mathcal{U}$ is given by

$$\frac{d}{dt} \equiv (dx_t^0, \ldots, dx_t^{m+1}).$$

In particular, the curve C is called a *regular curve* if $\frac{dC}{dt} \neq 0$ holds everywhere. A non-null regular curve C can be paramerized by arc length in the sense that $g(\frac{dC}{dt}, \frac{dC}{dt}) = \pm 1$ is valid everywhere. On the other hand, since each null vector has zero length, the usual arc length parameterization is not possible for a null curve. $(\mathbf{R}_1^{m+2}, g)$ is called *Minkowski space* defined by a *Minkowski metric*

$$g(x, y) = -x^0 y^0 + \sum_{a=1}^{m+1} x^a y^a.$$

The set of all null vectors of $\mathbf{R}_1^{m+2}$ forms what is called the *light cone* given by

$$(x^0)^2 = \sum_{a=1}^{m+1} (x^a)^2, \quad x^0 \neq 0.$$

Physically, $\mathbf{R}_1^4$ and $\mathbf{R}_1^3$ are important Minkowski spaces studied in general relativity. Here are two examples of non-null curves in $\mathbf{R}_1^3$.

Example 1. The hyperbola $(x^0)^2 = (x^1)^2 + 1$, $x^2 = 0$ is a spacelike curve. This can be parameterized by arc length using the parameter $C(t) = (\cosh t, \sinh t, 0)$ since $g(\frac{dC}{dt}, \frac{dC}{dt}) = 1$.

Example 2. The hyperbola $(x^0)^2 = (x^1)^2 - 1$, $x^2 = 0$ is timelike with an arc length parameter $C(t) = (\sinh t, \cosh t, 0)$ since $g(\frac{dC}{dt}, \frac{dC}{dt}) = -1$. The line

$C(t) = (t, t, 0)$, $t \neq 0$ lies entirely on the light cone.

We first assume that C is a non-null curve in M. Let TC be the 2-dimensional tangent bundle manifold of C. Then, its normal bundle manifold, defined by,

$$TC^{\perp} = \{X \in \Gamma(TM) : g(X, V) = 0\}, \quad V \equiv \frac{d}{dt}, \tag{1.3.1}$$

is a $2(m + 1)$-dimensional non-null bundle subspace of TM satisfying

$$TM = TC \perp TC^{\perp}, \quad TC \cap TC^{\perp} = \{\emptyset\}. \tag{1.3.2}$$

Along C, a vector field Y is said to be parallel, with respect to ∇, if $\nabla_V Y = 0$. Using this and (1.1.8), we conclude that Y is parallel along C if and only if

$$\frac{dY^k}{dt} + \Gamma^k_{ij} Y^i \frac{dx^j}{dt} = 0. \tag{1.3.3}$$

The curve C is called a *geodesic* if V is parallel along C, i.e., if $\nabla_V V = fV$ for some smooth function f along C. For a non-null curve C, it is possible to find an arc-length parameter s of C such that f is zero along C and then the geodesic equation $\nabla_V V = 0$ can be expressed, in local coordinate system (x^i), as

$$\frac{d^2 x^k}{ds^2} + \Gamma^k_{ji} \frac{dx^j}{ds} \frac{dx^i}{ds} = 0. \tag{1.3.4}$$

Two arc length parameters s_1 and s_2 are related by $s_2 = as_1 + b$, where a and b are constants. If the connection ∇ is smooth (or C^r), then the theory of differential equations certifies that, given a point x of M and a tangent vector X_x, there is a *maximal geodesic* $C(s)$ such that $C(0) = x$ and $\frac{dx^i}{ds}|_{s=0} = X^i_x$.

It is important to note that an arbitrary curve need not have causal structure, but, a geodesic always does since parallel translations preserve causal character of vectors. This is a very useful geometric property of geodesic curves.

Now we let the curve C be a *null curve* which preserves its causal character. Then, all its tangent vectors are null. Thus, C is a null curve if and only if at each point x of C we have

$$g\left(\frac{d}{dt}, \frac{d}{dt}\right) = 0.$$

The normal bundle manifold of TC, defined exactly as in (1.3.1), is given by

$$TC^{\perp} = \left\{X \in \Gamma(TM) : g\left(X, \frac{d}{dt}\right) = 0\right\}, \quad \dim(TC^{\perp})_x = m + 1. \tag{1.3.5}$$

However, null curves behave differently than the non-null curves as follows:

(1) $TC^{\perp}$ is also a null bundle subspace of TM.

(2) $TC \cap TC^{\perp} = TC \rightarrow TC \oplus TC^{\perp} \neq TM$.

Thus, contrary to the case of non-null curves, since the normal bundle manifold $TC^\perp$ contains the tangent bundle TC of C, (1.3.2) does not hold for any null curve as their sum is not the whole of the tangent bundle space TM. In other words, a vector of T_xM cannot be decomposed uniquely into a component tangent to C and a component perpendicular to C. Moreover, since the length of any arc of a null curve is zero, arc-length parameter makes no sense for null curves. Later on, the reader will see several other differences between non-null and null geometries. Thus, one can not use, in the usual way, the standard theory of non-null curves (in general, non-null submanifolds) in the study of the geometry of null curves (in general, lightlike submanifolds). Because of this anomaly, null curves (and, in general, lightlike submanifolds) have been studied by several ways corresponding to their use in a given problem. In this book we follow the technique first introduced by Bejancu [11] and then presented in a 1996 Duggal-Bejancu's book [28]. We also review the works of Honda-Inoguchi [57], Inoguchi-Lee [64] on null curves in $\mathbf{R}_1^3$, Ferrández-Giménez-Lucas [35, 36, 37, 38, 39, 40, 41] on null curves and its applications. In general, throughout the book, we suggest many references on null curves (such as Barros [6], Bonnor [18], Cartan [20], Graves[45], Ikawa [61] etc.), discuss the works of Gutiérrez et al. [49, 50, 51, 52] on conjugate points along null geodesics and highlight the significance of null geodesics in geometry and physics.

1.4 Screen and null transversal bundles

In order to develop the geometry of null curves (in line with the case of non-null curves) our objective of this section is to change the equation (1.3.2) such that the tangent bundle TM can be split into three non-intersecting complementary (but non-orthogonal) vector bundles. For this purpose consider a complementary vector bundle $S(TC^\perp)$ to TC in $TC^\perp$. This means that

$$TC^\perp = TC \oplus S(TC^\perp).$$

Following [28, chapter 1] we call $S(TC^\perp)$ a *screen vector bundle* of C in M which obviously is non-degenerate. Since we assume that M is paracompact, there always exist a screen bundle. Thus, along C we have the following decomposition

$$TM_{|C} = S(TC^\perp) \oplus_{orth} S(TC^\perp)^\perp, \tag{1.4.1}$$

where $S(TC^\perp)^\perp$ is a complementary orthogonal vector bundle to $S(TC^\perp)$ in $TM_{|C}$. Note that $S(TC^\perp)^\perp$ is of rank 2 and contains TC. Clearly, contrary to the non-null case, the equation (1.4.1) is not unique as it depends on the choice of a screen vector bundle which is not unique. We will examine such a dependence later on. On the other hand, given a $S(TC^\perp)$ for a null curve C, there exists a unique null vector bundle of rank 1 which plays a roll similar to the roll of unique normal vector bundle of a non-null curve and TM splits into three non-intersecting sub bundles. For this we state and prove the following theorem:

Theorem 4.1. *Let C be a null curve of a proper semi-Riemannian manifold (M, g) and $\pi : ntr(C) \to M$ be a sub bundle of a screen vector bundle $S(TC^\perp)^\perp$*

of C such that $S(TC^\perp)^\perp = TC \oplus ntr(C)$. Let $V \in \Gamma^\infty(\mathcal{U}, ntr(C))$ be a locally defined nowhere zero section, defined on the open subset $\mathcal{U} \subseteq M$. Then

(i) $g(\frac{d}{dt}, V) \neq 0$ everywhere on $\mathcal{U} \subseteq M$.

(ii) If we consider $N_V \in \Gamma^\infty(\mathcal{U}, S(TC^\perp)^\perp)$ given by

$$N_V = \frac{1}{g(\frac{d}{dt}, V)} \left\{ V - \frac{g(V, V)}{2g(\frac{d}{dt}, V)} \frac{d}{dt} \right\}, \tag{1.4.2}$$

then $ntr(C)$ is a unique vector bundle over C of rank 1 such that on each $\mathcal{U} \subset C$ there is a unique vector field $N \in \Gamma(ntr(TC)_{|\mathcal{U}})$ satisfying

$$g(N_V, N_V) = 0, \quad g\left(\frac{d}{dt}, N_V\right) = 1. \tag{1.4.3}$$

(iii) The tangent bundle TM splits into the following three bundle spaces:

$$TM_{|C} = TC \oplus ntr(C) \oplus_{orth} S(TC^\perp) = TC \oplus tr(TC). \tag{1.4.4}$$

Proof. Suppose $g(\frac{d}{dt}, V) = 0$ for some $x_0 \in \mathcal{U}$. Then g would be degenerate on TM at least at $x_0 \in \mathcal{U}$ which is a contradiction. Thus, (i) holds. Let V' be another nowhere zero section in ntr on $\mathcal{U}'$ with $\mathcal{U} \cap \mathcal{U}' \neq \emptyset$. Then $V' = \alpha V$ for some smooth function $\alpha \in \mathcal{U} \cap \mathcal{U}'$, and it follows from (1.4.2) that $N_V = N_{V'}$ on $\mathcal{U} \cap \mathcal{U}'$. This also shows that N depends neither on $ntr(C)$ nor on its local section V. Now, the relations of (1.4.3) easily follow from (1.3.2), which proves (ii). Finally, (1.4.4) of (iii) holds if we set $tr(C) = ntr(C) \oplus S(TC^\perp)$, which completes the proof.

According to the terminology in [28] we call $tr(C)$ and $ntr(C)$ the *transversal bundle* and the *null transversal bundle* with respect to $S(TC^\perp)$ and with respect to $\frac{d}{dt}$ of C in that order. It is easy to see that if C is a null curve of a semi-Riemannian manifold (M, g) of index q, then, its $S(TC^\perp)$ is also semi-Riemannian but of index $q - 1$. Thus, (1.4.4) replaces (1.3.2) such that TM splits into a sum of three non-intersecting complementary (but non-orthogonal) vector bundles.

Note. The statement and the proof of above theorem is an improved version of the one appeared in [28, page 53].

1.5   Quasi-orthonormal basis along a null curve

In this section, we show that, based on the decomposition equation (1.3.4) there exists a quasi-orthonormal basis along a null curve C of a proper semi-Riemannian manifold (M_q^{m+2}, g). Since g is of constant index q, at each point $x \in M$ there is an associated semi-Euclidean space $T_x M$ whose quadratic form is of type $(0, p, q)$ with $p \cdot q \neq 0$ and $p + q = m + 2$. Based on the convention (1.1.2), let $\{e_1, \ldots, e_{m+2}\}$ be an orthonormal basis of $T_x M$ such that $\{e_1, \ldots, e_q\}$ and $\{e_{q+1}, \ldots, e_{q+p}\}$ are unit timelike and spacelike vectors, respectively. To construct a basis including some null vectors we consider the following types.

Type I $(q < p)$. Construct vectors

$$\xi_i = \frac{1}{\sqrt{2}} \{e_{q+i} + e_i\}; \quad \xi_i^* = \frac{1}{\sqrt{2}} \{e_{q+i} - e_i\}, \quad i \in \{1, \ldots, q\},$$

which satisfy

$$g(\xi_i, \xi_j) = g(\xi_i^*, \xi_j^*) = 0,$$

and

$$g(\xi_i, \xi_j^*) = \delta_{ij}, \quad i, j \in \{1, \ldots, q\}.$$

Thus

$$\{\xi_1, \ldots, \xi_q, \xi_1^*, \ldots, \xi_q^*, e_{2q+1}, \ldots, e_{q+p}\}$$

is a basis of $T_x M$ with $2q$ null vectors and $p - q$ spacelike vectors. One can also take at least 2 null base vectors and all others non-null.

Type II $(p < q)$. In this case define

$$\xi_a = \frac{1}{\sqrt{2}} \{e_{q+a} + e_a\}, \quad \xi_a^* = \frac{1}{\sqrt{2}} \{e_{q+a} - e_a\}, \quad a \in \{1, \ldots, p\},$$

and, in a similar way, obtain a basis

$$\{\xi_1, \ldots, \xi_p, \xi_1^*, \ldots, \xi_p^*, e_{p+1}, \ldots, e_q\}$$

which contains $2p$ null vectors and $q - p$ timelike vectors.

Type III $(p = q)$. Similarly, we obtain a totally null basis

$$\{\xi_1, \ldots, \xi_q, \xi_1^*, \ldots, \xi_q^*\}.$$

The construction of above three types shows that, in general, there exists a basis

$$B = \{\xi_1, \ldots, \xi_r, \xi_1^*, \ldots, \xi_r^*, u_1, \ldots, u_t\}$$

of a proper semi-Euclidean space $T_x M$, at any point $x \in M$ if

$$\begin{aligned} g(\xi_i, \xi_j) &= g(\xi_i^*, \xi_j^*) = 0; \quad g(\xi_i, \xi_j^*) = \delta_{ij}, \\ g(u_\alpha, \xi_i) &= g(u_\alpha, \xi_i^*) = 0; \quad g(u_\alpha, u_\beta) = \epsilon_\alpha \delta_{\alpha\beta}, \end{aligned} \tag{1.5.1}$$

for any $i, j \in \{1, \ldots, r\}$ and $\alpha, \beta \in \{1, \ldots, t\}$. The basis B, satisfying above, is called a *quasi-orthonormal basis* of $T_x M$. In particular, let C be a null curve of a Lorent space M_1^{m+2}, then, a quasi-orthonormal basis of $T_x M$, along C, is given by

$$B = \{\xi, \xi^*, u_1, \ldots, u_m\}$$

such that $T_x C = Span\{\xi\}$. Also, based on the decomposition equation (1.4.4), we have $T_x(ntr(C)) = Span\{\xi^*\}$ and $T_x(S(TC^\perp)) = Span\{u_1, \ldots, u_m\}$.

Next, consider an n-dimensional lightlike subspace W of an $(m+2)$-dimensional proper semi-Euclidean space V. Then a quasi-orthonormal basis

$$B = \{f_1, \ldots, f_r, f_1^*, \ldots, f_r^*, u_1, \ldots, u_t\}$$

such that $W = Span\{f_1, \ldots, f_r, u_1, \ldots, u_s\}$, if $n = r + s$, $1 \le s \le t$, or

$$W = Span\{f_1, \ldots, f_n\}, \quad \text{if} \quad n \le r,$$

is called a quasi-orthonormal basis of V along W.

Proposition 5.1 ([28]). *There exists a quasi-orthonormal basis of V along W.*

Proof. First, suppose $null\, W = r < \min\{n, m + 2 - n\}$. Then we have

$$W = Rad\, W \perp W',$$

and

$$W^\perp = Rad\, W \perp W'',$$

where W' and W'' are some screen subspaces. We decompose V as follows

$$V = W' \perp (W')^\perp. \tag{1.5.2}$$

As W'' is a non-degenerate subspace of $(W')^\perp$ we obtain

$$(W')^\perp = W'' \perp (W'')^\perp, \tag{1.5.3}$$

where $(W'')^\perp$ is the complementary orthogonal subspace to W'' in $(W')^\perp$. It is easy to see that $Rad\, W$ is a subspace of $(W'')^\perp$. Denote by U a complementary subspace to $Rad\, W$ in $(W'')^\perp$. As $(W'')^\perp$ is of dimension $2r$ we may consider the basis $\{f_1, \ldots, f_r\}$ and $\{v_1, \ldots, v_r\}$ of $Rad\, W$ and U, respectively. Now, we look for $\{f_1^*, \ldots, f_r^*\}$ given by

$$f_i^* = A_i^j\, f_j + B_i^j\, u_j, \tag{1.5.4}$$

and satisfying the relations in the first line of (1.5.1). By direct calculations, one obtains that $g(f_i, f_k^*) = \delta_{ik}$ if and only if

$$B_k^j\, g(f_i, v_j) = \delta_{ik}. \tag{1.5.5}$$

As $\det[g(f_i, v_j)] \ne 0$, (otherwise $(W'')^\perp$ would be degenerate), the system (1.5.5) has a unique solution (B_k^j). Next, by using (1.5.4) and (1.5.5) one obtains $g(f_i^*, f_j^*) = 0$ if and only if

$$A_j^i + A_i^j + B_i^h\, B_j^k\, g(v_h, v_k) = 0$$

which proves the existence of A_i^j from (1.5.2). Finally, from (1.5.2) and (1.5.3), and taking into account of the above construction, we obtain the following decomposition

$$V = W' \perp W'' \perp (Rad\, W \oplus \{f_1^*, \ldots, f_r^*\}).$$

Hence we have a quasi-orthonormal basis of V along W given by

$$\{f_1, \ldots, f_r, f_1^*, \ldots, f_r^*, u_1, \ldots, u_{n-r}, w_1, \ldots, w_{m+2-n-r}\},$$

where $\{u_1, \ldots, u_{n-r}\}$ and $\{w_1, \ldots, w_{m-n-r}\}$ are two orthonormal basis of W' and W'', respectively. In this case

$$W = Span\{f_1, \ldots, f_r, u_1, \ldots, u_{n-r}\}.$$

In case $r = n < m + 2 - n$, it follows that $Rad\, W = W \subset W^\perp$. We put

$$W^\perp = W \perp W'',$$

where W'' is an arbitrary screen subspace of $W^\perp$. One obtains for V the orthogonal decomposition

$$V = W'' \perp (W'')^\perp,$$

where $(W'')^\perp$ is the complementary orthogonal subspace to W'' in V. Moreover $(W'')^\perp$ is of dimension $2n$ and contains W. In a similar way as in the first case we find the quasi-orthonormal basis of V along W

$$\{f_1, \ldots, f_n, f_1^*, \ldots, f_n^*, w_1, \ldots, w_{m-2n}\},$$

where $\{w_1, \ldots, w_{m-2n}\}$ is an orthonormal basis of W'' and

$$W = Span\{f_1, \ldots, f_n\}.$$

In case $r = m + 2 - n < n$, it follows that $Rad\, W = W^\perp \subset W$. Then we set

$$W = W^\perp \perp W',$$

where W' is a screen subspace of W. Thus V is decomposed as follows

$$V = W' \perp (W')^\perp,$$

where $(W')^\perp$ is the complementary orthogonal subspace to W' in V. It follows that $(W')^\perp$ is of dimension $2(m+2-n)$ and it contains $W^\perp$. Then the quasi-orthonormal basis of V along W is given by

$$\{f_1, \ldots, f_{m+2-n}, f_1^*, \ldots, f_{m-n}^*, u_1, \ldots, u_{2n-m+2}\},$$

where $\{u_1, \ldots, u_{2n-m+2}\}$ is an orthonormal basis of W'. In this case

$$W = Span\{f_1, \ldots, f_{m-n}, u_1, \ldots, u_{2n-m+2}\}.$$

Finally, if $r = n = \frac{m+2}{2}$, we get $Rad\, W = W = W^\perp$ and the decomposition

$$V = W \oplus Span\{f_1^*, \ldots, f_n^*\}.$$

Then the quasi-orthonormal basis of V along W is given by

$$\{f_1, \ldots, f_n, f_1^*, \ldots, f_n^*\},$$

where $\{f_1, \ldots, f_n\}$ is a basis of W.

For more details on the quasi-orthonormal basis, see [28, Chapter 1].

1.6   Brief notes and exercises

Serret-Frenet equations of a curve in $\mathbf{R}^3$. Let $C = C(s)$ be a curve in $\mathbf{R}^3$ with its Frenet frame $(\mathbf{t}, \mathbf{n}, \mathbf{b})$ consisting of unit tangent, normal and binormal vectors respectively. Then, the differential equations

$$
\begin{aligned}
\mathbf{t}' &= \kappa \mathbf{n} \\
\mathbf{n}' &= -\kappa \mathbf{t} + \tau \mathbf{b} \\
\mathbf{b}' &= -\tau \mathbf{n}
\end{aligned}
$$

are called *Serret-Frenet* (or briefly Frenet) equations of C. These equations are basic in the development of the theory of curves in $\mathbf{R}^3$. The functions κ and τ are called curvature and torsion functions respectively. The fundamental theorem of space curves establishes the existence and uniqueness of a curve with given Frenet frame. Unfortunately, in general, the unique existence of a null curve can not be established.

Null curves of $\mathbf{R}_1^3$. Consider a 3-dimensional *Minkowski space* $\mathbf{R}_1^3$ defined as a space to be the usual 3-dimensional vector space consisting of vectors $\{(x^0, x^1, x^2) | x^0, x^1, x^2 \in \mathbf{R}\}$, but with a linear connection ∇ corresponding to its Minkowski metric $\langle\,,\,\rangle$ given by

$$
\langle x, y \rangle = -x^0 y^0 + x^1 y^1 + x^2 y^2.
$$

The set of all null vectors of $\mathbf{R}_1^3$ forms the light cone in coordinates;

$$
\{(x^0, x^1, x^2) \mid (x^0)^2 = (x^1)^2 + (x^2)^2, \ x^0 \neq 0\}
$$

in $\mathbf{R}_1^3$, the rules of calculus are the same as in Euclidean space. Thus, we speak of immersions or regular or smooth curves just as in the Euclidean case. However, since any $\mathbf{R}_1^3$ can have a null curve (for example, any null vector can generate a null curve) as stated before the calculus of null curves is quite different than the non-null curves.

To derive the Frenet type equations for a null curve C, defined by $C : [a, b] \to \mathbf{R}_1^3$, Cartan [20] has shown that with respect to an affine parameter, say p, and a positively oriented set $\{C'(p), C''(p), C'''(p)\}, \forall p \in [a, b]$, there exists a local frame $F = \{\xi = C', N, W, \}$, called *Cartan frame* satisfying

$$
\begin{aligned}
\langle \xi, \xi \rangle &= \langle N, N \rangle = 0, \quad \langle \xi, N \rangle = 1, \\
\langle W, \xi \rangle &= \langle W, N \rangle = 0, \quad \langle W, W \rangle = 1,
\end{aligned}
$$

with the vector product $\times$ given by $\xi \times W = -\xi$, $\xi \times N = -W$ and $W \times N = -N$. The Cartan equations are given by

$$
\begin{aligned}
\xi' &= \nabla_\xi \xi = kW, \\
N' &= \nabla_\xi N = -\tau W, \\
W' &= \nabla_\xi W = -\tau \xi + kN,
\end{aligned} \tag{1.6.1}
$$

where k and τ are the curvature and torsion functions of C with respect to F. If $k = 0$ then the first Cartan equation reduces to

$$\frac{d^2 x^i}{dp^2} + \Gamma^i_{jk} \frac{dx^j}{dp} \frac{dx^k}{dp} = 0, \quad i \in \{0, 1, 2\},$$

where Γ^i_{jk} are the Christoffel symbols induced by ∇. Thus C is a *null geodesic* of $\mathbf{R}^3_1$ if and only if its curvature function k vanishes identically on C. This characterization result also holds for any null curve of a semi-Riemannian manifold. For $(k \neq 0)$, Honda-Inoguchi [57] have recently proved that there exists a Cartan frame for any non-geodesic null curve $C = C(t)$ in $\mathbf{R}^3_1$ such that the Cartan equations (1.6.1) hold for the original parameter t. In chapter 2 we show that one can take a canonical affine parameter for every non-geodesic null curve in a Lorentzian manifold M^{m+2}_1. Here we show that it is possible to express the Cartan frame F in terms of ξ and its covariant derivatives up to the order 3 as follows:

Suppose $k = \frac{1}{\sigma} \neq 0$ on $\mathcal{U}$ and denote $\sigma_1 = \xi(\sigma)$. From (1.6.1) we obtain

$$W = \sigma \xi', \quad N = -\sigma \left(\tau \xi + \sigma_1 \xi' + \sigma \xi'' \right).$$

Again using (1.6.1) and above two equations, we obtain

$$k = \pm \| \xi' \|, \quad \tau = \frac{\sigma}{2} \left\{ \sigma^2 \langle \xi'', \xi'' \rangle - \sigma^{-2} (\sigma_1)^2 \right\}.$$

Null tetrad of $\mathbf{R}^4_1$. Consider a Euclidean space $\mathbf{R}^4$ and a canonical basis $E = \{e_1 = (1, 0, 0, 0), \ldots, e_4 = (0, 0, 0, 1)\}$. Define on $\mathbf{R}^4$ the Minkowski metric:

$$ds^2 = -(dx^1)^2 + (dx^2)^2 + (dx^3)^2 + (dx^4)^2,$$

with respect to a local coordinate system (x^1, x^2, x^3, x^4). Then, we say that $\mathbf{R}^4_1$ is a *Minkowski* vector space with the above metric of Special Relativity, where the coordinate x^1 represents the time component. Using Type 1, we construct a quasi-orthonormal basis $B = \{\xi, N, e_3, e_4\}$ by the following transformation:

$$\xi = \frac{1}{\sqrt{2}} \{e_2 + e_1\}; \quad N = \frac{1}{\sqrt{2}} \{e_2 - e_1\},$$

$$g(\xi, \xi) = g(N, N) = 0, \quad g(\xi, N) = 1.$$

Let C be a null curve of $\mathbf{R}^4_1$, with the basis B, along C, such $T_x C = Span\{\xi\}$, $T_x(ntr(C)) = Span\{N\}$ and $T_x(S(TC^\perp)) = Span\{e_3, e_4\}$, for any $x \in \mathbf{R}^4_1$.

The *complexified vector space* of $\mathbf{R}^4_1$ is the complex vector space $(\mathbf{R}^4_1)^c$ of vectors $x + iy$, $x, y \in \mathbf{R}^4_1$, $i = \sqrt{-1}$. The scalar product g on $\mathbf{R}^4_1$ induces a scalar product g^c on $(\mathbf{R}^4_1)^c$ which is a symmetric non-degenerate **C**-bilinear mapping on $(\mathbf{R}^4_1)^c$. Then the quasi-orthonormal basis B of $\mathbf{R}^4_1$ induces the so called *null tetrad* of $(\mathbf{R}^4_1)^c$ (see Trautman et al. [107, Page 57])

$$T = \{\ell, n, m, \bar{m}\}; \quad \ell = \xi, \quad n = -N, \quad m = \frac{1}{\sqrt{2}} (e_3 + ie_4), \quad \bar{m} = \frac{1}{\sqrt{2}} (e_3 - ie_4)$$

consisting of two real null vectors and two conjugate complex null vectors with respect to g^c such that the only surviving scalar products are

$$g^c\left(\ell, n\right) = -1; \quad g^c\left(m, \bar{m}\right) = 1.$$

Denoting the dual basis of T by $T^* = \{\eta_1, \eta_2, \eta_3, \eta_4\}$ of the null tetrad T and obtain the associated quadratic form h^c of g^c by $h^c = 2\{\eta_3\,\eta_4 - \eta_1\,\eta_2\}$.

Using above, one can obtain a local null tetrad with respect to a coordinate neighborhood on a spacetime manifold, which has been a useful tool in relativity. Moreover, it is possible to express $\{n, e_3, e_4\}$ in terms of ℓ and its covariant derivatives up to the order 3. For details, see [28, pages 62, 63].

Using Cartan's approach [20], Bonnor [18], Cöken and Ciftci [22] and Ikawa [61] have done some work on null curves in a 4-dimensional Minkowski spacetime. See chapter 5 for fundamental existence and uniqueness theorems on null Cartan curves in Minkowski spaces and pseudo-Euclidean spaces of index two.

Exercises

(1) (a) Show that the circle $\mathbf{S}^1$ in $\mathbf{R}^3$ can be made into a smooth 1-manifold by constructing an atlas with two charts.

(b) Show that a one-chart does not exist. Thus, a circle is not homeomorphic to a line or interval.

(2) The n-sphere $\mathbf{S}^n$ is defined by the set of points (x^i) in $\mathbf{R}^{n+1}$ such that $\sum_{i+1}^{n+1} (x^i)^2 = a^2$ centered at the origin with radius a. A coordinate patch for a neighborhood of origin is:

$$x^1 = y^1, \ x^2 = y^2, \ \ldots, \ x^n = y^n, \ x^{n+1} = \sqrt{a^2 - (y^1)^2 - (y^2)^2 \ldots - (y^n)^2}$$

where the mapping is into the n-dimensional neighborhood $\sum_{i=1}^{n} (y^i)^2 < a^2$ (the interior of $\mathbf{S}^{n-1}$). Establish the analogous patches around the other [diametrically opposite] endpoints, and obtain an atlas of $2n + 2$ charts to see $\mathbf{S}^n$ as a manifold.

(3) Show that the hyperboloid of one sheet $4x^2 + 4y^2 + z^2 = 16$ is a smooth 2-manifold M, by the coordinatization $(k = 1, 2)$, where $((k - 2)\pi < \theta < h\pi)$ and

$$\mathbf{r}_k : x = 2\cos\theta\cosh\phi, \quad y = 2\sin\theta\cosh\phi, \quad z = 4\sinh\phi$$

with the neighborhoods $\mathcal{U}_{(1)} = \mathcal{U}_a$ and $a = (2, 0, 0)$, $\mathcal{U}_{(2)} = \mathcal{U}_b$ and $b = (-2, 0, 0)$.

(4) Find a convenient atlas showing that the set in $\mathbf{R}^4$ given by the equation

$$x^2 + y^2 + z^2 - t^2 = a^2$$

can be made into a smooth 3-manifold. [Hint: six charts will suffice.]

(5) Determine the curve whose curvature and torsion are given by

$$\kappa = (1/2as)^{\frac{1}{2}}, \quad \tau = 0.$$

(6) Show that the curve with $\kappa = \sqrt{2}\,\frac{1}{s^2+4} = \tau$ is a general helix on a cylinder whose cross section is a catenary.

(7) Consider a null Cartan curve C of $\mathbf{R}_1^3$ given by the equations

$$x^0 = \sinh p, \quad x^1 = p, \quad x^2 = \cosh p,$$

with a Cartan frame $F = \left\{\xi = \frac{d}{dp}, W, N\right\}$ as follows:

$$\xi = (\cosh p, 1, \sinh p), \quad W = (\sinh p, 0, \cosh p), \quad N = \frac{1}{2}(-\cosh p, 1, -\sinh p).$$

Then, using Cartan equations show that $k_1 = 1$ and $k_2 = -\frac{1}{2}$.

(8) Consider a curve C of $\mathbf{R}_1^4$ defined by

$$x^0 = 5t, \quad x^1 = 3\cos t, \quad x^2 = 3\sin t, \quad x^3 = 4t,$$

and $0 \le t \le 1$ is an affine parameter. Prove that C is a null curve.

(9) Show that a curve C given by

$$x^1 = \frac{1}{3}\{2t - 1\}, \quad x^2 = \frac{1}{2}t^2 - t, \quad x^3 = t\sin t + \cos t, \quad x^4 = \sin t - t\cos t,$$

with $t > 1$ is a null curve of a semi-Euclidean space $\mathbf{R}_2^4$ of index 2.

(10) Consider a curve in a semi-Euclidean space $\mathbf{R}_2^6$ given by

$$C : (\alpha t^2 - 2t, \ \beta t^2, \ \alpha t^3, \ \sqrt{2}t, \ \cos\sqrt{2}t, \ \sin\sqrt{2}t),$$

where α, β, $t \in \mathbf{R}$. Find a relation between α and β such that C is null.

Chapter 2

Null curves in Lorentzian manifolds M_1^{m+2}

In this chapter we first recall [28, chapter 3] the *Frenet equations* and their *Frenet frames* along a smooth null curve C of a real Lorentzian manifold (M, g), denoted by (M_1^{m+2}, g). We also introduce another new set of Frenet equations, called *Natural Frenet equations* (see Jin [71]), examine the dependence of Frenet frames on the coordinate neighborhood of C and the choice of a screen vector bundle. Since the length of any null curve is zero, we use an affine parameter, called, a *distinguished parameter* which plays the role similar to the role of an arc-length parameter for non-null curves. Then, we study the geometry of null curves in M_1^3, M_1^4, M_1^5 and M_1^6. Throughout we provide several solved examples.

2.1 Frenet frames along null curves

Let (M_1^{m+2}, g) be a real $(m + 2)$-dimensional Lorentzian manifold and C be a smooth null curve in M_1^{m+2} locally given by

$$x^i = x^i(t), \quad t \in I \subset \mathbf{R}, \qquad i \in \{0, 1, \ldots, (m + 1)\}$$

for a coordinate neighborhood $\mathcal{U}$ on C. For Lorentzian manifold M_1^{m+2}, it is clear that the screen vector bundle $S(TC^\perp)$ is a Riemannian vector bundle of rank m. Let ∇ be the Levi-Civita (metric) connection on M_1^{m+2} (see a note in section 2.6). Then, $\nabla g = 0$, that is,

$$(\nabla_X g)(Y, Z) = X(g(Y, Z)) - g(\nabla_X Y, Z) - g(Y, \nabla_X Z) = 0, \tag{2.1.1}$$

for any $X, Y, Z \in \Gamma(TM)$. We use this, the equations (1.4.3) through (1.4.4) and set $\frac{d}{dt} = \xi$. From $g(\xi, \xi) = 0$ and $g(\xi, N) = 1$ we have $g(\nabla_\xi \xi, N) = -g(\xi, \nabla_\xi N) = h$, where h is a smooth function on $\mathcal{U}$. All this and the fact that $S(TC^\perp)$ is Riemannian vector bundle implies

$$\nabla_\xi \xi = h\xi + k_1 W_1$$

where we denote $\kappa_1 = \|\nabla_\xi \xi\|$ the first curvature function and W_1 is a unit spacelike vector field along C. From $g(\nabla_\xi N, \xi) = -h$, $g(N, W_1) = 0$, we have $g(\nabla_\xi N, N) = 0$, $g(\nabla_\xi N, W_1) = \kappa_2$, where k_2 denotes the second curvature function. This implies that $\nabla_\xi N = -hN + \kappa_2 W_1 + T_1$, where T_1 is a spacelike vector perpendicular to ξ, N and W_1. Define $\kappa_3 = \|T_1\|$ the third curvature function so that $W_2 = T_1/\kappa_3$ is a spacelike unit vector. Then,

$$\nabla_\xi N = -hN + \kappa_2 W_1 + \kappa_3 W_2.$$

Also, since $g(\nabla_\xi W_1, \xi) = -\kappa_1$, if we put $g(\nabla_\xi W_1, W_2) = \kappa_4$, then,

$$\nabla_\xi W_1 = -\kappa_2 \xi - \kappa_1 N + \kappa_4 W_2 + T_2,$$

where k_4 denotes the fourth curvature function. Since T_2 is also a spacelike vector field, we define $\kappa_5 = \|T_2\|$ so that $W_3 = T_2/\kappa_5$. Then,

$$\nabla_\xi W_1 = -\kappa_2 \xi - \kappa_1 N + \kappa_4 W_2 + \kappa_5 W_3.$$

Repeating above process for all the m unit vectors of an orthonormal basis $\{W_1, \cdots, W_m\}$ of $\Gamma(S(TC^\perp))$, we obtains the following equations

$$
\begin{aligned}
\nabla_\xi \xi &= h\xi + \kappa_1 W_1, \\
\nabla_\xi N &= -hN + \kappa_2 W_1 + \kappa_3 W_2, \\
\nabla_\xi W_1 &= -\kappa_2 \xi - \kappa_1 N + \kappa_4 W_2 + \kappa_5 W_3, \\
\nabla_\xi W_2 &= -\kappa_3 \xi - \kappa_4 W_1 + \kappa_6 W_3 + \kappa_4 W_4, \\
\nabla_\xi W_3 &= -\kappa_5 W_1 - \kappa_6 W_2 + \kappa_8 W_4 + \kappa_9 W_5,
\end{aligned}
\qquad (2.1.2)
$$

$$
\cdots\cdots\cdots\cdots\cdots\cdots\cdots\cdots\cdots
$$

$$
\begin{aligned}
\nabla_\xi W_{m-2} &= -\kappa_{2m-5} W_{m-4} - \kappa_{2m-4} W_{m-3} + \kappa_{2m-2} W_{m-1} + \kappa_{2m-1} W_m, \\
\nabla_\xi W_{m-1} &= -\kappa_{2m-3} W_{m-3} - \kappa_{2m-2} W_{m-2} + \kappa_{2m} W_m, \\
\nabla_\xi W_m &= -\kappa_{2m-1} W_{m-2} - \kappa_{2m} W_{m-1},
\end{aligned}
$$

provided $m \geq 5$, where h and $\{\kappa_1, \ldots, \kappa_{2m}\}$ are smooth functions on $\mathcal{U}$. In general, for any $m > 0$ we call

$$F = \left\{ \frac{d}{dt} = \xi, N, W_1, \ldots, W_m \right\}$$

a *general Frenet frame* on M along C with respect to the screen vector bundle $S(TC^\perp)$ and the set $\{\kappa_1, \ldots, \kappa_{2m}\}$ the *curvature functions* of C with respect to F. The equations (2.1.2) are called the *general Frenet equations* with respect to F.

Example 1. Consider a null curve C in $\mathbf{R}_1^5$ given by the equation

$$x^1 = \sqrt{2}\sinh t, \quad x^2 = \sqrt{2}\cosh t, \quad x^3 = \cos t, \quad x^4 = \sin t, \quad x^5 = t, \quad t \in \mathbf{R}.$$

Then

$$\xi = \left(\sqrt{2}\cosh t, \ \sqrt{2}\sinh t, \ -\sin t, \ \cos t, \ 1 \right).$$

If we take $V = (0, 0, 0, 0, 1)$, then $g(\xi, V) = 1$ and $g(V, V) = 1$. Thus, by the equation (1.4.2) in Theorem 4.1 of chapter 1, we have

$$N = -\frac{1}{2}\left(\sqrt{2}\cosh t,\ \sqrt{2}\sinh t,\ -\sin t,\ \cos t,\ -1\right).$$

Since

$$\nabla_\xi \xi = \left(\sqrt{2}\sinh t,\ \sqrt{2}\cosh t,\ -\cos t,\ -\sin t,\ 0\right),$$

$h = g(\nabla_\xi \xi, N) = 0$ and $\kappa_1 = \|\nabla_\xi \xi\| = \sqrt{3}$, we let

$$W_1 = \frac{1}{\sqrt{3}}\left(\sqrt{2}\sinh t,\ \sqrt{2}\cosh t,\ -\cos t,\ -\sin t,\ 0\right),$$

then we have $\nabla_\xi \xi = \sqrt{3}\,W_1$. And, from

$$\nabla_\xi N = -\frac{1}{2}\left(\sqrt{2}\sinh t,\ \sqrt{2}\cosh t,\ -\cos t,\ -\sin t,\ 0\right)$$

and $\kappa_2 = g(\nabla_\xi N, W_1) = -\frac{\sqrt{3}}{2}$, we have $\nabla_\xi N = -\frac{\sqrt{3}}{2}W_1$ and $\kappa_3 = 0$. Choose $W_2 = \frac{1}{\sqrt{3}}\left(\sinh t,\ \cosh t,\ \sqrt{2}\cos t,\ \sqrt{2}\sin t,\ 0\right)$, then

$$\nabla_\xi W_1 = \frac{1}{\sqrt{3}}\left(\sqrt{2}\cosh t,\ \sqrt{2}\sinh t,\ \sin t,\ -\cos t,\ 0\right),$$

$\kappa_4 = g(\nabla_\xi W_1, W_2) = 0$ and the vector field $\nabla_\xi W_1 - \frac{\sqrt{3}}{2}\xi + \sqrt{3}\,N$ is given by

$$-\frac{2}{\sqrt{3}}(\sqrt{2}\cosh t,\ \sqrt{2}\sinh t,\ -2\sin t,\ 2\cos t,\ 0).$$

Since $\kappa_5 = \|\nabla_\xi W_1 - \frac{\sqrt{3}}{2}\xi + \sqrt{3}\,N\| = \frac{2\sqrt{2}}{\sqrt{3}}$, we have

$$W_3 = -\frac{1}{\sqrt{2}}(\sqrt{2}\cosh t,\ \sqrt{2}\sinh t,\ -2\sin t,\ 2\cos t,\ 0)$$

and $\nabla_\xi W_1 = \frac{\sqrt{3}}{2}\xi - \sqrt{3}\,N + \frac{2\sqrt{2}}{\sqrt{3}}W_3$. From

$$W_2 = \frac{1}{\sqrt{3}}\left(\cosh t,\ \sinh t,\ -\sqrt{2}\sin t,\ \sqrt{2}\cos t,\ 0\right),$$

and $\kappa_6 = g(\nabla_\xi W_2, W_3) = -\frac{1}{\sqrt{3}}$, we have

$$\nabla_\xi W_2 = -\frac{1}{\sqrt{3}}W_3,\qquad \nabla_\xi W_3 = -\frac{2\sqrt{2}}{\sqrt{3}}W_1 + \frac{1}{\sqrt{3}}W_2.$$

Example 2. Consider the null curve in $\mathbf{R}_1^3$ given by

$$x^1 = \frac{1}{3}t^3 + t,\quad x^2 = t^2,\quad x^3 = \frac{1}{3}t^3 - t,\qquad t \in \mathbf{R}.$$

Then,

$$\xi = \left(t^2 + 1,\ 2t,\ t^2 - 1 \right).$$

If we take $V = \left(0,\ \frac{1}{2t},\ 0 \right)$, then $g\left(\xi,\ V \right) = 1$ and $g(V,\ V) = \frac{1}{4t^2}$. Thus by (1.4.2) in Theorem 4.1 of chapter 1 we have

$$N = -\frac{1}{8t^2} \left(t^2 + 1,\ -2t,\ t^2 - 1 \right).$$

Since $\nabla_\xi \xi = 2\left(t,\ 1,\ t \right)$, $h = g\left(\nabla_\xi \xi,\ N \right) = \frac{1}{t}$, $\kappa_1 = ||\nabla_\xi \xi|| = 2$ and $\nabla_\xi \xi - h\xi = \frac{1}{t}\left(t^2 - 1,\ 0,\ t^2 + 1 \right)$, we have

$$W_1 = \frac{1}{2t} \left(t^2 - 1,\ 0,\ t^2 + 1 \right)$$

and $\nabla_\xi \xi = \frac{1}{t}\xi + 2W_1$. From $\nabla_\xi N = \frac{1}{4t^3}\left(1,\ -t,\ -1 \right)$, $k_2 = g\left(\nabla_{\frac{d}{dp}} N,\ W_1 \right) = -\frac{1}{4t^2}$, we have $\nabla_\xi N = -\frac{1}{t}N - \frac{1}{4t^2}W_1$ and $\nabla_\xi W_1 = \frac{1}{4t^2}\xi - 2N$.

Clearly, contrary to the Riemannian case, the above general Frenet frame and its general Frenet equations are not unique as they depend on the parameter on C and the screen vector bundle. For example, take another screen vector bundle $S(TC^\perp)$ with $N = -\frac{1}{2}(1, 0, 1)$, $W_1 = (t, 1, t)$ of the null curve C in example 2. Then, C has the general Frenet equations (2.1.2) such that $\kappa_1 = 2$ and $\kappa_2 = h = 0$. This Frenet frame and its Frenet equations are different from those in example 2. To deal with this non-uniqueness problem, Bonnor [18] was the first who introduced a unique Frenet frame (called the *Cartan frame*) along null curves in $\mathbf{R}_1^4$ with the minimum number of curvature functions. Bonnor's work was recently generalized by Ferrández-Giménez-Lucas [37] for null Cartan curves in M_1^{m+2} and they studied some classification theorems on null helices (see chapter 5 of this book). For further research on Cartan curves, in the following we introduce another form of Frenet equations (different from the general Frenet equations (2.1.2)) for null curves which includes Cartan equations as a special case. We first need the following result:

Let $F = \left\{ \frac{d}{dt},\ N,\ W_1,\ \dots,\ W_m \right\}$ and $\bar{F} = \left\{ \frac{d}{d\bar{t}},\ \bar{N},\ \bar{W}_1,\ \dots,\ \bar{W}_m \right\}$ be two Frenet frames with respect to $(t, S(TC^\perp), \mathcal{U})$ and $(\bar{t}, \bar{S}(TC^\perp), \bar{\mathcal{U}})$. The general transformations relating the elements of F and $\bar{F}$ on $\mathcal{U} \cap \bar{\mathcal{U}} \neq \emptyset$ are given by

$$\frac{d}{d\bar{t}} = \frac{dt}{d\bar{t}}\frac{d}{dt},$$

$$\bar{N} = -\frac{1}{2}\frac{dt}{d\bar{t}}\sum_{\alpha=1}^{m}(c_\alpha)^2\frac{d}{dt} + \frac{d\bar{t}}{dt}N + \sum_{\alpha=1}^{m}c_\alpha W_\alpha, \qquad (2.1.3)$$

$$\bar{W}_\alpha = \sum_{\beta=1}^{m} B_\alpha^\beta \left(W_\beta - \frac{dt}{d\bar{t}} c_\beta \frac{d}{dt} \right),$$

where c_α and B_α^β are smooth functions on $\mathcal{U} \cap \bar{\mathcal{U}} \neq \emptyset$ and the $m \times m$ matrix $\left[B_\alpha^\beta(x) \right]$ is an element of orthogonal group $O(m)$ for each $x \in \mathcal{U} \cap \bar{\mathcal{U}}$. We call (2.1.3) the transformation of screen vector bundle of C.

Theorem 1.1. *Let C be a null curve of a Lorentzian manifold (M_1^{m+1}, g) with a general Frenet frame F with respect to the screen vector bundle $S(TC^\perp)$ such that $\kappa_1 \neq 0$. Then there exists a screen vector bundle $\bar{S}(TC)$ which induces another Frenet frame $\bar{F}$ on $\bar{\mathcal{U}}$ such that $\bar\kappa_4 = \bar\kappa_5 = 0$ on $\mathcal{U} \cap \bar{\mathcal{U}}$.*

Proof. From the first equation of (2.1.2) and (2.1.3), we have

$$\bar\kappa_1 \, B_1^1 = \kappa_1 \left(\frac{dt}{d\bar{t}} \right)^2 ; \qquad \bar\kappa_1 \, B_1^\alpha = 0, \quad \alpha \in \{2, \cdots, m\}.$$

Since $\kappa_1 \neq 0$ on $\mathcal{U} \cap \bar{\mathcal{U}}$, we have $\bar\kappa_1 \neq 0$ on $\mathcal{U} \cap \bar{\mathcal{U}}$ and $B_1^2 = \cdots = B_1^m = 0$. Also $[B_\alpha^\beta(x)]$ being orthogonal, we infer that $B_1^1 = B_1 = \pm 1$ and $B_2^1 = \cdots = B_m^1 = 0$. Moreover, from the third equation of (2.1.2), we have

$$\bar\kappa_4 \, B_2^2 + \bar\kappa_5 \, B_3^2 = B_1^1 \left(\kappa_4 + \kappa_1 c_2 \frac{dt}{d\bar{t}} \right) \frac{dt}{d\bar{t}},$$

$$\bar\kappa_4 \, B_2^3 + \bar\kappa_5 \, B_3^3 = B_1^1 \left(\kappa_5 + \kappa_1 c_3 \frac{dt}{d\bar{t}} \right) \frac{dt}{d\bar{t}},$$

$$\bar\kappa_4 \, B_2^\alpha + \bar\kappa_5 \, B_3^\alpha = B_1^1 \kappa_1 c_\alpha \left(\frac{dt}{d\bar{t}} \right)^2, \quad \alpha \in \{4, \cdots, m\}.$$

Taking into account that

$$c_2 = -\frac{\kappa_4}{\kappa_1} \frac{d\bar{t}}{dt}; \quad c_3 = -\frac{\kappa_5}{\kappa_1} \frac{d\bar{t}}{dt}; \quad c_\alpha = 0, \quad \alpha \in \{4, \cdots, m\}$$

in the last equations, we have $\bar\kappa_4 = \bar\kappa_5 = 0$.

Remark. If we take $t = \bar{t}$ in Theorem 1.1, then $c_2 = -\frac{\kappa_4}{\kappa_1}$; $c_3 = -\frac{\kappa_5}{\kappa_1}$ and

$$\bar{N} = -\frac{1}{2} \left(\frac{\kappa_4^2 + \kappa_5^2}{\kappa_1^2} \right) \xi + N - \frac{\kappa_4}{\kappa_1} W_2 - \frac{\kappa_5}{\kappa_1} W_3,$$

$$\bar{W}_2 = W_2 + \frac{\kappa_4}{\kappa_1} \xi,$$

$$\bar{W}_3 = W_3 + \frac{\kappa_5}{\kappa_1} \xi,$$

$$\bar{W}_i = W_i, \quad i \in \{1, 4, \cdots, m\}.$$

Relabeling $\bar{N} = N$, $\bar{W}_1 = W_1$, $\bar{W}_2 = W_2$, $\bar\kappa_i = \kappa_i$, $i \in \{1, 2, 3\}$ and $\bar{S}(TC^\perp) = S(TC^\perp)$ and we take only the first four equations in (2.1.2) as follow:

$$\nabla_\xi \xi = h \, \xi + \kappa_1 \, W_1,$$

$$\nabla_\xi N = -h \, N + \kappa_2 \, W_1 + \kappa_3 \, W_2,$$

$$\nabla_\xi W_1 = -\kappa_2 \, \xi - \kappa_1 \, N,$$

$$\nabla_\xi W_2 = -\kappa_3 \, \xi + R_3,$$

where R_3 is a spacelike vector field in $\Gamma(S(TC^\perp))$ perpendicular to ξ, N, W_1 and W_2. Define the new fourth curvature function κ_4 by $\kappa_4 = \| R_3 \|$ and let $W_3 = \frac{R_3}{\kappa_4}$, then W_3 is also a unit spacelike vector field along C. Thus we have

$$\nabla_\xi W_2 = - \kappa_3\, \xi + \kappa_4\, W_3.$$

Repeating above process for all m unit vectors of an orthonormal basis $\{W_1, \ldots, W_m\}$ of $\Gamma(S(TC^\perp))$, and simplifying we obtain the following:

Theorem 1.2. Let C be a null curve of a Lorentzian manifold (M_1^{m+2}, g). Then there exists a Frenet frame $F = \{\xi, N, W_1, \ldots, W_m\}$ satisfying the following equations

$$
\begin{aligned}
\nabla_\xi \xi &= h\xi + \kappa_1 W_1, \\
\nabla_\xi N &= - hN + \kappa_2 W_1 + \kappa_3 W_2, \\
\nabla_\xi W_1 &= - \kappa_2 \xi - \kappa_1 N, \\
\nabla_\xi W_2 &= - \kappa_3 \xi + \kappa_4 W_3, \\
\nabla_\xi W_3 &= - \kappa_4 W_2 + \kappa_5 W_4,
\end{aligned}
\tag{2.1.4}
$$

$$
\begin{aligned}
\cdots\cdots\cdots\cdots \\
\nabla_\xi W_i &= - \kappa_{i+1} W_{i-1} + \kappa_{i+2} W_{i+1}, \quad i \in \{3, \ldots, m-1\}, \\
\nabla_\xi W_m &= - \kappa_{m+1} W_{m-1},
\end{aligned}
$$

where $\{\kappa_1, \ldots, \kappa_{m+1}\}$ are smooth functions on $\mathcal{U}$, $\{W_1, \ldots, W_m\}$ is a certain orthonormal basis of $\Gamma(S(TC^\perp)|_\mathcal{U})$.

We call the frame $F = \{\xi, N, W_1, \ldots, W_m\}$ in theorem 1.2 a *Natural Frenet frame* on M_1^{m+2} along C with respect to the given screen vector bundle $S(TC^\perp)$ and their corresponding equations (2.1.4) its *Natural Frenet equations* (see Jin [71]), with the curvature functions $\{\kappa_1, \ldots, \kappa_{m+1}\}$. Later on, we will show that a special case of this natural Frenet equations is the *null Cartan equations* used in the study of null helices and several other physical properties of null curves.

Example 3. Consider a null curve C in $\mathbf{R}_1^5$ given by the equation

$$C(t) = \left(\sqrt{2}\sinh t, \ \sqrt{2}\cosh t, \ \cos t, \ \sin t, \ t \right), \quad t \in \mathbf{R}.$$

We choose the Natural Frenet frame F as follows

$$\xi = \left(\sqrt{2}\cosh t, \ \sqrt{2}\sinh t, \ - \sin t, \ \cos t, \ 1 \right),$$

$$N = - \frac{1}{18} \left(15\sqrt{2}\cosh t, \ 15\sqrt{2}\sinh t, \ -7\sin t, \ 7\cos t, \ -1 \right),$$

$$W_1 = \frac{1}{\sqrt{3}} \left(\sqrt{2}\sinh t, \ \sqrt{2}\cosh t, \ - \cos t, \ - \sin t, \ 0 \right),$$

$$W_2 = \frac{1}{\sqrt{3}} \left(\sinh t, \ \cosh t, \ \sqrt{2}\cos t, \ \sqrt{2}\sin t, \ 0 \right),$$

$$W_3 = \frac{1}{3} (\cosh t, \ \sinh t, \ \sin t, \ - \cos t, \ 2\sqrt{2}).$$

Then the Natural Frenet equations (2.1.4) give

$$\nabla_\xi \xi = \sqrt{3}\, W_1, \qquad\qquad \nabla_\xi N = -\frac{37}{18\sqrt{3}}\, W_1 - \frac{4\sqrt{2}}{9\sqrt{3}}\, W_2,$$

$$\nabla_\xi W_1 = \frac{37}{18\sqrt{3}}\, \xi - \sqrt{3}\, N, \quad \nabla_\xi W_2 = \frac{4\sqrt{2}}{9\sqrt{3}}\, \xi - \frac{1}{3\sqrt{3}}(\sqrt{2}+1)\, W_3,$$

$$\nabla_\xi W_3 = \frac{1}{3\sqrt{3}}(\sqrt{2}+1)\, W_2.$$

2.2 Invariance of Frenet frames

In this section we examine the dependence of the Frenet frames and implicitly the Frenet equations (2.1.2) and (2.1.4) on both the transformations of the coordinate neighborhood and the screen vector bundle of C.

First, we fix a screen vector bundle $S(TC^\perp)$ and consider two Frenet frames F and F^* along $\mathcal{U}$ and $\mathcal{U}^*$ respectively with $\mathcal{U} \cap \mathcal{U}^* \neq \emptyset$. Then,

$$\frac{d}{dt^*} = \frac{dt}{dt^*}\frac{d}{dt}, \quad N^* = \frac{dt^*}{dt} N, \quad W_\alpha{}^* = \sum_{\alpha=1}^m A_\alpha^\beta W_\beta, \tag{2.2.1}$$

where $\beta \in \{1, \ldots, m\}$ and A_α^β are smooth functions an $\mathcal{U} \cap \mathcal{U}^*$ and the $m \times m$ matrix $[A_\alpha^\beta(x)]$ is an element of the orthogonal group $O(m)$ for each $x \in \mathcal{U} \cap \mathcal{U}^*$. We call (2.2.1) the *transformation of coordinate neighborhood* of C (the parameter on C). Using above and the first equations of Frenet equations (2.1.2) and (2.1.4) for Frenet frames F and F^*, we obtain

$$\frac{d^2 t}{dt^{*2}} + h\left(\frac{dt}{dt^*}\right)^2 = h^* \frac{dt}{dt^*}, \tag{2.2.2}$$

$$\kappa_1^* A_1^1 = \kappa_1 \left(\frac{dt}{dt^*}\right)^2, \tag{2.2.3}$$

$$\kappa_1^* A_1^2 = \cdots = \kappa_1^* A_1^m = 0. \tag{2.2.4}$$

Using the Natural Frenet equations (2.1.4), we have

Proposition 2.1. *Let C be a null curve of a Lorentz manifold M_1^{m+2} and F, F^* two Natural Frenet frames on $\mathcal{U}$ and $\mathcal{U}^*$ with curvature functions $\{\kappa_1, \ldots, \kappa_{m+1}\}$ and $\{\kappa_1^*, \ldots, \kappa_{m+1}^*\}$ respectively, induced by a fixed screen bundle $S(TC^\perp)$. Let $\mathcal{U} \cap \mathcal{U}^* \neq \emptyset$ and $\prod_{i=1}^{m+1} k_i \neq 0$ on $\mathcal{U} \cap \mathcal{U}^*$. At any point of $\mathcal{U} \cap \mathcal{U}^*$ we have*

$$\kappa_1^* = \kappa_1 A_1 \left(\frac{dt}{dt^*}\right)^2,$$

$$\kappa_2^* = \kappa_2 A_1, \quad \kappa_3^* = \kappa_3 A_2,$$

$$\kappa_\alpha^* = \kappa_\alpha A_{\alpha-1}\frac{dt}{dt^*}, \quad \alpha \in \{4, \ldots, m+1\}, \text{ where } A_i = \pm 1.$$

Hence, κ_2 and κ_3 are invariant functions up to a sign, with respect to the parameter transformations on C.

Proof. First, from (2.2.3) it follows that $\kappa_1^* \neq 0$ on $\mathcal{U} \cap \mathcal{U}^*$ and thus (2.2.4) implies $A_1^2 = \cdots = A_1^m = 0$. Since $\left[A_\alpha^\beta \right]$ is an orthogonal matrix we infer $A_1^1 = A_1 = \pm 1$ and $A_2^1 = \cdots = A_m^1 = 0$. Then, from the second equation of (2.1.4) with respect to F and F^*, and taking into account that $\kappa_3 \neq 0$, obtain $\kappa_3^* \neq 0$ on $\mathcal{U} \cap \mathcal{U}^*$ which implies $A_2^3 = A_2^2 = \cdots = A_2^m = A_m^2 = 0$ and $A_2^2 = A_2 = \pm 1$. Moreover, the second and third equations in proposition also follow. Further on, from the third equation in (2.1.4) we obtain $A_3^4 = A_4^3 = \cdots = A_3^m = A_m^3 = 0$ and $A_3^3 = A_3 = \pm 1$, since $\kappa_5^* \neq 0$ on $\mathcal{U} \cap \mathcal{U}^*$. Besides, it follows that

$$\kappa_4^* = \kappa_1 A_1 A_2 \frac{dt}{dt^*} = \kappa_4 A_3 \frac{dt}{dt^*} ; \qquad A_3 = A_1 A_2$$

and

$$\kappa_5^* = \kappa_5 A_1 A_3^3 \frac{dt}{dt^*} = \kappa_5 A_4 \frac{dt}{dt^*} ; \qquad A_4 = A_1 A_3^3,$$

that is, the first two relations of the last equation in this proposition. In a similar way, by using the remaining equations from (2.1.4) for F and F^* one obtains the other relations in the last equation of this proposition. Then, the last statement follows immediately which completes the proof.

Using (2.1.2) and the method of proposition 2.1, we have

Proposition 2.2. Let C be a null curve of a Lorentz manifold M_1^{m+2} and F, F^* two general Frenet frames on $\mathcal{U}$ and $\mathcal{U}^*$ with curvature functions $\{k_1, \ldots, k_{2m}\}$ and $\{k_1^*, \ldots, k_{2m}^*\}$ respectively, induced by a fixed screen bundle $S(TC^\perp)$. Suppose $\mathcal{U} \cap \mathcal{U}^* \neq \emptyset$ and $\prod_{i=1}^m k_{2i-1} \neq 0$ on $\mathcal{U} \cap \mathcal{U}^*$. Then at any point of $\mathcal{U} \cap \mathcal{U}^*$ we have

$$k_1^* = k_1 A_1 \left(\frac{dt}{dt^*} \right)^2 ,$$

$$k_2^* = k_2 A_1, \quad k_3^* = k_3 A_2,$$

$$k_\alpha^* = k_\alpha A_{\alpha-1} \frac{dt}{dt^*}, \quad \alpha \in \{4, \ldots, 2m\}, \text{ where } A_i = \pm 1.$$

Hence, k_2 and k_3 are invariant functions up to a sign, with respect to the parameter transformations on C.

Next, we study the effect of a different screen for $\bar{F}$ and prove the following:

Proposition 2.3. Let C be a null curve of a Lorentzian manifold M_1^{m+2} and F and $\bar{F}$ be two Natural Frenet frames on $\mathcal{U}$ and $\bar{\mathcal{U}}$ respectively. Suppose $\prod_{i=1}^{m+1} \kappa_i \neq 0$ on $\mathcal{U} \cap \bar{\mathcal{U}} \neq \phi$. Then, their curvature functions are related by

$$\bar{\kappa}_1 = \kappa_1 B_1 \left(\frac{dt}{d\bar{t}} \right)^2 ,$$

$$\bar{\kappa}_2 = \left\{ \kappa_2 + \bar{h}\,c_1 + \frac{dc_1}{d\bar{t}} - \frac{1}{2}\kappa_1\,c_1^2 \left(\frac{dt}{d\bar{t}}\right)^2 \right\} B_1,$$

$$\bar{\kappa}_3 = \kappa_3\,B_2,$$

$$\bar{\kappa}_\alpha = \kappa_\alpha\,B_{\alpha-1}\frac{dt}{d\bar{t}}, \quad \alpha \in \{4, \cdots, m\}, \text{ where } B_i = \pm 1,$$

and $c_2 = \cdots = c_m = 0$. Hence, κ_3 is invariant functions up to a sign, with respect to the transformations of the screen vector bundle of C.

Proof. From the third equation of (2.1.3) and (2.1.4), we have the first equation of this proposition and $\bar{\kappa}_1\,c_\alpha = 0\,(\alpha \neq 1)$ on $\mathcal{U} \cap \bar{\mathcal{U}}$. Thus we have $c_\alpha = 0\,(\alpha \neq 1)$. Also, from the second equation of (2.1.3) and (2.1.4), we have the second equation of proposition and $\bar{\kappa}_3\,B_2^2 = \kappa_3$; $\bar{\kappa}_3\,B_2^\alpha = 0\,(\alpha \geq 3)$. Thus we have $B_2 = B_2^2 = \pm 1$ and $B_2^\alpha = B_\alpha^2 = 0\,(\alpha \geq 3)$. Repeating this process for all other equations of (2.1.4) and set $B_\alpha = B_{\alpha-1}^{\alpha-1}B_\alpha^\alpha\,(\alpha \geq 3)$, we obtain all the relations in proposition, which completes the proof.

Proposition 2.4. *Let C be a null curve of a Lorentz manifold M_1^{m+2} and F, $\bar{F}$ two Frenet frames on $\mathcal{U}$ and $\bar{\mathcal{U}}$ induced by two screen bundle $S(TC^\perp)$ and $\bar{S}(TC^\perp)$ respectively. Then, the vanishing of the first curvature κ_1 on a neighborhood is independent of both the parameter transformations on C and the screen vector bundle transformations. Moreover, it is always possible to find a parameter on C such that C is a null geodesic of M.*

Proof. Using the first equation in (2.1.2) and (2.1.4) for F and $\bar{F}$ we obtain

$$\bar{h} = \frac{d^2t\,d\bar{t}}{d\bar{t}^2\,dt} + h\frac{dt}{d\bar{t}} + \kappa_1 c_1\left(\frac{dt}{d\bar{t}}\right)^2, \tag{2.2.5}$$

$$\bar{\kappa}_1 B_1^1 = \kappa_1\left(\frac{dt}{d\bar{t}}\right)^2, \tag{2.2.6}$$

$$\bar{\kappa}_1 B_1^2 = \cdots = \bar{\kappa}_1 B_1^m = 0. \tag{2.2.7}$$

If $\kappa_1 = 0$ on C, then, $\bar{\kappa}_1 = 0$ on C otherwise we have $B_1^1 = B_1^2 = \cdots = B_1^m = 0$, which is a contradiction as $[B_\alpha^\beta(x)] \in O(m)$. Hence, due to the first equations of the proposition 2.1, 2.2 and 2.3, we say that the vanishing of κ_1 is independent of both the parameter transformations on C and the screen vector bundle transformations which proves the first part. In case $\kappa_1 = 0$ the first Frenet equation becomes

$$\nabla_\xi \xi = h\xi.$$

It is easy to show (see [28, page 58]) that one can always find a special parameter on C such that $h = 0$ so that C is a null geodesic of M, which completes the proof.

Note. Following the terminology in [28], a special parameter, say p, is called a *distinguished parameter* for which $h = 0$ and C is a geodesic null curve in M

if $\kappa_1 = 0$. This distinguished (also called pseudo-arc [37]) parameter plays a role similar to the role of an arc-length parameter for non-null curves.

Theorem 2.5. *Let $C(p)$ be a non-geodesic null curve of a Lorentzian manifold (M_1^{m+2}, g), where p is a distinguished parameter on C. Then there exists a Natural Frenet frame $\{\xi, N, W_1, \ldots, W_m\}$ satisfying the following equations*

$$
\begin{aligned}
\nabla_\xi \xi &= \kappa_1 W_1, \\
\nabla_\xi N &= \kappa_2 W_1 + \kappa_3 W_2, \\
\nabla_\xi W_1 &= -\kappa_2 \xi - \kappa_1 N, \\
\nabla_\xi W_2 &= -\kappa_3 \xi + \kappa_4 W_3, \\
\nabla_\xi W_3 &= -\kappa_4 W_2 + \kappa_5 W_4,
\end{aligned}
\tag{2.2.8}
$$

$$\cdots\cdots\cdots\cdots$$

$$
\begin{aligned}
\nabla_\xi W_i &= -\kappa_{i+1} W_{i-1} + \kappa_{i+2} W_{i+1}, \quad i \in \{3, \ldots, m-1\}, \\
\nabla_\xi W_m &= -\kappa_{m+1} W_{m-1}.
\end{aligned}
$$

Note. Observe that, in particular, (2.2.8) are the Cartan equations if we consider a suitable basis such that, for any distinguished parameter p, the first curvature function $\kappa_1 = 1$. In next section we verify this fact in low dimensions. The condition $\kappa_1 = 1$ was first used by Bonnor [18] followed by others.

Example 4. Consider the null curve in $\mathbf{R}_1^4$ given by

$$C(p) = (\sinh p, \ \cosh p, \ \sin p, \ \cos p), \quad p \in \mathbf{R}.$$

Then the tangent vector $\frac{d}{dp}$ is given by

$$\frac{d}{dp} = (\cosh p, \ \sinh p, \ \cos p, \ -\sin p).$$

If we take $V = (0, \ 0, \ -\cos p, \ \sin p)$, then $g\left(\frac{d}{dp}, V\right) = -1$ and $g(V, V) = 1$. Thus, by (1.4.2) in Theorem 4.1 of chapter 1, we have

$$N = -\frac{1}{2}(\cosh p, \ \sinh p, \ -\cos p, \ \sin p).$$

Since

$$\nabla_{\frac{d}{dp}} \frac{d}{dp} = (\sinh p, \ \cosh p, \ -\sin p, \ -\cos p),$$

$h = g\left(\nabla_{\frac{d}{dp}} \frac{d}{dp}, N\right) = 0$ and $\|\nabla_{\frac{d}{dp}} \frac{d}{dp}\| = \sqrt{2}$, we let

$$W_1 = \frac{1}{\sqrt{2}}(\sinh p, \ \cosh p, \ -\sin p, \ -\cos p).$$

Then the first equation of (2.2.8) is given by $\nabla_{\frac{d}{dp}} \frac{d}{dp} = \sqrt{2}\, W_1$. Also, from

$$\nabla_{\frac{d}{dp}} N = (\sinh p, \ \cosh p, \ \sin p, \ \cos p)$$

and $\kappa_2 = g\left(\nabla_{\frac{d}{dp}} N, W_1\right) = 0$, we get $W_2 = \frac{1}{\sqrt{2}}(\sinh p, \cosh p, \sin p, \cos p)$ and $\nabla_{\frac{d}{dp}} N = -\frac{1}{\sqrt{2}} W_2$. Thus the parameter p is a distinguished parameter.

Example 5. Consider the null curve in $\mathbf{R}_1^4$ given by

$$C(p) = \left(\frac{1}{3}p^3 + 2p, \quad p^2, \quad \frac{1}{3}p^3, \quad 2p\right), \quad p \in \mathbf{R}.$$

Then

$$\frac{d}{dp} = \left(p^2 + 2, \quad 2p, \quad p^2, \quad 2\right).$$

If we take $V = \left(0, 0, 0, \frac{1}{2}\right)$, then $g\left(\frac{d}{dp}, V\right) = 1$ and $g(V, V) = \frac{1}{4}$. Thus,

$$N = -\frac{1}{8}\left(p^2 + 2, \quad 2p, \quad t^p, \quad -2\right).$$

Since $\nabla_{\frac{d}{dp}} \frac{d}{dp} = 2(p, 1, p, 0)$, $h = g\left(\nabla_{\frac{d}{dp}} \frac{d}{dp}, N\right) = 0$ and $\kappa_1 = \left\|\nabla_{\frac{d}{dp}} \frac{d}{dp}\right\| = 2$, we have $W_1 = (p, 1, p, 0)$ and $\nabla_{\frac{d}{dp}} \frac{d}{dp} = 2W_1$. From $\nabla_{\frac{d}{dp}} N = -\frac{1}{4}(p, 1, p, 0)$ and $\kappa_2 = g(\nabla_{\frac{d}{dp}} N, W_1) = -\frac{1}{4}$, we have $\nabla_{\frac{d}{dp}} N = -\frac{1}{4} W_1$ and $k_3 = 0$. Thus the parameter p is a distinguished parameter. Choose $W_2 = \frac{1}{2}(p^2, 2p, p^2 - 2, 0)$, then $\nabla_{\frac{d}{dp}} W_1 = \frac{1}{4}\frac{d}{dp} - 2N - W_2$ and $\nabla_{\frac{d}{dp}} W_2 = W_1$.

2.3   Null Cartan curves in M_1^3, M_1^4 and M_1^5

Consider a null curve C of M_1^{m+2}, together with a Frenet frame F along C with respect to a distinguished parameter p. We call the pair $(C(p), F)$ a *framed null curve* [64]. As we know from previous section, in general $(C(p), F)$ is not unique since it depends on both p and the screen vector bundle $S(TC^\perp)$. To deal with this problem of non-uniqueness, we follow Bonnor [18] and find a *Cartan Frenet frame* with the minimum number of curvature functions which are invariant under Lorentzian transformations. Let us start with the case of a framed null curve $(C(p), F)$ of a 3-dimensional Lorentz Manifold M_1^3, with $F = \left\{\xi = \frac{d}{dp}, N, W\right\}$ whose Frenet equations are

$$
\begin{aligned}
\nabla_\xi \xi &= \kappa_1 W, \\
\nabla_\xi N &= \kappa_2 W, \\
\nabla_\xi W &= -\kappa_2 \xi - \kappa_1 N.
\end{aligned}
\tag{2.3.1}
$$

We know from the proposition 2.4 that the vanishing of κ_1 on a neighborhood is independent of both the parameter transformation on C and $S(TC^\perp)$. According to Graves [45] a framed null curve, with respect to a distinguished parameter, is called a *generalized null cubic curve* if $\kappa_2 = 0$. Since, it follows from Proposition 2.1 and 2.2 that κ_2 is invariant up to a sign, with respect to the parameter transformations on C, the definition of generalized null cubic is independent of the choice of parameterization. If, in addition to $\kappa_2 = 0$, κ_1 is a non-zero constant, then, C is called a *null cubic curve*. Following is an example.

Example 6 ([87]). Let ϕ and ψ be functions satisfying $\phi' = (\psi')^2$. Then

$$C(p) = \left(\frac{1}{\sqrt{2}} \left(p + \frac{\phi(p)}{2} \right), \ \frac{1}{\sqrt{2}} \left(p - \frac{\phi(p)}{2} \right), \ \psi(p) \right)$$

is a null curve with respect to a distinguished parameter p with Frenet frame

$$F = \left\{ \xi = C'(p), \ N = \left(-\frac{1}{\sqrt{2}}, \ \frac{1}{\sqrt{2}}, \ 0 \right), \ W = \left(\frac{\psi'(p)}{\sqrt{2}}, \ -\frac{\psi'(p)}{\sqrt{2}}, \ 1 \right) \right\}.$$

Direct computation shows that the curvature κ and the torsion κ_1 of C are

$$\kappa(p) = \psi''(p), \quad \kappa_1 = 0.$$

Thus, C is a generalized null cubic. In particular, if κ is constant, then

$$\phi(p) = \frac{\kappa^2}{3} p^2 + b\kappa p^2 + b^2 p + c, \quad \psi(p) = \frac{\kappa}{2} p^2 + ap + b,$$

for some constants a, b, c, which shows that C is a cubic curve.

Suppose $\kappa_1 \neq 0$ on $\mathcal{U}$. Then, it follows from (2.2.5) that p is invariant for all $\bar{F}$ obtained from F by (2.1.3) with $c_1 = 0$, which we assume.

Now, consider another Frenet frame $\bar{F} = \left\{ \frac{d}{d\bar{p}}, \ \bar{N}, \ \bar{W}_1 \right\}$ induced by the screen vector bundle $\bar{S}(TC^\perp)$, with parameter $\bar{p}$ on $\bar{\mathcal{U}}$, $\mathcal{U} \cap \bar{\mathcal{U}} \neq \emptyset$. Then from (2.1.3) and $c_1 = 0$, it follows that

$$\frac{d}{d\bar{p}} = \frac{dp}{d\bar{p}} \frac{d}{dp}, \quad \bar{N} = \frac{d\bar{p}}{dt} N, \quad \bar{W}_1 = B_1 W_1,$$

where $B_1 = \pm 1$. By direct calculations, using (2.3.1) and above we obtain

$$\bar{\kappa}_1 = \kappa_1 B_1 \left(\frac{dt}{d\bar{p}} \right)^2, \quad \bar{\kappa}_2 = B_1 \kappa_2. \tag{2.3.2}$$

Consequently, to get a canonical parameter for a non-geodesic $C(p)$, it is logical to assume that the first curvature κ_1 is a non-zero constant. Without loss of generality, assume a parameter p such that $g(C'', C'') = \kappa_1 = 1$. Choose $\xi = C'$ and $W = C''$. It is easy to show that the unique null transversal bundle is generated by

$$N = -C^{(3)} - \kappa_2 C', \quad \text{where} \quad \kappa_2 = \frac{1}{2} g(C^{(3)}, C^{(3)}).$$

Thus, labeling $\kappa_1 = 1$ and $\kappa_2 = \tau$, we have proved the following:

Theorem 3.1. *Let $C(p)$ be a non-geodesic null curve of a Lorentz manifold (M_1^3, g) where p is a distinguished parameter on C. Consider a basis $E = \{C', C'', C^{(3)}\}$ of $T_{C(p)}M_1^3$ for all p, such that $g(C'', C'') = 1$. Then, there exists only one Frenet frame $F = \{\xi, N, W\}$ for which $(C(p), F)$ is a framed null curve with Frenet equations*

$$\begin{aligned}
\nabla_\xi \xi &= W, \\
\nabla_\xi N &= \tau W, \\
\nabla_\xi W &= -\tau \xi - N.
\end{aligned} \tag{2.3.3}$$

where F and E have the same positive orientation. Moreover, the torsion function τ is invariant upto a sign, under Lorentzian transformations.

Note. Following Bonnor [18] we call the equations (2.3.3) the *Cartan Frenet equations* and $(C(p), F)$ is its *null Cartan curve*.

Example 7. Let C be a null curve of $\mathbf{R}_1^3$ defined by

$$t = \frac{1}{2}p\sqrt{p^2+1} + \frac{1}{2}\ln\left|\sqrt{p^2+1}+p\right|, \quad x = \frac{1}{2}p^2, \quad y = p,$$

with respect to local coordinates (t, x, y) and a distinguished parameter p. Choose a Frenet frame $F = \{\xi, N, W\}$ as follows:

$$\xi = \left(\sqrt{p^2+1}, \; p, \; 1\right), \quad W = \left(\frac{p}{\sqrt{p^2+1}}, \; 1, \; 0\right),$$

$$N = \left(-\frac{1}{2}\sqrt{p^2+1}, \; \frac{1}{2}p, \; \frac{1}{2}\right).$$

Thus, using the Frenet equations (2.3.3), we obtain $\kappa = 1$ and $\tau = -\frac{1}{2}$. Therefore, $(C(p), F)$ is a null Cartan curve of $\mathbf{R}_1^3$.

The null Cartan curves satisfying theorem 3.1 are completely determined only by a single curvature (also called torsion) function and it is considered as a *canonical representation* for non-geodesic null curves in M_1^3, which has been very effective in the study of null curves. In particular, Honda-Inoguchi [57] have recently done some work on null curves in Minkowski 3-space $\mathbf{R}_1^3$, which we discuss in chapter 5 (section 2) of this book.

Now let $(C(p), F)$ be a framed null curve of a 4-dimensional Lorentzian manifold M_1^4, with a general Frenet frame $F = \left\{\xi = \frac{d}{dp}, N, W_1, W_2\right\}$ and Frenet equations

$$\begin{aligned}
\nabla_\xi \xi &= \kappa_1 W_1, \\
\nabla_\xi N &= \kappa_2 W_1 + \kappa_3 W_2, \\
\nabla_\xi W_1 &= -\kappa_2 \xi - \kappa_1 N + \kappa_4 W_2, \\
\nabla_\xi W_2 &= -\kappa_3 \xi - \kappa_4 W_1,
\end{aligned} \tag{2.3.4}$$

where p is a distinguished parameter. Recall from proposition 2.2 that κ_2 and κ_3 are invariant up to a sign, with respect to the parameter transformations on C. Suppose $\kappa_1 \neq 0$ on $\mathcal{U}$. Then, as per (2.2.5), p is invariant for all $\bar{F}$ obtained from F by (2.1.3) with $c_1 = 0$ which we assume for this case also.

Now consider another general Frenet frame $\bar{F} = \left\{ \frac{d}{d\bar{p}}, \bar{N}, \bar{W}_1, \bar{W}_2 \right\}$ induced by the screen vector bundle $\bar{S}(TC^\perp)$, with parameter $\bar{p}$ on $\bar{\mathcal{U}}$, $\mathcal{U} \cap \bar{\mathcal{U}} \neq \emptyset$. Then, it follows from (2.2.6) and (2.2.7) that $B_1^2 = B_2^1 = 0$; $B_1^1 = B_1 = \pm 1$; $B_2^2 = B_2 = \pm 1$. Therefore, by (2.1.3) and $c_1 = 0$, we obtain

$$\frac{d}{d\bar{p}} = \frac{dp}{d\bar{p}} \frac{d}{dp}$$

$$\bar{N} = -\frac{1}{2} \frac{dp}{d\bar{p}} (c_2)^2 \frac{d}{dp} + \frac{d\bar{p}}{dp} N + c_2 W_2$$

$$\bar{W}_1 = B_1 W_1; \quad \bar{W}_2 = B_2 \left(W_2 - \frac{dp}{d\bar{p}} c_2 \frac{d}{dp} \right).$$

By direct calculations, using (2.3.4) for both frames and above we obtain

$$\bar{\kappa}_1 = B_1 \left(\frac{dp}{d\bar{p}} \right)^2 \kappa_1$$

$$\bar{\kappa}_2 = B_1 \left\{ \kappa_2 - \frac{\kappa_1}{2} (c_2)^2 \frac{dp}{d\bar{p}} - c_2 \frac{dp}{d\bar{p}} \kappa_4 \right\}$$

$$\bar{\kappa}_3 = B_2 \left\{ \kappa_3 + \frac{dc_2}{d\bar{p}} \right\} \tag{2.3.5}$$

$$\bar{\kappa}_4 = B_1 B_2 \left\{ \kappa_4 + c_2 \frac{dp}{d\bar{p}} \kappa_1 \right\} \frac{dp}{d\bar{p}}.$$

Proposition 3.1. *Let $(C(p), F)$ be a non-geodesic framed null curve of a Lorentzian manifold (M_1^4, g) where p is a distinguished parameter on C and $F = \left\{ \frac{d}{dp}, N, W_1, W_2 \right\}$ is a general Frenet frame on $\mathcal{U} \subset C$ with respect to $S(TC^\perp)$. Then there exists a screen vector bundle $\bar{S}(TC^\perp)$ which induces a Frenet frame $\bar{F} = \left\{ \frac{d}{dp}, \bar{N}, \bar{W}_1, \bar{W}_2 \right\}$ on $\mathcal{U}$ such that $\bar{\kappa}_4 = 0$ on $\mathcal{U}$.*

Proof. Define the following vector fields in terms of the elements of F on $\mathcal{U}$:

$$\bar{N} = -\frac{1}{2} \left(\frac{\kappa_4}{\kappa_1} \right)^2 \frac{d}{dp} + N - \frac{\kappa_4}{\kappa_1} W_2$$

$$\bar{W}_1 = W_1; \quad \bar{W}_2 = W_2 + \frac{\kappa_4}{\kappa_1} \frac{d}{dp}. \tag{2.3.6}$$

Let $\mathcal{U}^* \subset C$ be another coordinate neighborhood with parameter p^* on C and $\mathcal{U} \cap \mathcal{U}^* \neq \emptyset$. Using (2.3.5) we obtain

$$\kappa_1^* = \kappa_1 A_1 \left(\frac{dp}{dp^*} \right)^2 ; \quad \kappa_4^* = \kappa_4 A_3 \frac{dp}{dp^*}$$

$$W_1^* = A_1 W_1 ; \qquad W_2^* = A_2 W_2 , \tag{2.3.7}$$

on $\mathcal{U} \cap \mathcal{U}^* \neq \emptyset$. Define $\{\bar{N}^*, W_1^*, \bar{W}_2^*\}$ by (2.3.6) but on $\mathcal{U}^*$ and with respect to $F^* = \{\frac{d}{dp^*}, N^*, W_1^*, W_2^*\}$, induced by the same $S(TC^\perp)$ on $\mathcal{U}^*$. Then by using (2.3.6) and (2.3.7) we obtain

$$\bar{N}^* = \frac{dp^*}{dp} \bar{N}; \quad \bar{W}_1^* = A_1 \bar{W}_1; \quad \bar{W}_2^* = A_2 W_2.$$

Hence there exists a vector bundle $\bar{S}(TC^\perp)$ spanned on $\mathcal{U}$ by $\{\bar{W}_1, \bar{W}_2\}$ given by (2.3.6). Moreover, it is easy to check that $\bar{S}(TC^\perp)$ is complementary to TC in $TC^\perp$. The null transversal vector bundle constructed in theorem 4.1 of chapter 1, but with respect to $\bar{S}(TC^\perp)$, is locally represented by $\bar{N}$ from (2.3.6). Finally, taking into account that $\bar{p} = p$ and $c_2 = -\frac{\kappa_4}{\kappa_1}$ in the last formula of (2.3.5), we obtain $\bar{\kappa}_4 = 0$, which completes the proof.

Note. Proposition 3.1 also holds for any null curve of M_1^4 with respect to a general parameter t (see [28, page 61]).

Consequently, as in the case of M_1^3, to get a canonical parameter for a non-geodesic $C(p)$, we assume $k_1 = 1$ so that if $\xi = C'$, then, $g(C'', C'') = 1$. Choose $W = C''$. It is easy to show that the unique null transversal bundle is given by

$$N = -C^{(3)} - \kappa_2 C', \quad \text{where} \quad \kappa_2 = \frac{1}{2} g(C^{(3)}, C^{(3)}).$$

Thus, labeling $\kappa_2 = \kappa_1$ and $\kappa_3 = \kappa_2$ we have proved the following:

Theorem 3.2. *Let $C(p)$ be a non-geodesic null curve of a Lorentzian manifold (M_1^4, g) where p is a distinguished parameter on C. Consider a basis $E = \{C', C'', C^{(3)}, C^{(4)}\}$ of $T_{C(p)} M_1^4$ for all p, such that $g(C'', C'') = 1$. Then, there exists only one Frenet frame $F = \{\xi, N, W_1, W_2\}$ for which $(C(p), F)$ is a null Cartan curve with Cartan Frenet equations*

$$\begin{aligned}
\nabla_\xi \xi &= W_1, \\
\nabla_\xi N &= \kappa_1 W_1 + \kappa_2 W_2, \\
\nabla_\xi W_1 &= -\kappa_1 \xi - N, \\
\nabla_\xi W_2 &= -\kappa_2 \xi,
\end{aligned} \tag{2.3.8}$$

where F and E have the same positive orientation. Moreover, the curvature functions κ_1 and κ_2 is invariant upto a sign, under Lorentzian transformations.

Example 8. Consider a null curve C in $\mathbf{R}_1^4$ given by the equation

$$C(p) = \left(\frac{1}{6}p^3 + p, \ \frac{1}{2}p^2, \ \frac{1}{6}p^3, \ p \right), \quad p \in \mathbf{R}.$$

Then

$$\xi = \left(\frac{1}{2}p^2 + 1, \ p, \ \frac{1}{2}p^2, \ 1 \right).$$

Let $N = -(1, 0, 1, 0)$, $W_1 = (p, 1, p, 0)$ and $W_2 = (1, 0, 1, 1)$, then the Frenet equations (2.3.8) give $\kappa_1 = \kappa_2 = 0$, that is,

$$\nabla_\xi \xi = W_1, \quad \nabla_\xi N = 0, \quad \nabla_\xi W_1 = -N, \quad \nabla_\xi W_2 = 0.$$

Example 9. Consider a null curve C and its Frenet frames in $\mathbf{R}_1^4$ given by

$$C(p) = \frac{1}{\sqrt{2}}(\sinh p, \quad \cosh p, \quad \sin p, \quad \cos p), \quad p \in \mathbf{R};$$

$$\xi = \frac{1}{\sqrt{2}}(\cosh p, \quad \sinh p, \quad \cos p, \quad -\sin p),$$

$$N = \frac{1}{\sqrt{2}}(-\cosh p, \quad -\sinh p, \quad \cos p, \quad -\sin p),$$

$$W_1 = \frac{1}{\sqrt{2}}(\sinh p, \quad \cosh p, \quad -\sin p, \quad -\cos p),$$

$$W_2 = \frac{1}{\sqrt{2}}(\sinh p, \quad \cosh p, \quad \sin p, \quad \cos p),$$

then $\nabla_\xi \xi = W_1, \quad \nabla_\xi N = W_2, \quad \nabla_\xi W_1 = -N, \quad \nabla_\xi W_2 = -\xi; \quad \kappa_1 = 0.$

In particular, Bonnor [18] studied the fundamental existence and congruence theorems of null Cartan curves in a Minkowski 4-space. Recently, Cöken-Ciftci [22] have done some work on null Cartan curves (see chapter 5).

For the case of framed null curves in M_1^5, with a distinguished parameter p and $\kappa_1 = 1$, using a similar argument one can show that $\kappa_4 = \kappa_5 = 0$. Then, labeling $\kappa_2 = \kappa_1$, $\kappa_3 = \kappa_2$ and $\kappa_6 = \kappa_3$ one can prove the following:

Theorem 3.3. Let $C(p)$ be a non-geodesic null curve of a Lorentzian manifold (M_1^5, g). Consider a basis $E = \{C', C'', C^{(3)}, C^{(4)}, C^{(5)}\}$ of $T_{C(p)} M_1^5$ for all p, such that $g(C'', C'') = 1$. Then, there exists only one Frenet frame $F = \{\xi, N, W_1, W_2, W_3\}$ for which $(C(p), F)$ is a null Cartan curve with Cartan Frenet equations

$$
\begin{aligned}
\nabla_\xi \xi &= W_1, \\
\nabla_\xi N &= \kappa_1 W_1 + \kappa_2 W_2, \\
\nabla_\xi W_1 &= -\kappa_1 \xi - N, \\
\nabla_\xi W_2 &= -\kappa_2 \xi + \kappa_3 W_3, \\
\nabla_\xi W_3 &= -\kappa_3 W_2,
\end{aligned}
\tag{2.3.9}
$$

where F and E have the same positive orientation. Moreover, the curvature functions $\{\kappa_1, \kappa_2, \kappa_3\}$ are invariant upto a sign, under Lorentzian transformations.

Example 10. Consider a null curve C and its Frenet frames in $\mathbf{R}_1^5$ given by

$$C(p) = \frac{1}{2\sqrt{2}}\left(\frac{1}{3}p^3 + 2p, \quad p^2, \quad \frac{1}{3}p^3, \quad 2\cos p, \quad 2\sin p\right), \quad p \in \mathbf{R};$$

$$\xi = \frac{1}{2\sqrt{2}}\left(p^2 + 2,\ 2p,\ p^2,\ -2\sin p,\ 2\cos p\right),$$

$$N = -\frac{1}{8\sqrt{2}}\left(p^2 + 10,\ 2p,\ p^2 + 8,\ 6\sin p,\ -6\cos p\right),$$

$$W_1 = \frac{1}{\sqrt{2}}\left(p,\ 1,\ p,\ -\cos p,\ -\sin p\right),$$

$$W_2 = -\frac{1}{\sqrt{2}}\left(p,\ 1,\ p,\ \cos p,\ \sin p\right),$$

$$W_3 = \frac{1}{4}\left(p^2 - 2,\ 2p,\ p^2 - 4,\ 2\sin p,\ -2\cos p\right).$$

Then the Frenet equations (2.3.9) give

$$\nabla_\xi \xi = W_1,\quad \nabla_\xi N = \frac{1}{4}W_1 + \frac{1}{2}W_2,\quad \nabla_\xi W_1 = -\frac{1}{4}\xi - N,$$

$$\nabla_\xi W_2 = -\frac{1}{2}\xi + \frac{1}{\sqrt{2}}W_3,\quad \nabla_\xi W_3 = -\frac{1}{\sqrt{2}}W_2.$$

Example 11. Consider a null curve C and its Frenet frames in $\mathbf{R}_1^5$ given by

$$C(p) = \frac{1}{\sqrt{3}}\left(\sqrt{2}\sinh p,\ \sqrt{2}\cosh p,\ \sin p,\ \cos p,\ p\right),\quad p \in \mathbf{R};$$

$$\xi = \frac{1}{\sqrt{3}}\left(\sqrt{2}\cosh p,\ \sqrt{2}\sinh p,\ \cos p,\ -\sin p,\ 1\right),$$

$$N = -\frac{1}{6\sqrt{3}}\left(5\sqrt{2}\cosh p,\ 5\sqrt{2}\sinh p,\ -7\cos p,\ 7\sin p,\ -1\right),$$

$$W_1 = \frac{1}{\sqrt{3}}\left(\sqrt{2}\sinh p,\ \sqrt{2}\cosh p,\ -\sin p,\ -\cos p,\ 0\right),$$

$$W_2 = \frac{1}{\sqrt{6}}\left(\sqrt{2}\sinh p,\ \sqrt{2}\cosh p,\ 2\sin p,\ 2\cos p,\ 0\right),$$

$$W_3 = -\frac{1}{3\sqrt{2}}\left(\sqrt{2}\cosh p,\ \sqrt{2}\sinh p,\ -2\cos p,\ 2\sin p,\ 4\right).$$

Then,

$$\nabla_\xi \xi = W_1,\quad \nabla_\xi N = -\frac{1}{6}W_1 - \frac{4}{3\sqrt{2}}W_2,\quad \nabla_\xi W_1 = \frac{1}{6}\xi - N,$$

$$\nabla_\xi W_2 = \frac{4}{3\sqrt{2}}\xi + \frac{\sqrt{2}}{3}W_3,\quad \nabla_\xi W_3 = -\frac{\sqrt{2}}{3}W_2.$$

2.4 Null Cartan curves in M_1^6

Consider a null curve C $(\kappa_1 \neq 0)$ of a 6-dimensional Lorentz manifold M_1^6. Let $F = \{\frac{d}{dt},\ N,\ W_1,\ W_2,\ W_3,\ W_4\}$ and $\bar{F} = \{\frac{d}{dt},\ \bar{N},\ \bar{W}_1,\ \bar{W}_2,\ \bar{W}_3,\ \bar{W}_4\}$ be two

Frenet frames with respect to $(t, S(TC^\perp), \mathcal{U})$ and $(\bar{t}, \bar{S}(TC^\perp), \bar{\mathcal{U}})$ respectively, where $\mathcal{U} \cap \bar{\mathcal{U}} \neq \emptyset$. Then, following are the transformation equations:

$$\frac{d}{d\bar{t}} = \frac{dt}{d\bar{t}} \frac{d}{dt},$$

$$\bar{N} = -\frac{1}{2} \frac{d}{d\bar{t}} \sum_{\alpha=1}^{4} (c_\alpha)^2 \frac{d}{dt} + \frac{d\bar{t}}{dt} N + \sum_{\alpha=1}^{4} c_\alpha W_\alpha, \qquad (2.4.1)$$

$$\bar{W}_\alpha = B_\alpha \left(W_\alpha - \frac{dt}{d\bar{t}} c_\alpha \frac{d}{dt} \right), \quad B_\alpha = B_\alpha^\alpha = \pm 1,$$

where $\alpha \in \{1, 2, 3, 4\}$. We call this transformation as *diagonal transformation*.

Lemma 4.1. Let C be a null curve of M_1^6 and F and $\bar{F}$ be two general Frenet frames of related by (2.4.1) such that $k_1 \neq 0$. Then $c_4 = 0$ and the curvature functions with respect to F and $\bar{F}$ are related by

$$\bar{\kappa}_1 = B_1 \kappa_1 \left(\frac{dt}{d\bar{t}} \right)^2,$$

$$\bar{\kappa}_2 = B_1 \left\{ \kappa_2 + \bar{h}c_1 + \frac{dc_1}{d\bar{t}} - \frac{1}{2} \kappa_1 \left(\frac{dt}{d\bar{t}} \right)^2 \sum_{\alpha=1}^{4} (c_\alpha)^2 - (c_2\kappa_4 + c_3\kappa_5) \frac{dt}{d\bar{t}} \right\},$$

$$\bar{\kappa}_3 = B_2 \left\{ \kappa_3 + \bar{h}c_2 + \frac{dc_2}{d\bar{t}} + (c_1\kappa_4 - c_3\kappa_6 - c_4\kappa_7) \frac{dt}{d\bar{t}} \right\},$$

$$\bar{\kappa}_4 = B_1 B_2 \left\{ \kappa_4 + c_2 \frac{dt}{d\bar{t}} \kappa_1 \right\} \frac{dt}{d\bar{t}}, \qquad (2.4.2)$$

$$\bar{\kappa}_5 = B_1 B_3 \left\{ \kappa_5 + c_3 \frac{dt}{d\bar{t}} \kappa_1 \right\} \frac{dt}{d\bar{t}},$$

$$\bar{\kappa}_\alpha = B_{\alpha-1} \kappa_\alpha \frac{dt}{d\bar{t}}, \quad \alpha \in \{6, 7, 8\},$$

where $B_5 = B_2 B_3$, $B_6 = B_2 B_4$, $B_7 = B_3 B_4$ and $B_\alpha = \pm 1$.

Proof. The relations (2.4.2) follow by straightforward calculations from (2.1.2) when written for both general Frenet frames and by the use of (2.4.1).

Theorem 4.1. Let C be a null curve of M_1^6 with general Frenet frame satisfying $\kappa_1 \neq 0$. Then there exist a 2-surface S which is invariant with respect to both the parameter transformations on C and the diagonal screen vector bundle transformations.

Proof. Let $I(C)$ be an integral curve of the vector field W_4. Since $c_4 = 0$ the surface $S = C \times I(C)$ is always invariant with respect to the diagonal screen vector bundle transformations.

Remark. Every tangent space T_xS, $x \in S$ in above theorem always has basis which are made up of one null and other spacelike. This means $T_xS \cap \Lambda = L - \{0\}$, where L is a 1-dimensional space and Λ the null cone of M_1^6. Hence each T_xS, $x \in S$ is a lightlike vector subspace of Lorentz vector space [82, p. 42].

Theorem 4.2. *Let C be a null curve of M_1^6 and $S(TC^\perp)$ be a screen vector bundle of C. Suppose F is a general Frenet frame on $\mathcal{U} \subset C$ with respect to $S(TC^\perp)$ such that $\kappa_1 \neq 0$ on $\mathcal{U}$. Then there exists a screen vector bundle $\bar{S}(TC^\perp)$ which induces a Frenet frame $\bar{F}$ on $\mathcal{U}$ such that $\bar{\kappa}_4 = 0$ or $\bar{\kappa}_5 = 0$.*

Proof. Define the following vector fields in terms of elements of F on $\mathcal{U}$:

$$\bar{N} = -\frac{1}{2}\left(\frac{\kappa_4}{\kappa_1}\right)^2 \frac{d}{dt} + N - \frac{\kappa_4}{\kappa_1}W_2$$

$$\bar{W}_2 = W_2 + \frac{\kappa_4}{\kappa_1}\frac{d}{dt} \qquad\qquad (2.4.3)$$

$$\bar{W}_i = W_i, \ \ i \in \{1, 3, 4\}.$$

Or, alternatively,

$$\bar{N} = -\frac{1}{2}\left(\frac{\kappa_5}{\kappa_1}\right)^2 \frac{d}{dt} + N - \frac{\kappa_5}{\kappa_1}W_3$$

$$\bar{W}_3 = W_3 + \frac{\kappa_5}{\kappa_1}\frac{d}{dt} \qquad\qquad (2.4.4)$$

$$\bar{W}_i = W_i, \ \ i \in \{1, 2, 4\}.$$

Let $\mathcal{U}^* \subset C$ be another coordinate neighborhood with parameter t^* on C and $\mathcal{U} \cap \mathcal{U}^*$. By Proposition 2.2 it follows

$$\kappa_1{}^* = \kappa_1 A_1 \left(\frac{dt}{dt^*}\right)^2 ;$$

$$\kappa_4{}^* = \kappa_4 A_3 \frac{dt}{dt^*} \left[\text{or } \kappa_5{}^* = \kappa_5 A_4 \frac{dt}{dt^*}\right] ; \qquad (2.4.5)$$

$$W_i{}^* = A_i W_i, \ \ i \in \{1, 2, 3, 4\}$$

on $\mathcal{U} \cap \mathcal{U}^*$. Define $\{\bar{N}^*, \bar{W}_1^*, \bar{W}_2^*, \bar{W}_3^*, \bar{W}_4^*\}$ by (2.4.3) [or (2.4.4)] but on $\mathcal{U}^*$ with respect to $F^* = \{\frac{d}{dt^*}, N^*, W_1{}^*, W_2{}^*, W_3{}^*, W_4{}^*\}$, induced by the same $S(TC^\perp)$ on $\mathcal{U}^*$. Then by using (2.2.1), (2.4.3) [or (2.4.4)] and (2.4.5) we obtain

$$\bar{N}^* = \frac{dt^*}{dt}\bar{N}, \ \ \bar{W}_i^* = A_i\bar{W}_i, \ \ i \in \{1, 2, 3, 4\}.$$

Hence there exists a vector bundle $\bar{S}(TC^\perp)$ spanned by $\{\bar{W}_1, \bar{W}_2, \bar{W}_3, \bar{W}_4\}$ on $\mathcal{U}$ given by (2.4.3) [or (2.4.4)]. Moreover, it is easy to check that $\bar{S}(TC^\perp)$ is complementary to TC in $TC^\perp$. The null transversal vector bundle constructed with respect to $(TC^{\perp})$, is locally represented by $\bar{N}$ from (2.4.3) [or (2.4.4)]. Finally, taking into account that $t = \bar{t}$ and $c_2 = -\frac{\kappa_4}{\kappa_1}$ [or $c_3 = -\frac{\kappa_5}{\kappa_1}$] in the fourth [or fifth] formula of (2.4.2), one obtains $\bar{\kappa}_4 = 0$ [or $\bar{\kappa}_5 = 0$].

Next, we consider the general case such that the transformation of Frenet fields of frames (2.4.1) becomes

$$\frac{d}{d\bar{t}} = \frac{dt}{d\bar{t}}\frac{d}{dt},$$

$$\bar{N} = -\frac{1}{2}\frac{d}{d\bar{t}}\sum_{\alpha=1}^{4}(c_\alpha)^2\frac{d}{dt} + \frac{d\bar{t}}{dt}N + \sum_{\alpha=1}^{4}c_\alpha W_\alpha,$$

$$\bar{W}_1 = B_1\left(W_1 - \frac{d}{d\bar{t}}c_1\frac{d}{dt}\right),$$ (2.4.6)

$$\bar{W}_\alpha = \sum_{\beta=2}^{4}B_\alpha^\beta\left(W_\beta - \frac{dt}{d\bar{t}}c_\beta\frac{d}{dt}\right), \quad \alpha, \beta \in \{2, 3, 4\}.$$

By the method of lemma 4.1 we have

Lemma 4.2. *Let C be a null curve of M_2^6 and F and $\bar{F}$ be two general Frenet frames of related by (2.4.6) such that $\kappa_1 \neq 0$. Then the curvature functions with respect to F and $\bar{F}$ are related by*

$$\bar{\kappa}_1 = \kappa_1 B_1\left(\frac{dt}{d\bar{t}}\right)^2,$$

$$\bar{\kappa}_2 = B_1\left\{\kappa_2 + \bar{h}c_1 + \frac{dk_1}{d\bar{t}} - \frac{1}{2}\kappa_1\left(\frac{dt}{d\bar{t}}\right)^2\sum_{\alpha=1}^{4}(c_\alpha)^2 - (c_2\kappa_4 + c_3\kappa_5)\frac{dt}{d\bar{t}}\right\},$$

$$\bar{\kappa}_3 = \kappa_3 B_2^2 + \bar{h}\sum_{\alpha=2}^{4}B_2^\alpha c_\alpha + \sum_{\alpha=2}^{4}B_2^\alpha\frac{dc_\alpha}{dt} + (c_1\kappa_4 - c_3\kappa_6 - c_4\kappa_7)B_2^2\frac{dt}{d\bar{t}}$$

$$+ \quad (c_1\kappa_5 + c_2\kappa_6 - c_4\kappa_8)B_2^3\frac{dt}{d\bar{t}} + (c_2\kappa_7 + c_3\kappa_8)B_2^4\frac{dt}{d\bar{t}},$$

$$\bar{\kappa}_4 = B_1\left\{B_2^2\left(\kappa_4 + \kappa_1\frac{dt}{d\bar{t}}c_2\right) + B_2^3\left(\kappa_5 + \kappa_1\frac{dt}{d\bar{t}}c_3\right) + B_2^4\kappa_1\frac{dt}{d\bar{t}}c_4\right\}\frac{dt}{d\bar{t}},$$

$$\bar{\kappa}_5 = B_1\left\{B_3^2\left(\kappa_4 + \kappa_1\frac{dt}{d\bar{t}}c_2\right) + B_3^3\left(\kappa_5 + \kappa_1\frac{dt}{d\bar{t}}c_3\right) + B_3^4\kappa_1\frac{dt}{d\bar{t}}c_4\right\}\frac{dt}{d\bar{t}},$$

$$\bar{\kappa}_6 = \begin{vmatrix} B_2^2 & B_2^3 & k_8 \\ B_2^3 & B_3^3 & -\kappa_7 \\ B_2^4 & B_3^4 & \kappa_6 \end{vmatrix}\frac{dt}{d\bar{t}} + \sum_{\alpha=2}^{4}B_3^\alpha\frac{dB_2^\alpha}{d\bar{t}},$$

$$\bar{\kappa}_7 = \begin{vmatrix} B_2^2 & -\kappa_8 & B_4^2 \\ B_2^3 & \kappa_7 & B_4^3 \\ B_2^4 & -\kappa_6 & B_4^4 \end{vmatrix}\frac{dt}{d\bar{t}} + \sum_{\alpha=2}^{4}B_4^\alpha\frac{dB_2^\alpha}{d\bar{t}},$$

$$\bar{\kappa}_8 = \left| \left(\begin{array}{ccc} \kappa_8 & B_3^2 & B_4^2 \\ -\kappa_7 & B_3^3 & B_4^3 \\ \kappa_6 & B_3^4 & B_4^4 \end{array} \right) \right| \frac{dt}{d\bar{t}} + \sum_{\alpha=2}^{4} B_4^\alpha \frac{dB_3^\alpha}{d\bar{t}}.$$

By the method of theorem 4.2 and using the following equations

$$\bar{N} = -\frac{1}{2}\left(\frac{\kappa_4^2 + \kappa_5^2}{\kappa_1^2}\right)\frac{d}{dt} + N - \frac{\kappa_4}{\kappa_1}W_2 - \frac{\kappa_5}{\kappa_1}W_3,$$

$$\bar{W}_1 = W_1, \quad \bar{W}_2 = W_2 + \frac{\kappa_4}{\kappa_1}\frac{d}{dt},$$

$$\bar{W}_3 = W_3 + \frac{\kappa_5}{\kappa_1}\frac{d}{dt}, \quad \bar{W}_4 = W_4,$$

instead of (2.4.3) and (2.4.4) in theorem 4.2, we obtain

Theorem 4.3. *Let C be a null curve of M_1^6 and $S(TC^\perp)$ be a screen vector bundle of C. Suppose F is a general Frenet frame on $\mathcal{U} \subset C$ with respect to $S(TC^\perp)$ such that $\kappa_1 \neq 0$. Then there exists a screen vector bundle $\bar{S}(TC^\perp)$ which induces a Frenet frame $\bar{F}$ on $\mathcal{U}$ such that $\bar{\kappa}_4 = \bar{\kappa}_5 = 0$.*

Example 12. Consider a null curve C in $\mathbf{R}_1^6$ given by the equation

$$C(t) = \left(\frac{1}{3}t^3 + 2t, \ t^2, \ \frac{1}{3}t^3, \ \sqrt{2}t, \ \cos\sqrt{2}t, \ \sin\sqrt{2}t \right), \quad t \in \mathbf{R}.$$

Then,

$$\frac{d}{dt} = \left(t^2 + 2, \ 2t, \ t^2, \ \sqrt{2}, \ -\sqrt{2}\sin\sqrt{2}t, \ \sqrt{2}\cos\sqrt{2}t \right).$$

If we take $V = \left(0, \ 0, \ 0, \ 0, \ -\frac{1}{\sqrt{2}}\sin\sqrt{2}t, \ \frac{1}{\sqrt{2}}\cos\sqrt{2}t \right)$, then, $g\left(\frac{d}{dt}, V\right) = 1$ and $g(V, V) = \frac{1}{2}$. Thus, by the equation (1.4.2) in theorem 4.1 of chapter 1, we have

$$N = -\frac{1}{4}\left(t^2 + 2, \ 2t, \ t^2, \ \sqrt{2}, \ \sqrt{2}\sin\sqrt{2}t, \ -\sqrt{2}\cos\sqrt{2}t \right).$$

Since

$$\nabla_{\frac{d}{dt}}\frac{d}{dt} = 2\left(t, \ 1, \ t, \ 0, \ -\cos\sqrt{2}t, \ -\sin\sqrt{2}t \right),$$

$h = g\left(\nabla_{\frac{d}{dt}}\frac{d}{dt}, N\right) = 0$ and $\kappa_1 = \|\nabla_{\frac{d}{dt}}\frac{d}{dt}\| = 2\sqrt{2}$, we let

$$W_1 = \frac{1}{\sqrt{2}}\left(t, \ 1, \ t, \ 0, \ -\cos\sqrt{2}t, \ -\sin\sqrt{2}t \right).$$

Then we have $\nabla_{\frac{d}{dt}}\frac{d}{dt} = 2\sqrt{2}W_1$. From

$$\nabla_{\frac{d}{dt}}N = -\frac{1}{2}\left(t, \ 1, \ t, \ 0, \ \cos\sqrt{2}t, \ \sin\sqrt{2}t \right)$$

and $\kappa_2 = g(\nabla_{\frac{d}{dt}} N, W_1) = 0$, let

$$W_2 = -\frac{1}{\sqrt{2}}\left(t,\ 1,\ t,\ 0,\ \cos\sqrt{2}t,\ \sin\sqrt{2}t \right).$$

Then we have $\nabla_{\frac{d}{dt}} N = \frac{1}{\sqrt{2}} W_2$ and $\kappa_3 = \frac{1}{\sqrt{2}}$. Also, from

$$\nabla_{\frac{d}{dt}} W_1 = \frac{1}{\sqrt{2}}\left(1,\ 0,\ 1,\ 0,\ \sqrt{2}\sin\sqrt{2}t,\ -\sqrt{2}\cos\sqrt{2}t \right),$$

$\kappa_4 = g(\nabla_{\frac{d}{dt}} W_1, W_2) = 0$ and

$$\nabla_{\frac{d}{dt}} W_1 + 2\sqrt{2}\,N = -\frac{1}{\sqrt{2}}\left(t^2 + 1,\ 2t,\ t^2 - 1,\ \sqrt{2},\ 0,\ 0 \right),$$

we let

$$W_3 = -\frac{1}{\sqrt{2}}\left(t^2 + 1,\ 2t,\ t^2 - 1,\ \sqrt{2},\ 0,\ 0 \right).$$

Then we have $\nabla_{\frac{d}{dt}} W_1 = 2\sqrt{2}\,N + W_3$ and $\kappa_5 = 1$. Since

$$\nabla_{\frac{d}{dt}} W_2 = -\frac{1}{\sqrt{2}}\left(1,\ 0,\ 1,\ 0,\ -\sqrt{2}\sin\sqrt{2}t,\ \sqrt{2}\cos\sqrt{2}t \right)$$

and $\kappa_6 - g(\nabla_{\frac{d}{dt}} W_2, W_3) = -1$, then $\nabla_{\frac{d}{dt}} W_2 + \frac{1}{\sqrt{2}}\frac{d}{dt} + W_3 = 0$ and $\kappa_7 = 0$.

By the same method, from

$$\nabla_{\frac{d}{dt}} W_3 = -\sqrt{2}\,(t,\ 1,\ t,\ 0,\ 0,\ 0),$$

we have $\nabla_{\frac{d}{dt}} W_3 + W_1 - W_2 = 0$ and $\kappa_8 = 0$. Choose

$$W_4 = \frac{1}{\sqrt{2}}\left(1,\ 0,\ 1,\ \sqrt{2},\ 0,\ 0 \right).$$

Then $\{W_1, W_2, W_3, W_4\}$ is an orthonormal basis of $S(TC^{\perp})$ and $\nabla_{\frac{d}{dt}} W_4 = 0$.

Next since $\kappa_1 = 2\sqrt{2}$, $\kappa_4 = 0$ and $\kappa_5 = 1$, let

$$\bar{N} = -\frac{1}{16}\frac{d}{dt} + N - \frac{1}{2\sqrt{2}} W_3, \quad \bar{W}_3 = W_3 + \frac{1}{2\sqrt{2}}\frac{d}{dt},$$

$$\bar{W}_1 = W_1, \qquad \bar{W}_2 = W_2, \qquad \bar{W}_4 = W_4,$$

that is,

$$\bar{N} = -\frac{1}{16}\left(t^2 + 6,\ 2t,\ t^2 + 4,\ \sqrt{2},\ 3\sqrt{2}\sin\sqrt{2}t,\ -3\sqrt{2}\cos\sqrt{2}t \right),$$

$$\bar{W}_1 = \frac{1}{\sqrt{2}}\left(t,\ 1,\ t,\ 0,\ -\cos\sqrt{2}t,\ -\sin\sqrt{2}t \right),$$

$$\bar{W}_2 = -\frac{1}{\sqrt{2}}\left(t,\ 1,\ t,\ 0,\ \cos\sqrt{2}t,\ \sin\sqrt{2}t \right),$$

$$\bar{W}_3 = -\frac{1}{2\sqrt{2}}\left(t^2,\ 2t,\ t^2 - 2,\ \sqrt{2},\ \sqrt{2}\sin\sqrt{2}t,\ -\sqrt{2}\cos\sqrt{2}t \right),$$

$$\bar{W}_4 = \frac{1}{\sqrt{2}}\left(1,\ 0,\ 1,\ \sqrt{2},\ 0,\ 0 \right),$$

then $\bar{\kappa}_4 = g\left(\nabla_{\frac{d}{dt}} \bar{W}_1, \bar{W}_2\right) = 0$ and $\bar{\kappa}_5 = g\left(\nabla_{\frac{d}{dt}} \bar{W}_1, \bar{W}_3\right) = 0$. In this case the Natural Frenet equations (2.1.4) is given by

$$\nabla_{\frac{d}{dt}} \frac{d}{dt} = 2\sqrt{2}\,\bar{W}_1, \qquad\qquad \nabla_{\frac{d}{dt}} \bar{N} = \frac{1}{4\sqrt{2}}\bar{W}_1 + \frac{1}{2\sqrt{2}}\bar{W}_2,$$

$$\nabla_{\frac{d}{dt}} \bar{W}_1 = -\frac{1}{4\sqrt{2}}\frac{d}{dt} - 2\sqrt{2}\,\bar{N}, \quad \nabla_{\frac{d}{dt}} \bar{W}_2 = -\frac{1}{\sqrt{2}}\frac{d}{dt} - \bar{W}_3,$$

$$\nabla_{\frac{d}{dt}} \bar{W}_3 = \bar{W}_2, \qquad\qquad \nabla_{\frac{d}{dt}} \bar{W}_4 = 0.$$

Finally, for a distinguished parameter p and $\kappa_1 = 1$, by a procedure similar to the one in previous section one can show that $\kappa_4 = \kappa_5 = \kappa_7 = 0$. Then, labeling $\kappa_2 = \kappa_1$, $\kappa_3 = \kappa_2$, $\kappa_6 = \kappa_3$ and $\kappa_8 = \kappa_4$ we have the following:

Theorem 4.4. *Let $C(p)$ be a non-geodesic null curve of a Lorentzian manifold (M_1^6, g). Consider a basis $E = \{C', C'', C^{(3)}, C^{(4)}, C^{(5)}, C^{(6)}\}$ of $T_{C(p)}M_1^6$ for all p, such that $g(C'', C'') = 1$. Then, there exists only one Frenet frame $F = \{\xi, N, W_1, W_2, W_3, W_4\}$ for which $(C(p), F)$ is a null Cartan curve with Cartan Frenet equations*

$$\begin{aligned}
\nabla_\xi \xi &= W_1, \\
\nabla_\xi N &= \kappa_1 W_1 + k_2 W_2, \\
\nabla_\xi W_1 &= -\kappa_1 \xi - N, \\
\nabla_\xi W_2 &= -\kappa_2 \xi + \kappa_3 W_3, \\
\nabla_\xi W_3 &= -\kappa_3 W_2 + \kappa_4 W_4 \\
\nabla_\xi W_4 &= -\kappa_4 W_3
\end{aligned} \qquad (2.4.7)$$

where F and E have the same positive orientation. Moreover, the curvature functions $\{\kappa_1, \kappa_2, \kappa_3, \kappa_4\}$ are invariant up to a sign, under Lorentzian transformations.

Example 13. Consider a null curve C and its Frenet frames in $\mathbf{R}_1^6$ given by

$$C(p) = \frac{1}{\sqrt{2}}\left(\frac{1}{18}p^3 + 2p,\ \frac{1}{2}p^2,\ \frac{1}{18}p^3 - p,\ \sqrt{2}p,\ \cos p,\ \sin p\right), \quad p \in \mathbf{R};$$

$$\xi = \frac{1}{\sqrt{2}}\left(\frac{1}{6}p^2 + 2,\ p,\ \frac{1}{6}p^2 - 1,\ \sqrt{2},\ -\sin p,\ \cos p\right),$$

$$N = -\frac{1}{4\sqrt{2}}\left(\frac{1}{6}p^2 + \frac{10}{3},\ p,\ \frac{1}{6}p^2 + \frac{1}{3},\ \sqrt{2},\ 3\sin p,\ -3\cos p\right),$$

$$W_1 = \frac{1}{\sqrt{2}}\left(\frac{1}{3}p,\ 1,\ \frac{1}{3}p,\ 0,\ -\cos p,\ -\sin p\right),$$

$$W_2 = -\frac{1}{\sqrt{2}}\left(\frac{1}{3}p,\ 1,\ \frac{1}{3}p,\ 0,\ \cos p,\ \sin p\right),$$

$$W_3 = -\frac{1}{2}\left(\frac{1}{6}p^2 + \frac{4}{3},\ p,\ \frac{1}{6}p^2 - \frac{5}{3},\ \sqrt{2},\ \sin p,\ -\cos p\right).$$

Then the Natural Frenet equations (2.4.6) give

$$\nabla_\xi \xi = W_1, \qquad \nabla_\xi N = \frac{1}{4} W_1 + \frac{1}{2} W_2,$$

$$\nabla_\xi W_1 = -\frac{1}{4}\xi - N, \quad \nabla_\xi W_2 = -\frac{1}{2}\xi - \frac{1}{\sqrt{2}} W_3,$$

$$\nabla_\xi W_3 = \frac{1}{\sqrt{2}} W_2, \qquad \nabla_\xi W_4 = 0.$$

2.5   Fundamental theorem of null curves in $\mathbf{R}_1^{m+2}$

In this section we reproduce two results (taken from [28, chapter 3]) on the geometry of null curves, namely, a theorem on the reduction of codimension and the *fundamental existence and uniqueness theorem for null curves* in $\mathbf{R}_1^{m+2}$.

Let $\mathbf{R}_1^{m+2}$ be an $(m+2)$-dimensional Minkowski space with the metric

$$g(x, y) = -x^0 y^0 + \sum_{\alpha=1}^{m+1} x^\alpha y^\alpha.$$

Suppose C is a null curve in $\mathbf{R}_1^{m+2}$ locally given by the equations

$$x^i = x^i(t), \quad t \in I \subset \mathbf{R}, \quad i \in \{0, 1, \dots, m+1\}. \tag{2.5.1}$$

Theorem 5.1 ([28]). *Let C be a null curve of $\mathbf{R}_1^{m+2}$, $m > 2$. Suppose there exists a screen vector bundle $S(TC^\perp)$ which induces on each coordinate neighborhood $\mathcal{U} \subset C$ a Frenet frame $F = \{\xi, N, W_1, \dots, W_m\}$ such that, $\{\kappa_\alpha\}$, $\alpha \in \{1, \dots, (2n-4)\}$, $n \in \{3, \dots, m\}$ are nowhere zero and $\kappa_{2n-3}, \kappa_{2n-2}, \kappa_{2n-1}$ are everywhere zero on $\mathcal{U}$. Then C lies in an n-dimensional Lorentz plane of $\mathbf{R}_1^{m+2}$. In case κ_{2m-1} and κ_{2m} are everywhere zero on $\mathcal{U}$, C lies in a Lorentz hyperplane of $\mathbf{R}_1^{m+2}$.*

Proof. Let $\triangle(t)$ be the n-dimensional Lorentz subspace of $\mathbf{R}_1^{m+2}$ spanned by $\{\xi, N(t), W_1(t), \dots, W_{n-2}(t)\}$. We first prove $\triangle(t)$, $t \in I$, are parallel n-dimensional planes. To this end, note that $\nabla_\xi X$ is just $X'(t)$ for any vector field X defined on $\mathcal{U}$. Then from the Frenet equations (2.1.2) and taking into account that $k_{2n-3} = k_{2n-2} = k_{2n-1} = 0$, obtain

$$W_i'(t) = \sum_{j=1}^n A_{ij}(t) W_j(t), \quad i \in \{1, 2, \dots, n\}, \tag{2.5.2}$$

where $W_{n-1} = N$ and $W_n = \frac{d}{dt}$. As $\triangle(t)$ is a Lorentz subspace (containing a hyperbolic plane) it follows that its orthogonal complement $\triangle(t)^\perp$ in $T_{x(t)}\mathbf{R}_1^{m+2}$ is a spacelike subspace. Thus for a fixed t_0 there exist constant spacelike vector fields V_α, $\alpha \in \{1, \cdots m+2-n\}$, such that

$$g(W_i(t_0), V_\alpha) = 0, \quad \forall i \in \{1, 2, \ldots, n\}. \tag{2.5.3}$$

Furthermore, by using (2.5.2) and taking into account that ∇ is a metric connection we obtain the system of differential equations

$$\frac{d}{dt}(g(W_i(t), V_\alpha)) = g(W_i'(t), V_\alpha) = \sum_{j=1}^{n} A_{ij}(t)g(W_j(t), V_\alpha),$$

with initial conditions (2.5.2). By the uniqueness of solutions of the last equation we infer $g(W_i(t), V_\alpha) = 0$. Hence all n-planes $\triangle(t)$ are parallel with the initial one $\triangle(t_0)$. Now we recall the following general result from Spivak [105, p. 39].

Lemma 5.1. *Let C be an immersed curve of $\mathbf{R}^{m+2}$ such that $\xi \in \triangle(t)$ for all t, where $\triangle(t)$ are parallel n-dimensional planes of $\mathbf{R}^{m+2}$. Then C is a curve in some n-dimensional plane of $\mathbf{R}^{m+2}$.*

Thus by using lemma 5.1 one obtains that any $\mathcal{U}$ lies in an n-dimensional Lorentz plane V of $\mathbf{R}_1^{m+2}$. Consider another $\mathcal{U}^* \subset C$ such that $\mathcal{U} \cap \mathcal{U}^* \neq \emptyset$. Then from the above proof it follows that $\mathcal{U}^*$ lies in the same V, which implies that C lies in V. In a similar way the second assertion of the theorem can be proved.

The proof of next theorem follows main lines of the above proof but using the following general Frenet equations with $m = 2$.

$$\begin{array}{rcl}
\nabla_\xi \xi & = & h\xi + k_1 W_1 \\
\nabla_\xi N & = & -hN + k_2 W_1 + k_3 W_2 \\
\nabla_\xi W_1 & = & -k_2 \xi - k_1 N + k_4 W_2 \\
\nabla_\xi W_2 & = & -k_3 \xi - k_4 W_1 .
\end{array}$$

Theorem 5.2 ([28]). *Let C be a null curve of $\mathbf{R}_1^4$. Suppose there exists a screen vector bundle $S(TC^\perp)$ which induces on each coordinate neighborhood $\mathcal{U} \subset C$ a Frenet frame $F = \{\xi, N, W_1, W_2\}$ such that $\kappa_1 \neq 0$ and $\kappa_2 = \kappa_3 = \kappa_4 = 0$ on $\mathcal{U}$. Then C lies in a Lorentz hyperplane of $\mathbf{R}_1^4$.*

Next, we define in $\mathbf{R}_1^{m+2}$ the quasi-orthonormal basis

$$W_0^0 = \left(-\frac{1}{\sqrt{2}}, \frac{1}{\sqrt{2}}, 0, \ldots, 0\right), \quad W_1^0 = \left(\frac{1}{\sqrt{2}}, \frac{1}{\sqrt{2}}, 0, \ldots, 0\right)$$
$$W_2^0 = (0, 0, 1, 0, \ldots, 0), \ldots, \quad W_{m+1}^0 = (0, 0, \ldots, 0, 1), \tag{2.5.4}$$

where $\{W_0^0, W_1^0\}$ are null vectors such that

$$g(W_0^0, W_1^0) = 1$$

and $\{W_2^0, \ldots, W_{m+1}^0\}$ are orthonormal spacelike vectors. It is easy to see that

$$W_0^{0i}W_1^{0j} + W_0^{0j}W_1^{0i} + \sum_{\alpha=2}^{m+1} W_\alpha^{0i}W_\alpha^{0j} = h^{ij}, \qquad (2.5.5)$$

for any $i, j \in \{0, \ldots, m+1\}$, where we put

$$h^{ij} = \begin{cases} -1, & i = j = 0 \\ 1, & i = j \neq 0 \\ 0, & i \neq j. \end{cases}$$

Theorem 5.3 ([12]). *Let $\kappa_1, \ldots, \kappa_{2m} : [-\varepsilon, \varepsilon] \to \mathbf{R}$ be everywhere continuous functions, $x_o = (x_o^i)$ be a fixed point of $\mathbf{R}_1^{m+2}$ and let $W_0^o, \ldots, W_{m+1}^o$ be the quasi-orthonormal basis in (2.5.4). Then there exist a unique null curve C of $\mathbf{R}_1^{m+2}$ given by the equations $x^i = x^i(p)$, $p \in [-\varepsilon, \varepsilon]$, where p is a distinguished parameter on C, such that $x_0^i = x^i(0)$ and $\kappa_1, \ldots, \kappa_{2m}$ are the curvature functions of C, with respect to a general Frenet frame*

$$F = \left\{ \frac{d}{dp}, N, W_2, \ldots, W_{m+1} \right\}$$

that satisfies

$$\frac{d}{dp}(0) = W_0^o, \ N(0) = W_1^o, \ W_\alpha(0) = W_\alpha^o, \ \alpha \in \{2, \ldots, (m+1)\}.$$

Proof. Using the first and second equations of (2.1.2) with $h = 0$ we consider the system of differential equations

$$W_0'(p) = \kappa_1 W_2$$
$$W_1'(p) = \kappa_2 W_2 + \kappa_3 W_3$$
$$W_2'(p) = -\kappa_2 W_0 - \kappa_1 W_1 + \kappa_4 W_3 + \kappa_5 W_4$$
$$W_3'(p) = -\kappa_3 W_0 - \kappa_4 W_2 + \kappa_6 W_4 + \kappa_7 W_5 \qquad (2.5.6)$$

$$\cdots\cdots\cdots\cdots\cdots\cdots\cdots\cdots$$

$$W_{m-1}'(p) = -\kappa_{2m-5}W_{m-3} - \kappa_{2m-4}W_{m-2} + \kappa_{2m-3}W_m + \kappa_{2m-1}W_{m+1}$$
$$W_m'(p) = -\kappa_{2m-4}W_{m-2} - \kappa_{2m-2}W_{m-1} + \kappa_{2m}W_{m+1}$$
$$W_{m+1}'(p) = -\kappa_{2m-1}W_{m-1} - \kappa_{2m}W_m.$$

Then there exists a unique solution $\{W_0, \ldots, W_{m+1}\}$ satisfying the initial conditions $W_i(0) = W_i^0$, $i \in \{0, \ldots, m+1\}$. Now we claim that

$$\{W_0(p), \ldots, W_{m+1}(p)\}$$

is a quasi-orthonormal basis such that $\{W_0(p), W_1(p)\}$ is null vector fields and $\{W_2(p), \ldots, W_{m+1}(p)\}$ is spacelike vector fields respectively, for each $p \in [-\varepsilon, \varepsilon]$. To this end, by direct calculations using (2.5.6), we obtain

$$\frac{d}{dp}\left\{ W_0^i(p)W_1^j(p) + W_0^j(p)W_1^i(p) + \sum_{\alpha=2}^{m+1} W_\alpha^i(p)W_\alpha^j(p) \right\} = 0.$$

As for $p = 0$ we have (2.5.5), from the last equation it follows that

$$W_0^i(p)W_1^j(p) + W_0^j(p)W_1^i(p) + \sum_{\alpha=2}^{m+1} W_\alpha^i(p)W_\alpha^j(p) = h^{ij}. \qquad (2.5.7)$$

Further on, construct the field of frames

$$V_0 = \frac{1}{\sqrt{2}}(W_1 - W_0), \quad V_1 = \frac{1}{\sqrt{2}}(W_1 + W_0),$$
$$V_\alpha = W_\alpha, \quad \alpha \in \{2, \ldots, m+1\}. \qquad (2.5.8)$$

Then (2.5.7) becomes

$$\sum_{a=1}^{m+1} V_a^i V_a^j - V_0^i V_0^j = h^{ij}. \qquad (2.5.9)$$

Following Bonnor([18]), we define for each $p \in [-\varepsilon, \varepsilon]$ the matrix $D(p) = [d^{ij}(p)]$ such that

$$d^{00} = V_0^0; \ d^{0a} = \sqrt{-1}\, V_a^0; \ d^{a0} = -\sqrt{-1}\, V_0^a;$$

$$d^{ab} = V_b^a, \quad a, b \in \{1, \ldots, (m+1)\}.$$

By using (2.5.9) it is easy to check that $D(p) \times D(p)^t = I_{m+2}$, which implies that $\{V_i(p)\}$, $i \in \{0, \ldots, m+1\}$ is an orthonormal basis for any $p \in [-\varepsilon, \varepsilon]$. Then from (2.5.8) we conclude that $\{W_i(p)\}$, $i \in \{0, \ldots, m+1\}$ is a quasi-orthonormal basis for any $p \in [-\varepsilon, \varepsilon]$. The null curve is obtained by integrating the system

$$\frac{dx^i}{dp} = W_0^i(p), \quad x^i(0) = x_0^i.$$

Taking into account of (2.5.6) we see that

$$F = \left\{ \frac{d}{dp} = W_0(p),\ N(p) = W_1(p),\ W_2(p),\ \ldots,\ W_{m+1}(p) \right\}$$

is a Frenet frame for C with curvature functions $\{\kappa_1, \ldots, k_{2m}\}$ and p is the distinguished parameter on C, which completes the proof.

2.6 Brief notes and exercises

(A) Levi-Civita connection. Let (x^i) be a coordinate system on $\mathcal{U} \subset (M, g)$. Then, components of the metric tensor g on $\mathcal{U}$ are given by $g_{ij} = \langle \partial_i, \partial_j \rangle$ $(1 \le i, j \le m+2)$. Thus for any vector fields $X = \sum X^i \partial_i$ and $Y = \sum Y^j \partial_j$,

$$g(X, Y) = \langle X, Y \rangle = \sum g_{ij} X^i Y^j.$$

It is well-known that there exists a unique metric (also called Levi-Civita) connection ∇ on M with respect to which the covariant derivative of g is zero, that is, $g_{ij;k} = 0$. Using this we find the following identity for covariant constant g:

$$X(g(Y,\,Z)) = \nabla_X(g(Y,\,Z)) = (\nabla_X g)(Y,\,Z) + g(\nabla_X Y,\,Z) + g(Y,\,\nabla_X Z)$$

from which we obtain the equation (2.1.1). Also it is easy to show that

$$\begin{aligned} 2g(\nabla_X Y,\,Z) &= X(g(Y,\,Z)) + Y(g(X,\,Z)) - Z(g(X,\,Y)) \\ &+ g([X,\,Y],\,Z) + g([Z,\,X],\,Y) - g([Y,\,Z],\,X). \end{aligned}$$

(B) Orthogonal projection of a null curve. Let C be a null curve of $\mathbf{R}_1^{m+2}$ given by

$$x^i = x^i(t), \quad t \in I \subset \mathbf{R}, \qquad i \in \{0, 1, \ldots, (m+1)\}.$$

As $\frac{d}{dt} \neq 0$ it follows $\frac{dx^o}{dt} \neq 0$ at any point of C, no matter which parameter t is taken on C. Hence, locally, we may suppose $\frac{dx^o}{dt} > 0$. Then consider the orthogonal projection C^* of C on the Euclidean space $\mathbf{R}^{m+1}$, i.e., C^* is given by the equations

$$x^a = x^a(t), \quad t \in I, \quad a \in \{1, \ldots, m+1\}. \tag{2.6.1}$$

Then by using the formula for the arc length s of C^*, obtain

$$s = \int_u^t \left\{ \sum_{a=1}^{m+1} \left(\frac{dx^a}{dt}\right)^2 \right\}^{\frac{1}{2}} dt = \int_u^t \frac{dx^o}{dt}\,dt = x^o(t) - u^o. \tag{2.6.2}$$

Fix a point $A(u^i)$ of C, $u^i = x^i(u)$, $u \in I$. As $g(\frac{d}{dt}, \frac{d}{dt}) = 0$, we have

$$\left(\frac{dx^o}{dt}\right)^2 = \sum_{a=1}^{m+1}\left(\frac{dx^a}{dt}\right)^2. \tag{2.6.3}$$

Consider the cone $\Lambda_1^{m+1}(A)$ of $\mathbf{R}_1^{m+2}$ with vertex A, i.e.,

$$\Lambda_1^{m+1}(A) = \{x \in \mathbf{R}_1^{m+2}; \ g(xA,\,xA) = 0\}.$$

Observe that the light cone Λ_1^{m+1} of $\mathbf{R}_1^{m+2}$ is just $\Lambda_1^{m+1}(0) - \{0\}$. In case $g(x_o - A, x_o - A) \leq 0$, we say that x_o does not lie in the exterior of the cone $\Lambda_1^{m+1}(A)$. Since the length of C^* is not less than the Euclidean distance between points $(u^1, \ldots, u^{m+1})$ and $(x^1(t), \ldots, x^{m+1}(t))$ of $\mathbf{R}^{m+1}$, we infer

$$(x^o(t) - u^o)^2 \geq \sum_{a=1}^{m+1}(x^a(t) - u^a)^2,$$

from which we have the following result:

Proposition 1 ([28, page 72]). *The null curve C, locally, does not lie in the exterior of the cone $\Lambda_1^{m+1}(A)$.*

Since for null curves an arc length is not possible, we show that for a null curve C of $\mathbf{R}_1^{m+2}$ one can choose the arc length parameter of the orthogonal projection C^* of C on a Euclidean space $\mathbf{R}^{m+1}$ as a parameter for C. Suppose that $\frac{dx^0}{dt} > o$ and choose a coordinate system in $\mathbf{R}_1^{m+2}$ such that $u^o = 0$. Then, from (2.6.2) we get $x^o(t) = s$, which implies $t = t(s)$. Hence one can take the arc-length s of C^* as parameter a on C and obtain for C the local equations

$$x^o = s, \quad x^a = \bar{x}^a(s), \quad a \in \{1, \dots, m+1\}, \quad s \in [0, L] .$$

Since $\frac{d}{ds} = \left(1, \frac{d\bar{x}^1}{ds}, \dots, \frac{d\bar{x}^{m+1}}{ds}\right)$ is a null vector field, we get

$$\left(\frac{d\bar{x}^1}{ds}\right)^2 + \dots + \left(\frac{d\bar{x}^{m+1}}{ds}\right)^2 = 1 .$$

Then by using the natural parameterization of the sphere $S^m(1)$ of the Euclidean space $\mathbf{R}^{m+1}$, we state the following:

Proposition 2 ([28, page 73]). *A null curve C of $\mathbf{R}_1^{m+2}$ is given locally by the equations of the following form*

$$x^o = s, \quad x^1 = \int_o^s \cos b_1(s)ds + c_1, \quad x^\alpha = \int_o^s \cos b_\alpha(s)\Pi_{k=1}^{\alpha-1} \sin b_k(s)ds + c_\alpha ,$$

$$\alpha \in \{2, \dots, m\}, \quad x^{m+1} = \int_o^s \Pi_{k=1}^m \sin b_k(s)ds + c_{m+1} ,$$

where $c_k \in \mathbf{R}$, b_k are smooth functions for any $k \in \{1, \dots, m\}$, and s the arc length of the orthogonal projection C^* of C on $\mathbf{R}^{m+1}$ given by (2.6.1).

It is easy to see that $\nabla_{\frac{d}{ds}}\frac{d}{ds}$ is a spacelike vector field. Thus, locally one can choose W_1 as a unit vector field in the direction of $\nabla_{\frac{d}{ds}}\frac{d}{ds}$. As s is independent of the parameter on C, it follows that there is a spacelike distribution D along C, spanned locally by W_1. Choose $S(TC^\perp)$ such that it contains D and obtain that s is a distinguished parameter of C with respect to $S(TC^\perp)$.

In particular, for $m = 1$, we have the following general equations

$$x^o = s, \ x^1 = \int_o^s \cos b_1(s)ds + c_1, \quad x^2 = \int_o^s \sin b_1(s)ds + c_2 . \qquad (2.6.4)$$

of a null curve C in $\mathbf{R}_1^3$, where

$$\frac{d}{ds} = (1, \cos b_1(s), \sin b_1(s)) ; \quad \nabla_{\frac{d}{ds}}\frac{d}{ds} = b_1'(s)(0, -\sin b_1(s), \cos b_1(s)) .$$

Choose W_1 as a unit vector field in the direction of $\nabla_{\frac{d}{ds}}\frac{d}{ds}$ and by using the first equation in (2.3.1) wherein the distinguished parameter p is replaced by s, we obtain

$k_1 = b'_1(s)$. As $S(TC^\perp)$ is spanned by $W_1 = (0, -\sin b_1(s), \cos b_1(s))$, consider $V = (0, \cos b_1(s), \sin b_1(s)) \in S(TC^\perp)^\perp$ and by using (1.4.2) of chapter 1, where t is replaced by s, we obtain

$$N = (1/2)\,(-1,\ \cos b_1(s),\ \sin b_1(s))\ .$$

Finally, the second equation in (2.3.1) is replaced by

$$k_2 = \frac{b'_1(s)}{2} = \frac{k_1}{2}.$$

Therefore the arc-length s on C^* induces a Frenet frame for which the calculations are much simplified as compared with any other Frenet frame. In support of this assertion we have

Proposition 3 ([28, page 74]). *Let C be a null curve of $\mathbf{R}^3_1$ given by (2.6.4). Then k_1 is a non-zero constant if and only if there exist real constants $a, b \neq 0, c, d$ such that C is given by*

$$x^0 = s, \quad x^1 = \frac{1}{b}\sin(bs+a) + c, \quad x^2 = -\frac{1}{b}\cos(bs+a) + d. \qquad (2.6.5)$$

Proof. If C is given by (2.6.5) we get $b_1(s) = bs + a$ and so $k_1 = b'_1(s) = b \neq 0$. Conversely, if $k_1 = b \neq 0$, then, we get (2.6.5) by integrating in (2.6.4).

As curves given by (2.6.5) are known as circular helices, the proposition 3 gives a complete characterization of *null circular helices* by means of curvature k_1 with respect to a Cartan frame and parameter s.

Exercises

Find the Frenet frame and the curvature functions of

(1) A null curve C in $\mathbf{R}^3_1$ given by

$$x^0 = \sinh t, \quad x^1 = t, \quad x^2 = \cosh t, \quad t \in \mathbf{R}.$$

(2) A null curve C in $\mathbf{R}^3_1$ given by

$$x^0 = \frac{4}{3}t^3 + t, \quad x^1 = 2t^2, \quad x^2 = \frac{4}{3}t^3 - t, \quad t \in \mathbf{R}.$$

(3) A null curve C in $\mathbf{R}^4_1$ given by

$$x^0 = \frac{1}{\sqrt{2}}\sinh t, \ x^1 = \frac{1}{\sqrt{2}}\cosh t, \ x^2 = \frac{1}{\sqrt{2}}\sin t, \ x^3 = \frac{1}{\sqrt{2}}\cos t, \ t \in \mathbf{R}.$$

(4) A null curve C in $\mathbf{R}_1^4$ given by

$$x^0 = \log(\sec t + \tan t)\,,\ x^1 = -\cos t\,,\ x^2 = \sin t\,,\ x^3 = \log \sec t\,,\ t \in \mathbf{R}\,.$$

(5) A null curve C in $\mathbf{R}_1^4$ given by

$$x^0 = \frac{1}{6}t^3 + t\,,\ x^1 = \frac{1}{2}t^2\,,\ x^2 = \frac{1}{6}t^3\,,\ x^3 = t\,,\ t \in \mathbf{R}\,.$$

(6) A null curve C in $\mathbf{R}_1^6$ given by

$$\left(\frac{1}{2}t^3 + 2t,\ \frac{3}{2}t^2,\ \frac{1}{2}t^3 - t,\ \sqrt{2}t,\ \cos t,\ \sin t\right),\quad t \in \mathbf{R}\,.$$

(7) Using note (B), find a relation between the arc length of the orthogonal projection C^* of the null curves C taken from exercises (1) and (2).

(8) Consider a null curve C of $\mathbf{R}_1^4$ given by the equations

$$x^0 = s\,,\quad x^1 = \int_o^s \cos b_1(s) + c_1\,,\quad x^2 = \int_o^s \sin b_1(s)\cos b_2(s)ds + c_2\,,$$

$$x^3 = \int_0^s \sin b_1(s)\sin b_2(s)ds + c_3\,,$$

where s is the arc length parameter of the orthogonal projection C^* of C on $\mathbf{R}^3$. Find its Frenet frame and curvature functions.

(9) Let $C(p)$ be a non-geodesic Cartan null curve of a Lorentzian manifold (M_1^4, g) where p is a distinguished parameter on C, with its Frenet frame $F = \{\xi, N, W_1, W_2\}$ and Frenet equations (2.3.8) as given in theorem 3.2. Find an explicit expression of N, W_1 and W_2 in terms of ξ and its derivatives up to the order 3.

Chapter 3

Null curves in M_2^{m+2}

In this chapter we present results on the general study of a null curve C of semi-Riemannian manifold (M, g) with index $q = 2$, denoted by M_2^{m+2}. Since, as per chapter 1, any screen distribution $S(TC^\perp)$ will be of index 1, any of its base vectors might change its causal character on $\mathcal{U} \subset C$. To illustrate this we give the following two examples.

Example 1. Let C be the null curve in $\mathbf{R}_2^6$ given by

$$C : (\cosh t, \ At^3 + 2t, \ Bt^2, \ At^3, \ \sqrt{3}t, \ \sinh t)$$

where $B^2 = 3A$ and $A, \ B, \ t \in \mathbf{R}$. Then,

$$\xi = (\sinh t, \ 3At^2 + 2, \ 2Bt, \ 3At^2, \ \sqrt{3}, \ \cosh t),$$
$$\nabla_\xi \xi = (\cosh t, \ 6At, \ 2Bt, \ 6At, \ 0, \ \sinh t).$$

We need to know the causal character of the vector field

$$H(t) = \nabla_\xi \xi - h \, \xi$$

along C. By direct calculations we obtain

$$g(H(t), \ H(t)) \ = \ 1 - 4B^2.$$

Hence $H(t)$ is spacelike, timelike or null according as $\{ B < -\frac{1}{2} \, ; \ B > \frac{1}{2} \}$, $\{ -\frac{1}{2} < B < \frac{1}{2} \}$ or $\{ B^2 = \pm \frac{1}{2} \}$ respectively.

Example 2. Let C be the null curve of $\mathbf{R}_2^4$ given by

$$C : \left(t^2 + t, \ \sqrt{3}t \cosh t, \ \sqrt{3}t \sinh t, \ \frac{1}{2}t^2 + 2t \right).$$

In this case, since

$$\xi = \left(2t + 1, \ \sqrt{3}\cosh t + \sqrt{3}t \sinh t, \ \sqrt{3}\sinh t + \sqrt{3}t \cosh t, \ t + 2 \right),$$
$$\nabla_\xi \xi = \left(2, \ 2\sqrt{3}\sinh t + \sqrt{3}t \cosh t, \ 2\sqrt{3}\cosh t + \sqrt{3}t \sinh t, \ 1 \right),$$

we have

$$g\left(H(t),\, H(t)\right) = 3(3 - t^2).$$

In this case also as Example 1, $H(t)$ is spacelike, timelike or null according as $t \in (-\sqrt{3},\, \sqrt{3})$, $t \in (-\infty,\, -\sqrt{3}) \cup (\sqrt{3},\, \infty)$ or $t = \pm\sqrt{3}$, respectively.

Thus, above two examples suggest that the construction of a Frenet frame for null curves of semi-Riemannian manifolds of index greater than one should be done subject to some restrictive conditions on such curves. Due to this reason, Duggal and Jin [30] constructed two types of Frenet equations suitable for a null curve of a semi-Riemannian manifold of index 2 (Ferrández et. al [36] have also done some work, in particular, on pseudo-Euclidean spaces of index 2 (see chapter 5)). In this chapter we also introduce another two sets of new Frenet equations, called *Natural Frenet equations*. We present those two Frenet equations, study their invariant properties and discuss the geometry of null curves in 4 and 6 dimensional semi-Riemannian manifolds of index 2.

3.1   Type 1 general Frenet frames

Let C be a null curve of M_2^{m+2} such that $C = span\,\{\xi\}$ and N be its unique null transversal vector field (see Theorem 4.1 of chapter 1). It follows that any screen distribution $S(TC^\perp)$ of C will be Lorentz. In this section we study a class of null curves C whose Frenet frame is made up of two null vectors ξ and N, one timelike and $(m - 1)$ spacelike vector fields. Let us denote by *Type 1* the Frenet frame and Frenet equations of this class of C.

From $g\left(\nabla_\xi\,\xi,\, \xi\right) = 0$ and $g\left(\nabla_\xi\,\xi,\, N\right) = h$, we have

$$\nabla_\xi\,\xi = h\,\xi + R_1,$$

where $R_1 \in \Gamma(S(TC^\perp))$. Since $S(TC^\perp)$ is a Lorentz vector bundle, there are three cases (timelike, spacelike and null) by the causality of the vector field R_1. In this section we assume that R_1 is non-null. Based on this restriction, denote $\rho_1 = \|R_1\|$ and define the first curvature function κ_1 by $\kappa_1 = \epsilon_1\rho_1$ where $\epsilon_1 = +1$ or -1 according as R_1 is spacelike or timelike respectively. Now, if ρ_1 is non-zero for any t, we set $\bar{W}_1 = R_1/\rho_1$, then $\bar{W}_1$ is a unit vector field with the same causality as R_1 along C. Thus the above equation becomes

$$\nabla_\xi\,\xi = h\,\xi + \kappa_1\,\epsilon_1\,\bar{W}_1.$$

Also, from $g(\nabla_\xi\,N,\, \xi) = -h$, $g(\nabla_\xi\,N,\, N) = 0$ and $g(\nabla_\xi\,N,\, \bar{W}_1) = \kappa_2$, where κ_2 denotes the second curvature function, we also have

$$\nabla_\xi\,N = -h\,N + \kappa_2\,\epsilon_1\,\bar{W}_1 + R_2$$

where $R_2 \in \Gamma(S(TC^\perp))$. Thus R_2 is a vector field perpendicular to ξ, N and $\bar{W}_1$. Assume that R_2 is non-null. Denote $\rho_3 = \|R_2\|$ and define the third curvature function κ_3 by $\kappa_3 = \epsilon_2\,\rho_3$ where $\epsilon_2 = +1$ or -1 according as R_2 is spacelike or timelike respectively. If ρ_2 is non-zero for any t, we set $\bar{W}_2 = R_2/\rho_3$, then $\bar{W}_2$ is a non-null unit vector field with the same causality as R_2 along C. Thus we have

$$\nabla_\xi N = -h N + \kappa_2 \,\epsilon_1 \,\bar{W}_1 + \kappa_3 \,\epsilon_2 \,\bar{W}_2.$$

Repeating above process for all the m unit vectors of an orthonormal basis $\{\bar{W}_1, \ldots, \bar{W}_m\}$ of $\Gamma(S(TC^\perp))$ and then setting

$$W_i = \epsilon_i \,\bar{W}_i, \qquad i \in \{1, \ldots, m\}$$

and after some simplification we obtain the following equations

$$
\begin{aligned}
\nabla_\xi \xi &= h\xi + \kappa_1 W_1, \\
\nabla_\xi N &= -h N + \kappa_2 W_1 + \kappa_3 W_2, \\
\epsilon_1 \nabla_\xi W_1 &= -\kappa_2 \xi - \kappa_1 N + \kappa_4 W_2 + \kappa_5 W_3, \\
\epsilon_2 \nabla_\xi W_2 &= -\kappa_3 \xi - \kappa_4 W_1 + \kappa_6 W_3 + \kappa_7 W_4, \\
\epsilon_3 \nabla_\xi W_3 &= -\kappa_5 W_1 - \kappa_6 W_2 + \kappa_8 W_4 + \kappa_9 W_5,
\end{aligned}
\tag{3.1.1}
$$

$$\cdots\cdots\cdots\cdots\cdots\cdots$$

$$
\begin{aligned}
\epsilon_{m-1} \nabla_\xi W_{m-1} &= -\kappa_{2m-3} W_{m-3} - \kappa_{2m-2} W_{m-2} + \kappa_{2m} W_m, \\
\epsilon_m \nabla_\xi W_m &= -\kappa_{2m-1} W_{m-2} - \kappa_{2m} W_{m-1},
\end{aligned}
$$

where h and $\{\kappa_1, \ldots, \kappa_{2m}\}$ are smooth functions on $\mathcal{U}$, $\{W_1, \ldots, W_m\}$ is an orthonormal basis of $\Gamma(S(TC^\perp)_{|\mathcal{U}})$ and $(\epsilon_1, \ldots, \epsilon_m)$ is the signature of the manifold M_2^{m+2} such that $\epsilon_i \delta_{ij} = g(W_i, W_j)$. We call

$$F_1 = \{\xi, N, W_1, \ldots, W_m\} \tag{3.1.2}$$

a *general Frenet frame of Type 1* on M_2^{m+2} along C, with respect to a given screen vector bundle $S(TC^\perp)$ and the equations (3.1.1) its *general Frenet equations of Type 1*. The functions $\{\kappa_1, \ldots, \kappa_{2m}\}$ are called *curvature functions* of C with respect to F_1. Since the screen vector bundle is Lorentz, this implies that only one of W_i is timelike and all others are spacelike vector fields.

Example 3. Let C be the null curve in $\mathbf{R}_2^6$ given by

$$C : \left(\frac{1}{12}t^3 + t, \quad \sinh t, \quad \frac{1}{12}t^3 - t, \quad \frac{1}{2}t^2, \quad \cosh t, \quad t \right), \qquad t \in \mathbf{R}.$$

Then,

$$\xi = \left(\frac{1}{4}t^2 + 1, \quad \cosh t, \quad \frac{1}{4}t^2 - 1, \quad t, \quad \sinh t, \quad 1 \right).$$

Let $V = (0, \cosh t, 0, 0, \sinh t, 0)$, then $g(\xi, V) = -1$ and $g(V, V) = -1$. Thus, by equation (1.4.2) in Theorem 4.1 of chapter 1, we have

$$N = \frac{1}{2} \left(\frac{1}{4}t^2 + 1, \quad \cosh t, \quad \frac{1}{4}t^2 - 1, \quad t, \quad -\sinh t, \quad 1 \right).$$

Since
$$\nabla_\xi \xi = \left(\frac{t}{2}, \; \sinh t, \; \frac{t}{2}, \; 1, \; \cosh t, \; 0 \right),$$

$h = g\left(\nabla_\xi \xi, \, N\right) = 0$ and $\kappa_1 = \| \nabla_\xi \xi \| = \sqrt{2}$, we let

$$W_1 \;=\; \frac{1}{\sqrt{2}} \left(\frac{t}{2}, \; \sinh t, \; \frac{t}{2}, \; 1, \; \cosh t, \; 0 \right).$$

Then W_1 is spacelike and perpendicular to ξ and N; $\nabla_\xi \xi = \sqrt{2}\, W_1$. From

$$\nabla_\xi N = \frac{1}{2} \left(\frac{t}{2}, \; -\sinh t, \; \frac{t}{2}, \; 1, \; -\cosh t, \; 0 \right),$$

$\kappa_2 = g\left(\nabla_\xi N, \, W_1\right) = 0$ and $\kappa_3 = \| \nabla_\xi N \| = \frac{1}{\sqrt{2}}$, we let

$$W_2 = \frac{1}{\sqrt{2}} \left(\frac{t}{2}, \; -\sinh t, \; \frac{t}{2}, \; 1, \; -\cosh t, \; 0 \right).$$

Then W_2 is spacelike and perpendicular to ξ, N and W_1; $\nabla_\xi N = \frac{1}{\sqrt{2}} W_2$. From

$$\nabla_\xi W_1 = \frac{1}{\sqrt{2}} \left(\frac{1}{2}, \; \cosh t, \; \frac{1}{2}, \; 0, \; \sinh t, \; 0 \right), \quad \kappa_4 = g(\nabla_\xi W_1, W_2) = 0$$

and
$$\nabla_\xi W_1 + \sqrt{2}\, N = \frac{1}{\sqrt{2}} \left(\frac{1}{4}t^2 + \frac{3}{2}, \; 0, \; \frac{1}{4}t^2 - \frac{1}{2}, \; t, \; 0, \; 1 \right),$$

we let
$$W_3 = \left(\frac{1}{4}t^2 + \frac{3}{2}, \; 0, \; \frac{1}{4}t^2 - \frac{1}{2}, \; t, \; 0, \; 1 \right).$$

Then W_3 is a timelike vector field and $\nabla_\xi W_1 = -\sqrt{2}\, N + \frac{1}{\sqrt{2}} W_3$. Since

$$\nabla_\xi W_2 = \frac{1}{\sqrt{2}} \left(\frac{1}{2}, \; -\cosh t, \; \frac{1}{2}, \; 0, \; -\sinh t, \; 0 \right),$$

$\kappa_6 = g(\nabla_\xi W_2, W_3) = \frac{1}{\sqrt{2}}$, then $\nabla_\xi W_2 = -\frac{1}{\sqrt{2}}\xi + \frac{1}{\sqrt{2}}W_3$ and $\kappa_7 = 0$. Also, we have $\nabla_\xi W_3 = \frac{1}{\sqrt{2}}W_1 + \frac{1}{\sqrt{2}}W_2$ and $\kappa_8 = 0$. Let

$$W_4 = \left(\frac{1}{2}, \; 0, \; \frac{1}{2}, \; 0, \; 0, \; 1 \right).$$

Then $\{W_1, W_2, W_3, W_4\}$ is an orthonormal basis of $S(TC^\perp)$ and $\nabla_\xi W_4 = 0$.

Example 4. Let C be a null curve in $\mathbf{R}_2^6$ given by

$$C : \left(\cosh t, \; \frac{1}{27}t^3 + 2t, \; \frac{1}{3}t^2, \; \frac{1}{27}t^3, \; \sqrt{3}t, \; \sinh t \right), \qquad t \in \mathbf{R}$$

which is the curve in Example 1 such that $B = \frac{1}{3}$, therefore, $A = \frac{1}{27}$. Then this curve has the general Frenet frame of Type 1. We choose the general Frenet frame $F = \{\xi, N, W_1, W_2, W_3, W_4\}$ as follows;

$$\xi = \left(\sinh t, \ \frac{t^2}{9} + 2, \ \frac{2}{3}t, \ \frac{t^2}{9}, \ \sqrt{3}, \ \cosh t\right),$$

$$N = \frac{1}{2}\left(\sinh t, \ -\frac{t^2}{9} - 2, \ -\frac{2}{3}t, \ -\frac{t^2}{9}, \ -\sqrt{3}, \ \cosh t\right),$$

$$W_1 = -\frac{3}{\sqrt{5}}\left(\cosh t, \ \frac{2}{9}t, \ \frac{2}{3}, \ \frac{2}{9}t, \ 0, \ \sinh t\right),$$

$$W_2 = -\frac{1}{\sqrt{5}}\left(2\cosh t, \ t, \ 3, \ t, \ 0, \ 2\sinh t\right),$$

$$W_3 = -\left(0, \ \frac{t^2}{9} + \frac{3}{2}, \ \frac{2}{3}t, \ \frac{t^2}{9} - \frac{1}{2}, \ \sqrt{3}, \ 0\right),$$

$$W_4 = \left(0, \ \frac{\sqrt{3}}{2}, \ 0, \ \frac{\sqrt{3}}{2}, \ 1, \ 0\right).$$

The general Frenet equations (3.1.1) give

$$\nabla_\xi \xi = -\frac{\sqrt{5}}{3}W_1, \qquad\qquad \nabla_\xi N = -\frac{13}{6\sqrt{5}}W_1 + \frac{2}{\sqrt{5}}W_2,$$

$$\epsilon_1 \nabla_\xi W_1 = \frac{13}{6\sqrt{5}}\xi + \frac{\sqrt{5}}{3}N + \frac{4}{3\sqrt{5}}W_3, \ \epsilon_2 \nabla_\xi W_2 = -\frac{2}{\sqrt{5}}\xi - \frac{2}{\sqrt{5}}W_3,$$

$$\epsilon_3 \nabla_\xi W_3 = -\frac{4}{3\sqrt{5}}W_1 + \frac{2}{\sqrt{5}}W_2, \qquad\qquad \epsilon_3 \nabla_\xi W_4 = 0,$$

that is, $h = 0$ and

$$\kappa_1 = -\frac{\sqrt{5}}{3}, \quad \kappa_2 = -\frac{13}{6\sqrt{5}}, \quad \kappa_3 = \frac{2}{\sqrt{5}}, \quad \kappa_4 = 0,$$

$$\kappa_5 = \frac{4}{3\sqrt{5}}, \quad \kappa_6 = -\frac{2}{\sqrt{5}}, \quad \kappa_7 = 0, \quad \kappa_8 = 0.$$

3.2 Type 2 general Frenet frames

Now we study a class of null curves C whose Frenet frame is generated by a quasi-orthonormal basis consisting of the two null vector fields ξ and N and another two null vector fields L_i and L_{i+1} such that $g(L_i, L_{i+1}) = 1$ and $(m - 2)$ spacelike vector fields $\{W_\alpha\}$. In this case we will have Frenet frame of the form

$$F_2 = \{\xi, N, W_1, \cdots, L_i, L_{i+1}, \cdots, W_m\}.$$

We denote the Frenet frame and Frenet equations of this particular class of C by
Type 2. There are $(m-1)$ choices for L_i. On the other hand, if we set

$$W_i = \frac{L_i - L_{i+1}}{\sqrt{2}}, \qquad W_{i+1} = \frac{L_i + L_{i+1}}{\sqrt{2}}, \qquad (3.2.1)$$

then W_i and W_{i+1} are timelike and spacelike vector fields respectively. Then we
will have Frenet frame of the form

$$F_2 = \{\xi, N, W_1, \cdots, W_i, W_{i+1}, \cdots, W_m\}$$

instead of the above Frenet frame. We also denote the Frenet frame and Frenet
equations of this class of C by Type 2. To choose $\{L_1, L_2\}$, we let the vector field
$\nabla_\xi \xi - h\xi$ be null and define the curvature function $K_1(=1)$ by

$$\nabla_\xi \xi - h\xi = K_1 L_1,$$

where $L_1 \in \Gamma(S(TC^\perp))$. Thus L_1 is a null vector field everywhere perpendicular
to ξ and N along C. Since $S(TC^\perp)$ is Lorentz vector bundle, we can take a null
vector field L_2 in $\Gamma(S(TC^\perp))$ along C such that $g(L_1, L_2) = 1$. Set this case so
that the equation (3.2.1) holds for $i = 1$. Therefore W_1 and W_2 are perpendicular
to ξ and N, then we have

$$\nabla_\xi \xi = h\xi + \kappa_1 W_1 + \tau_1 W_2,$$

where $\kappa_1 = \tau_1 = \frac{1}{\sqrt{2}} K_1$. From $g(\nabla_\xi N, \xi) = -h$ and $g(\nabla_\xi N, N) = 0$ and
defining the second and third curvature functions by

$$g(\nabla_\xi N, W_1) = -\kappa_2, \qquad g(\nabla_\xi N, W_2) = \kappa_3$$

respectively, we obtain

$$\nabla_\xi N = -hN + \kappa_2 W_1 + \kappa_3 W_2 + R_3$$

where R_3 is a spacelike vector field perpendicular to ξ, N, W_1 and W_2. Now define
a torsion function τ_2 by $\tau_2 = \| R_3 \|$ which is non-zero for any t and set $W_3 = R_3/\tau_2$.
Clearly, W_3 is a unit spacelike vector field along C which is also perpendicular to
ξ, N, W_1 and W_2. Thus we obtain

$$\nabla_\xi N = -hN + \kappa_2 W_1 + \kappa_3 W_2 + \tau_2 W_3.$$

Also from the following results

$$g(\nabla_\xi W_1, \xi) = \kappa_1, \quad g(\nabla_\xi W_1, N) = \kappa_2, \quad g(\nabla_\xi W_1, W_1) = 0,$$

and defining the fourth and fifth curvature functions by

$$g(\nabla_\xi W_1, W_2) = -\kappa_4, \quad g(\nabla_\xi W_1, W_3) = -\kappa_5$$

respectively, we obtain

$$\nabla_\xi W_1 = \kappa_2\,\xi + \kappa_1\,N - \kappa_4\,W_2 - \kappa_5\,W_3 - R_4$$

where R_4 is a spacelike vector field perpendicular to ξ, N, W_1, W_2 and W_3. Now we define another torsion function τ_3 by $\tau_3 = \|\,R_4\,\|$ which is non-zero for any t and set $W_4 = R_4/\tau_3$. Then we obtain

$$\nabla_\xi W_1 = \kappa_2\,\xi + \kappa_1\,N - \kappa_4\,W_2 - \kappa_5\,W_3 - \tau_3\,W_4. \tag{3.2.2}$$

In a similar way we get

$$\nabla_\xi W_2 = -\,\kappa_3\,\xi - \tau_1\,N - \kappa_4\,W_1 + \kappa_6\,W_3 + \kappa_7\,W_4 + \tau_4\,W_5, \tag{3.2.3}$$

where κ_6 and κ_7 are the curvature functions, τ_4 is a torsion function and W_5 is a unit spacelike vector field which is perpendicular to ξ, N and $W_1, \cdots, W_4$.

Repeating above process for all the m unit vectors of an orthonormal basis $\{W_1, \cdots, W_m\}$ of $\Gamma(S(TC^\perp))$, we also obtain the following equations

$$
\begin{aligned}
\nabla_\xi \xi &= \ \ h\,\xi + \kappa_1\,W_1 + \tau_1\,W_2, \\
\nabla_\xi N &= -\,h\,N + \kappa_2\,W_1 + \kappa_3\,W_2 + \tau_2\,W_3, \\
\nabla_\xi W_1 &= \ \ \kappa_2\,\xi + \kappa_1\,N - \kappa_4\,W_2 - \kappa_5\,W_3 - \tau_3\,W_4, \\
\nabla_\xi W_2 &= -\,\kappa_3\,\xi - \tau_1\,N - \kappa_4\,W_1 + \kappa_6\,W_3 + \kappa_7\,W_4 + \tau_4\,W_5, \\
\nabla_\xi W_3 &= -\,\tau_2\,\xi - \kappa_5\,W_1 - \kappa_6\,W_2 + \kappa_8\,W_4 + \kappa_9\,W_5 + \tau_5\,W_6, \\
\nabla_\xi W_4 &= -\,\tau_3\,W_1 - \kappa_7\,W_2 - \kappa_8\,W_3 + \kappa_{10}\,W_5 + \kappa_{11}\,W_6 + \tau_6\,W_7
\end{aligned}
\tag{3.2.4}
$$

$$\cdots\cdots\cdots\cdots\cdots\cdots\cdots\cdots\cdots$$

$$
\begin{aligned}
\nabla_\xi W_{m-1} &= -\tau_{m-2}\,W_{m-4} - k_{2m-3}\,W_{m-3} - k_{2m-2}\,W_{m-2} + k_{2m}\,W_m, \\
\nabla_\xi W_m &= -\tau_{m-1}\,W_{m-3} - k_{2m-1}\,W_{m-2} - k_{2m}\,W_{m-1}.
\end{aligned}
$$

In the above case, we call

$$F_2 = \{\,\xi,\ N,\ W_1,\ \cdots,\ W_m\,\} \tag{3.2.5}$$

a *general Frenet frame of Type 2* on M_2^{m+2} along C, with respect to a given screen vector bundle $S(TC^\perp)$ and (3.2.4) its *general Frenet equations of Type 2*.

On the other hand, using the relations

$$g(\nabla_\xi N,\ \xi) = -\,h, \quad g(\nabla_\xi N,\ L_1) = K_3, \quad g(\nabla_\xi N,\ L_2) = K_2,$$

we can write

$$\nabla_\xi N = -\,h\,\xi + K_2\,L_1 + K_3\,L_2 + R_5$$

where R_5 is perpendicular to ξ, N, L_1, and L_2 along C. From the equations

$$
\begin{aligned}
-\,\kappa_2 &= g(\nabla_\xi N,\ W_1) = \frac{1}{\sqrt{2}}\{g(\nabla_\xi N,\ L_1 - L_2)\} = \frac{K_3 - K_2}{\sqrt{2}}, \\
\kappa_3 &= g(\nabla_\xi N,\ W_2) = \frac{1}{\sqrt{2}}\{g(\nabla_\xi N,\ L_1 + L_2)\} = \frac{K_3 + K_2}{\sqrt{2}},
\end{aligned}
$$

and

$$\nabla_\xi N \quad + \quad h\,\xi - K_2\,L_1 - K_3\,L_2$$
$$= \quad \nabla_\xi N + h\,\xi - \frac{K_2}{\sqrt{2}}\,(W_2 + W_1) - \frac{K_3}{\sqrt{2}}\,(W_2 - W_1)$$
$$= \quad \nabla_\xi N + h\,\xi + \left(\frac{K_3 - K_2}{\sqrt{2}}\right)W_1 - \left(\frac{K_3 + K_2}{\sqrt{2}}\right)W_2$$
$$= \quad \nabla_\xi N + h\,\xi - \kappa_2\,W_1 - \kappa_3\,W_2,$$

we conclude that $R_5 = \tau_2\,W_3$. Therefore,

$$\nabla_\xi N = -\,h\,\xi + K_2\,L_1 + K_3\,L_2 + \tau_2\,W_3.$$

Next, by the transformations (3.2.1) for $i = 1$ we have

$$\nabla_\xi L_1 \quad = \quad \frac{1}{\sqrt{2}}\,(\nabla_\xi W_2 + \nabla_\xi W_1),$$
$$\nabla_\xi L_2 \quad = \quad \frac{1}{\sqrt{2}}\,(\nabla_\xi W_2 - \nabla_\xi W_1).$$

Using (3.2.2) and (3.2.3) in above equation and the following results

$$K_4 = g(\nabla_\xi L_1, L_2) = \frac{1}{2}\,g(\nabla_\xi W_2 + \nabla_\xi W_1, W_2 - W_1) = -\kappa_4,$$
$$K_5 = g(\nabla_\xi L_1, W_3) = \frac{1}{\sqrt{2}}\,g(\nabla_\xi W_2 + \nabla_\xi W_1, W_3) = \frac{1}{\sqrt{2}}\,(\kappa_6 - \kappa_5),$$
$$K_7 = g(\nabla_\xi L_2, W_3) = \frac{1}{\sqrt{2}}\,g(\nabla_\xi W_2 - \nabla_\xi W_1, W_3) = \frac{1}{\sqrt{2}}\,(\kappa_6 + \kappa_5),$$

we obtain

$$\nabla_\xi L_1 = -\,K_3\,\xi + K_4\,L_1 + K_5\,W_3 + K_6\,W_4 + \ell\,W_5,$$
$$\nabla_\xi L_2 = -\,K_2\,\xi - K_1\,N - K_4\,L_2 + K_7\,W_3 + K_8\,W_4 + \ell\,W_5,$$

where $\ell = \frac{1}{\sqrt{2}}\tau_4$, $K_6 = \frac{1}{\sqrt{2}}(\kappa_7 - \tau_3)$ and $K_8 = \frac{1}{\sqrt{2}}(\kappa_7 + \tau_3)$.

Repeating above process and after some simplifications we get the following equations

$$\nabla_\xi \xi \quad = \quad h\,\xi + K_1\,L_1,$$
$$\nabla_\xi N = -\,h\,N + K_2\,L_1 + K_3\,L_2 + \tau_2\,W_3,$$
$$\nabla_\xi L_1 = -\,K_3\,\xi + K_4\,L_1 + K_5\,W_3 + K_6\,W_4 + \ell\,W_5,$$
$$\nabla_\xi L_2 = -\,K_2\,\xi - K_1\,N - K_4\,L_2 + K_7\,W_3 + K_8\,W_4 + \ell\,W_5,$$
$$\nabla_\xi W_3 = -\,\tau_2\,\xi - K_7\,L_1 - K_5\,L_2 + \kappa_8\,W_4 + \kappa_9\,W_5 + \tau_5\,W_6 \qquad (3.2.6)$$
$$\nabla_\xi W_4 = -\,K_8\,L_1 - K_6\,L_2 - \kappa_8\,W_3 + \kappa_{10}\,W_5 + \kappa_{11}\,W_6 + \tau_6\,W_7,$$

$$\cdots\cdots\cdots\cdots\cdots\cdots$$

$$\nabla_\xi W_{m-1} = -\tau_{m-2}\,W_{m-4} - \kappa_{2m-3}\,W_{m-3} - \kappa_{2m-2}\,W_{m-2} + \kappa_{2m}\,W_m,$$
$$\nabla_\xi W_m = -\tau_{m-1}\,W_{m-3} - \kappa_{2m-1}\,W_{m-2} - \kappa_{2m}\,W_{m-1}.$$

In this case, we also call

$$F_2 = \{\, \xi,\, N,\, L_1,\, L_2,\, W_3, \ldots, W_m \,\}$$

a *general Frenet frame of Type 2* on M_2^{m+2} along C, with respect to a given screen vector bundle $S(TC^{\perp})$ and the (3.2.6) its *general Frenet equations of Type 2*.

In the next cases, to choose $\{L_2, L_3\}$, $\{L_3, L_4\}$, $\cdots$, $\{L_{m-2}, L_{m-1}\}$ or $\{L_{m-1}, L_m\}$, let the vector fields

$$\nabla_\xi N + h\, N - \kappa_2\, W_1,$$
$$\nabla_\xi W_1 + \kappa_2\, \xi + \kappa_1\, N - \kappa_4\, W_2,$$

$$\cdots\cdots\cdots\cdots\cdots\cdots$$

$$\nabla_\xi W_{m-3} + \kappa_{2m-7}\, W_{m-5} + \kappa_{2m-6}\, W_{m-4} - \kappa_{2m-4}\, W_{m-2}$$

be null in turn. Then using a procedure same as above for each such cases, we finally obtain the following general equations

$$\nabla_\xi \xi \;=\; h\, \xi + \kappa_1\, W_1,$$
$$\nabla_\xi N = -h\, N + \kappa_2\, W_1 + \kappa_3\, W_2,$$
$$\nabla_\xi W_1 = -\kappa_2\, \xi - \kappa_1\, N + \kappa_4\, W_2 + \kappa_5\, W_3,$$

$$\cdots\cdots\cdots\cdots\cdots\cdots \qquad\qquad (3.2.7)$$

$$\nabla_\xi L_i \;=\; -K_{2i+1}\, W_{i-1} + K_{2i+3}\, L_i + K_{2i+4}\, W_{i+2} + K_{2i+5}\, W_{i+3} + \ell\, W_{i+2},$$
$$\nabla_\xi L_{i+1} = -K_{2i-1}\, W_{i-2} - K_{2i}\, W_{i-1} - K_{2i+3}\, L_{i+1} + K_{2i+6}\, W_{i+2}$$
$$\qquad\qquad + K_{2i+7}\, W_{i+3} + \ell\, W_{i+2}, \qquad 2 \le i \le m-1,$$

$$\cdots\cdots\cdots\cdots\cdots\cdots$$

$$\nabla_\xi W_{m-1} = -\tau_{m-2}\, W_{m-4} - \kappa_{2m-3}\, W_{m-3} - \kappa_{2m-2}\, W_{m-2} + \kappa_{2m}\, W_m,$$
$$\nabla_\xi W_m = -\tau_{m-1}\, W_{m-3} - \kappa_{2m-1}\, W_{m-2} - \kappa_{2m}\, W_{m-1}.$$

In the above case also the equations (3.2.7) are the general Frenet equations of *Type 2* with the general Frenet frame

$$F_2 = \{\, \xi,\, N,\, W_1,\, \ldots,\, L_i,\, L_{i+1}, \ldots, W_m \,\}. \qquad\qquad (3.2.8)$$

Also, by replacing L_i and L_{i+1} in (3.2.7) with its values in terms of W_i and W_{i+1} (see relations (3.2.1)), we get another set of general Frenet equations in terms of one timelike and all others spacelike basis of its Frenet frame. Proceeding as before, following are the general Frenet equations of Type 2:

$$\nabla_\xi \xi \;=\; h\, \xi + \kappa_1\, W_1 + \tau_1\, W_2,$$
$$\nabla_\xi N \;=\; -h\, N + \kappa_2\, W_1 + \kappa_3\, W_2 + \tau_2\, W_3,$$
$$\epsilon_1 \nabla_\xi W_1 = -\kappa_2\, \xi - \kappa_1\, N + \kappa_4\, W_2 + \kappa_5\, W_3 + \tau_3\, W_4,$$
$$\epsilon_2 \nabla_\xi W_2 = -\kappa_3\, \xi - \tau_1\, N - \kappa_4\, W_1 + \kappa_6\, W_3 + \kappa_7\, W_4 + \tau_4\, W_5, \qquad (3.2.9)$$
$$\epsilon_3 \nabla_\xi W_3 = -\tau_2\, \xi - \kappa_5\, W_1 - \kappa_6\, W_2 + \kappa_8\, W_4 + \kappa_9\, W_5 + \tau_5\, W_6,$$

$$\epsilon_4 \nabla_\xi W_4 = -\tau_3 W_1 - \kappa_7 W_2 - \kappa_8 W_3 + \kappa_{10} W_5 + \kappa_{11} W_6 + \tau_6 W_7$$

$$\cdots\cdots\cdots\cdots\cdots\cdots\cdots\cdots\cdots$$

$$\epsilon_{m-1} \nabla_\xi W_{m-1} = -\tau_{m-2} W_{m-4} - \kappa_{2m-3} W_{m-3} - \kappa_{2m-2} W_{m-2} + \kappa_{2m} W_m,$$
$$\epsilon_m \nabla_\xi W_m = -\tau_{m-1} W_{m-3} - \kappa_{2m-1} W_{m-2} - \kappa_{2m} W_{m-1},$$

whose Frenet frame is given by (3.2.5). We call the functions $\{\kappa_1, \ldots, k_{2m}\}$ and $\{\tau_1, \ldots, \tau_{m-1}\}$ as the *curvature* and the *torsion functions* of C for the general Frenet equations (3.2.9) respectively.

Remark. One can verify that the general Frenet equations (3.2.9) include all m-th different general Frenet equations of Type 1 and all $(m-1)$-th different general Frenet equations of Type 2. Also, for the dim $M = 4$ the general Frenet frame of Type 2 consists of all four null vector fields $\{\xi, N, L_1, L_2\}$.

Example 5. Consider a null curve in $\mathbf{R}_2^6$ given by

$$C : \left(At^3 - 2t, \quad Bt^2, \quad At^3, \quad \sqrt{2}\,t, \quad \cos \sqrt{2}\,t, \quad \sin \sqrt{2}\,t \right)$$

where $B^2 = 3A$ and $A, B, t \in \mathbf{R}$. Then, we have

$$\xi = \left(3At^2 - 2, \quad 2Bt, \quad 3At^2, \quad \sqrt{2}, \quad -\sqrt{2} \sin \sqrt{2}\,t, \quad \sqrt{2} \cos \sqrt{2}\,t \right),$$
$$\nabla_\xi \xi = \left(6At, \quad 2B, \quad 6At, \quad 0, \quad -2 \cos \sqrt{2}\,t, \quad -2 \sin \sqrt{2}\,t \right).$$

We need to know the causal character of the vector field $H(t)$ along C. By direct calculations we obtain $g(H(t), H(t)) = 4(1 - B^2)$. Hence $H(t)$ is spacelike, timelike or null according as $B^2 < 1$, $B^2 > 1$ and $B^2 = 1$ respectively.

Choose $B = 2$ for which $A = \frac{4}{3}$ and $H(t)$ is timelike. Therefore, the curve falls in the Type 1 with the following general Frenet frame

$$F = \{\xi, N, W_1, W_2, W_3, W_4\}, \quad \text{where}$$

$$\xi = \left(4t^2 - 2, 4t, 4t^2, \sqrt{2}, -\sqrt{2} \sin \sqrt{2}\,t, \sqrt{2} \cos \sqrt{2}\,t \right),$$

$$N = -\frac{1}{8} \left(4t^2 - 2, 4t, 4t^2, -\sqrt{2}, \sqrt{2} \sin \sqrt{2}\,t, -\sqrt{2} \cos \sqrt{2}\,t \right),$$

$$W_1 = -\frac{1}{\sqrt{3}} \left(4t, 2, 4t, 0, -\cos \sqrt{2}\,t, -\sin \sqrt{2}\,t \right),$$

$$W_2 = \frac{1}{\sqrt{3}} \left(2t, 1, 2t, 0, -2 \cos \sqrt{2}\,t, -2 \sin \sqrt{2}\,t \right),$$

$$W_3 = \frac{1}{\sqrt{17}} \left(8t^2, 8t, 8t^2 + 4, \frac{1}{\sqrt{2}}, \frac{1}{\sqrt{2}} \sin \sqrt{2}\,t, -\frac{1}{\sqrt{2}} \cos \sqrt{2}\,t \right),$$

$$W_4 = -\frac{1}{\sqrt{17}} \left(2t^2, 2t, 2t^2 + 1, -2\sqrt{2}, -\sqrt{2} \sin \sqrt{2}\,t, 2\sqrt{2} \cos \sqrt{2}\,t \right).$$

The general Frenet equations (3.1.1) give

$$h = 0, \; \kappa_1 = -2\sqrt{3}, \; \kappa_2 = \frac{5}{4\sqrt{3}}, \; \kappa_3 = \frac{1}{\sqrt{3}}, \; \kappa_4 = 0,$$

$$\kappa_5 = \sqrt{\frac{17}{3}}, \; \kappa_6 = \frac{10}{\sqrt{51}}, \; \kappa_7 = 2\sqrt{\frac{3}{17}}, \; \kappa_8 = 0.$$

Now set $B = \frac{1}{2}$ so $A = \frac{1}{12}$. The curve falls in Type 1, W_1 is spacelike and

$$\nabla_\xi N + h N - \kappa_2 W_1 = W_2$$

is timelike with the general Frenet frame

$$\xi = \left(\frac{t^2}{4} - 2, \, t, \, \frac{t^2}{4}, \, \sqrt{2}, \, -\sqrt{2} \sin \sqrt{2}\,t, \, \sqrt{2} \cos \sqrt{2}\,t \right),$$

$$N = -\frac{1}{8} \left(\frac{t^2}{4} - 2, \, t, \, \frac{t^2}{4}, \, -\sqrt{2}, \, \sqrt{2} \sin \sqrt{2}\,t, \, -\sqrt{2} \cos \sqrt{2}\,t \right),$$

$$W_1 = \frac{1}{\sqrt{3}} \left(\frac{t}{2}, \, 1, \, \frac{t}{2}, \, 0, \, -2 \cos \sqrt{2}\,t, \, -2 \sin \sqrt{2}\,t \right),$$

$$W_2 = \frac{1}{\sqrt{3}} \left(t, \, 2, \, t, \, 0, \, -\cos \sqrt{2}\,t, \, -\sin \sqrt{2}\,t \right),$$

$$W_3 = \frac{1}{\sqrt{17}} \left(\frac{t^2}{8}, \, \frac{t}{2}, \, \frac{t^2}{8} + 1, \, 2\sqrt{2}, \, 2\sqrt{2} \sin \sqrt{2}\,t, \, -2\sqrt{2} \cos \sqrt{2}\,t \right),$$

$$W_4 = -\frac{1}{\sqrt{17}} \left(\frac{t^2}{2}, \, 2t, \, \frac{t^2}{2} + 4, \, -\frac{1}{\sqrt{2}}, \, -\frac{1}{\sqrt{2}} \sin \sqrt{2}\,t, \, \frac{1}{\sqrt{2}} \cos \sqrt{2}\,t \right).$$

Use of above in the general Frenet equations (3.1.1) gives

$$h = 0, \; \kappa_1 = \sqrt{3}, \; \kappa_2 = \frac{5}{8\sqrt{3}}, \; \kappa_3 = -\frac{1}{2\sqrt{3}}, \; \kappa_4 = 0,$$

$$\kappa_5 = \frac{1}{2}\sqrt{\frac{17}{3}}, \; \kappa_6 = -\frac{5}{\sqrt{51}}, \; \kappa_7 = \sqrt{\frac{3}{17}}, \; \kappa_8 = 0.$$

Finally set $B = 1$ for which $A = \frac{1}{3}$ and a Type 2 general Frenet frame

$$F = \{ \xi, \, N, \, L_1, \, L_2, \, W_3, \, W_4 \}, \quad \text{where}$$

$$\xi = \left(t^2 - 2, \, 2t, \, t^2, \, \sqrt{2}, \, -\sqrt{2} \sin \sqrt{2}\,t, \, \sqrt{2} \cos \sqrt{2}\,t \right),$$

$$N = -\frac{1}{8} \left(t^2 - 2, \, 2t, \, t^2, \, -\sqrt{2}, \, \sqrt{2} \sin \sqrt{2}\,t, \, -\sqrt{2} \cos \sqrt{2}\,t \right),$$

$$L_1 = \left(-t, \, -1, \, -t, \, 0, \, \cos \sqrt{2}\,t, \, \sin \sqrt{2}\,t \right),$$

$$L_2 = \frac{1}{2} \left(t, \, 1, \, t, \, 0, \, \cos \sqrt{2}\,t, \, \sin \sqrt{2}\,t \right),$$

$$W_3 = \frac{1}{\sqrt{2}} \left(0, 0, 0, 1, \sin\sqrt{2}t, -\cos\sqrt{2}t \right),$$

$$W_4 = \frac{1}{2} \left(t^2, 2t, t^2 + 2, 0, 0, 0 \right).$$

Using the general Frenet equations (3.2.6) we get

$$h = 0, \quad K_1 = -2, \quad K_2 = 0, \quad K_3 = -\frac{1}{2}, \quad \tau_2 = 0, \quad K_4 = 0,$$

$$K_5 = -1, \quad K_6 = -1, \quad K_7 = -\frac{1}{2}, \quad K_8 = \frac{1}{2}, \quad \kappa_8 = 0.$$

Example 6. Let C be the null curve in $\mathbf{R}_2^6$ given by

$$C : \left(\cosh t, \ \frac{1}{12}t^3 + 2t, \ \frac{1}{2}t^2, \ \frac{1}{12}t^3, \ \sqrt{3}t, \ \sinh t \right), \quad t \in \mathbf{R},$$

which is the curve in Example 1 such that $B = \frac{1}{2}$, therefore, $A = \frac{1}{12}$. Then the curve have the general Frenet frame of Type 2. We choose the general Frenet frame $F = \{\xi, N, L_1, L_2, W_3, W_4\}$ as follows;

$$\xi = \left(\sinh t, \ \frac{t^2}{4} + 2, \ t, \ \frac{t^2}{4}, \ \sqrt{3}, \ \cosh t \right),$$

$$N = \frac{1}{2} \left(\sinh t, \ -\frac{t^2}{4} - 2, \ -t, \ -\frac{t^2}{4}, \ -\sqrt{3}, \ \cosh t \right),$$

$$L_1 = \frac{1}{\sqrt{2}} \left(\cosh t, \ \frac{t}{2}, \ 1, \ \frac{t}{2}, \ 0, \ \sinh t \right),$$

$$L_2 = \frac{1}{\sqrt{2}} \left(-\cosh t, \ \frac{t}{2}, \ 1, \ \frac{t}{2}, \ 0, \ -\sinh t \right),$$

$$W_3 = -\left(0, \ \frac{t^2}{4} + \frac{3}{2}, \ t, \ \frac{t^2}{4} - \frac{1}{2}, \ \sqrt{3}, \ 0 \right),$$

$$W_4 = \left(0, \ \frac{\sqrt{3}}{2}, \ 0, \ \frac{\sqrt{3}}{2}, \ 1, \ 0 \right).$$

The Frenet equations (3.2.6) with respect to $F = \{\xi, N, L_1, L_2, W_3, W_4\}$ give

$$h = 0, \quad K_1 = \sqrt{2}, \quad K_2 = 0, \quad K_3 = -\frac{1}{\sqrt{2}}, \quad \tau_2 = 0, \quad \kappa_4 = 0,$$

$$K_5 = \frac{1}{\sqrt{2}}, \quad K_6 = 0, \quad K_7 = \frac{1}{\sqrt{2}}, \quad K_8 = 0, \quad \kappa_8 = 0.$$

3.3   Natural Frenet equations

We consider two Frenet frames F and F^* along neighborhoods $\mathcal{U}$ and $\mathcal{U}^*$ respectively with respect to a given screen vector bundle $S(TC^{\perp})$ such that $F \cap F^* \neq \emptyset$. Then the parameter transformations are given by

$$\xi^* = \frac{dt}{dt^*}\xi, \quad N^* = \frac{dt^*}{dt}N, \quad W_\alpha^* = \sum_{\beta=1}^{m} A_\alpha^\beta W_\beta, \quad 1 \leq \alpha \leq m, \qquad (3.3.1)$$

where A_α^β are smooth functions on $\mathcal{U} \cap \mathcal{U}^*$ and the matrix $[A_\alpha^\beta(x)]$ is an element of the Lorentzian group $O(1, m-1)$ for any $x \in \mathcal{U} \cap \mathcal{U}^*$.

Let F and $\bar{F}$ be two Frenet frames with respect to $(t, S(TC^{\perp}), \mathcal{U})$ and $(\bar{t}, \bar{S}(TC^{\perp}), \bar{\mathcal{U}})$ respectively. Then the transformations of the screen vector bundle that relate elements of F and $\bar{F}$ on $\mathcal{U} \cap \bar{\mathcal{U}} \neq \emptyset$ are given by

$$\begin{aligned}
\bar{\xi} &= \frac{dt}{d\bar{t}}\xi, \\
\bar{N} &= -\frac{1}{2}\frac{dt}{d\bar{t}}\sum_{\alpha=1}^{m}\epsilon_\alpha(c_\alpha)^2\xi + \frac{d\bar{t}}{dt}N + \sum_{\alpha=1}^{m}c_\alpha W_\alpha, \\
\bar{W}_\alpha &= \sum_{\beta=1}^{m}B_\alpha^\beta\left(W_\beta - \epsilon_\beta\frac{dt}{d\bar{t}}c_\beta\xi\right), \qquad 1 \leq \alpha \leq m,
\end{aligned} \qquad (3.3.2)$$

where c_α and B_α^β are smooth functions on $\mathcal{U} \cap \bar{\mathcal{U}}$ and the $m \times m$ matrix $[B_\alpha^\beta(x)]$ is also an element of the Lorentzian group $O(1, m-1)$ for each $x \in \mathcal{U} \cap \bar{\mathcal{U}}$.

Using the parameter transformations (3.3.1) and the first equation of the general Frenet equations (3.2.9) for both F and F^*, we obtain

$$(\kappa_1^*, \tau_1^*, 0, \cdots, 0)\left[A_\alpha^\beta(x)\right] = (\kappa_1, \tau_1, 0, \cdots, 0)\left(\frac{dt}{dt^*}\right)^2. \qquad (3.3.3)$$

Also, using the transformations of screen vector bundle (3.3.2) and the first equation of (3.2.9) for both F and $\bar{F}$, we obtain

$$(\bar{\kappa}_1, \bar{\tau}_1, 0, \cdots, 0)\left[B_\alpha^\beta(x)\right] = (\kappa_1, \tau_1, 0, \cdots, 0)\left(\frac{dt}{d\bar{t}}\right)^2. \qquad (3.3.4)$$

First, we show that each type of the general Frenet frames always transform to the same type by the transformations of the coordinate neighborhood and the screen vector bundle of C. We denote $\widetilde{F} = F^*$ or $\bar{F}$ and $P_j^i = A_j^i$ or B_j^i.

Proposition 3.1. *Let C be a null curve of a semi-Riemannian manifold M_2^{m+2}. Then the type of general Frenet equations is invariant to the transformations of the coordinate neighborhood and the screen vector bundle of C.*

Proof. In the first case suppose $\widetilde{F} = \widetilde{F}_2$ and $F = F_1$. Then we have $\widetilde{\tau}_1 = \widetilde{\kappa}_1$ and $\tau_1 = 0$. This means from equations (3.3.3) and (3.3.4) that $P_1^2 = P_2^2$.

Since $\widetilde{W}_1$ and $\widetilde{W}_2$ are the timelike and spacelike vector fields respectively, the first row $(P_1^1, P_1^2, 0, \cdots, 0)$ of $[P_\alpha^\beta(x)]$ is timelike vector field and the second row $(P_2^1, P_2^2, 0, \cdots, 0)$ is a spacelike vector field of $\mathbf{R}_1^m$ and these vectors are perpendicular to each other. Thus, we have

$$(P_1^1)^2 - 1 = (P_2^1)^2 + 1 = P_1^1 P_2^1.$$

This means $P_1^1 = P_2^1$ which is a contradiction so is not possible.

Conversely, if $\widetilde{F} = \widetilde{F}_1$ and $F = F_2$, then $\widetilde{\tau}_1 = 0$ and $\tau_1 = -\kappa_1$. From the equations (3.3.3) and (3.3.4) we have $P_1^1 = -P_1^2$, This means that the first row $(P_1^1, P_1^2, 0, \cdots, 0)$ of the matrix $[P_\alpha^\beta(x)]$ is a null vector. Hence, $\widetilde{W}_1$ is null, so this case is also not possible, which complete the proof.

We introduce another set of Frenet equations of Type 1 (different from the general Frenet equations (3.1.1) of Type 1) for null curves as chapter 2. Using the method of theorem 1.1 and 1.2 of chapter 2, we have

Theorem 3.1. *Let C be a null curve of M_2^{m+2} with a general Frenet frame F_1 of Type 1 with respect to the screen vector bundle $S(TC^\perp)$ such that $\kappa_1 \neq 0$. Then there exists a screen vector bundle $\bar{S}(TC)$ which induces another Frenet frame $\bar{F}$ on $\bar{\mathcal{U}}$ such that $\bar{\kappa}_4 = \bar{\kappa}_5 = 0$ on $\mathcal{U} \cap \bar{\mathcal{U}}$.*

Theorem 3.2. *Let C be a Type 1 null curve M_2^{m+2}. Then there exists a frame $F_1 = \{\xi, N, W_1, \ldots, W_m\}$ satisfying the equations*

$$\begin{aligned}
\nabla_\xi \xi &= h\xi + \kappa_1 W_1, \\
\nabla_\xi N &= -hN + \kappa_2 W_1 + \kappa_3 W_2, \\
\epsilon_1 \nabla_\xi W_1 &= -\kappa_2 \xi - \kappa_1 N, \\
\epsilon_2 \nabla_\xi W_2 &= -\kappa_3 \xi + \kappa_4 W_3, \\
\epsilon_3 \nabla_\xi W_3 &= -\kappa_4 W_2 + \kappa_5 W_4,
\end{aligned} \qquad (3.3.5)$$

$$\cdots\cdots\cdots\cdots\cdots$$

$$\begin{aligned}
\epsilon_i \nabla_\xi W_i &= -\kappa_{i+1} W_{i-1} + \kappa_{i+2} W_{i+1}, \quad 3 \leq i \leq m-1, \\
\epsilon_m \nabla_\xi W_m &= -\kappa_{m+1} W_{m-1},
\end{aligned}$$

where $\{\kappa_1, \ldots, \kappa_{m+1}\}$ are smooth functions on $\mathcal{U}$, $\{W_1, \ldots, W_m\}$ is a certain orthonormal basis of $\Gamma(S(TC^\perp)|_\mathcal{U})$ and $(\epsilon_1, \ldots, \epsilon_m)$ is the signature of M_2^{m+1}.

We call the frame $F_1 = \{\xi, N, W_1, \ldots, W_m\}$ in theorem 3.2 a *Natural Frenet frame of Type 1* on M_2^{m+2} along C with respect to the given screen vector bundle $S(TC^\perp)$ and the equations (3.3.5) are called its *Natural Frenet equations of Type 1*. Finally, the functions $\{\kappa_1, \ldots, \kappa_{m+1}\}$ are called *curvature functions* of C with respect to the frame F_1.

Example 7. Consider the null curve C in $\mathbf{R}_2^6$ presented in Example 3 in Section 3.1. This curve have the Frenet frame of Type 1. If we take the Natural Frenet frames $F_1 = \{\, \xi,\ N,\ W_1,\ W_2,\ W_3,\ W_4\,\}$ given by

$$\xi = \left(\frac{1}{4}t^2 + 1,\ \cosh t,\ \frac{1}{4}t^2 - 1,\ t,\ \sinh t,\ 1\right),$$

$$N = \frac{1}{8}\left(\frac{7}{4}t^2 + 9,\ -5\cosh t,\ \frac{7}{4}t^2 - 5,\ 7t,\ -5\sinh t,\ 7\right),$$

$$W_1 = \frac{1}{\sqrt{2}}\left(\frac{t}{2},\ \sinh t,\ \frac{t}{2},\ 1,\ \cosh t,\ 0\right),$$

$$W_2 = \frac{1}{\sqrt{2}}\left(\frac{t}{2},\ -\sinh t,\ \frac{t}{2},\ 1,\ -\cosh t,\ 0\right),$$

$$W_3 = \frac{1}{2}\left(\frac{1}{4}t^2 + 2,\ -\cosh t,\ \frac{1}{4}t^2,\ t,\ -\sinh t,\ 1\right),$$

$$W_4 = \left(\frac{1}{2},\ 0,\ \frac{1}{2},\ 0,\ 0,\ 1\right).$$

Then the Natural Frenet equations of Type 1 with respect to the above frame are

$$\nabla_\xi \xi = \sqrt{2}\,W_1, \qquad\qquad \nabla_\xi N = \frac{1}{4\sqrt{2}}W_1 + \frac{3}{2\sqrt{2}}W_2,$$

$$\nabla_\xi W_1 = -\frac{1}{4\sqrt{2}}\xi - \sqrt{2}\,N, \quad \nabla_\xi W_2 = -\frac{3}{2\sqrt{2}}\xi + \frac{1}{\sqrt{2}}W_3,$$

$$\nabla_\xi W_3 = -\frac{1}{\sqrt{2}}W_2, \qquad\qquad \nabla_\xi W_4 = 0.$$

Example 8. Consider the null curve C in $\mathbf{R}_2^6$ presented in Example 4 in Section 3.1. This curve have also the Frenet frame of Type 1. If we take the Natural Frenet frames $F_1 = \{\, \xi,\ N,\ W_1,\ W_2,\ W_3,\ W_4\,\}$ as follows;

$$\xi = \left(\sinh t,\ \frac{t^2}{9} + 2,\ \frac{2}{3}t,\ \frac{t^2}{9},\ \sqrt{3},\ \cosh t\right),$$

$$N = \frac{9}{50}\left(\sinh t,\ -t^2 - \frac{142}{9},\ -6t,\ -t^2 + \frac{20}{9},\ -9\sqrt{3},\ \cosh t\right),$$

$$W_1 = -\frac{3}{\sqrt{5}}\left(\cosh t,\ \frac{2}{9}t,\ \frac{2}{3},\ \frac{2}{9}t,\ 0,\ \sinh t\right),$$

$$W_2 = -\frac{1}{\sqrt{5}}\left(2\cosh t,\ t,\ 3,\ t,\ 0,\ 2\sinh t\right),$$

$$W_3 = -\frac{4}{5}\left(\sinh t,\ \frac{t^2}{4} + \frac{31}{8},\ \frac{3}{2}t,\ \frac{t^2}{4},\ \frac{9}{4}\sqrt{3},\ \cosh t\right),$$

$$W_4 = \left(0,\ \frac{\sqrt{3}}{2},\ 0,\ \frac{\sqrt{3}}{2},\ 1,\ 0\right).$$

The Natural Frenet equations (3.3.5) of Type 1 give

$$\nabla_\xi \xi = -\frac{\sqrt{5}}{3}W_1, \qquad\qquad \nabla_\xi N = \frac{27}{10\sqrt{5}}W_1 + \frac{18}{5\sqrt{5}}W_2,$$

$$\nabla_\xi W_1 \;=\; -\frac{27}{10\sqrt{5}}\,\xi - \frac{\sqrt{5}}{3}\,N, \quad \nabla_\xi W_2 \;=\; -\frac{18}{5\sqrt{5}}\,\xi + \frac{3}{2\sqrt{5}}\,W_3,$$

$$\nabla_\xi W_3 \;=\; -\frac{3}{2\sqrt{5}}\,W_2, \qquad\qquad \nabla_\xi W_4 \;=\; 0.$$

We introduce another form of Frenet equations of Type 2 different from the general Frenet equations of Type 2.

Let $\tau_1 \neq 0$. Then, from the equation (3.3.3) and (3.3.4), we have

$$P_1^1 + P_2^1 \;=\; P_1^2 + P_2^2, \qquad P_1^\alpha \;=\; -P_2^\alpha, \quad \alpha \in \{3, \cdots, m\}, \qquad (3.3.6)$$

as $\widetilde{\tau}_1 = \widetilde{\kappa}_1$ and $\tau_1 = \kappa_1$. Using (3.3.1), (3.3.6) and $L_1 = \frac{1}{\sqrt{2}}(W_1 + W_2)$, we get

$$L_1^* \;=\; \left(A_1^1 + A_2^1 \right) L_1, \qquad (3.3.7)$$

where $A_1^1 + A_2^1 \neq 0$, otherwise the matrix $[\,A_\alpha^\beta(x)\,]$ is singular. Using (3.3.2), (3.3.6) and $L_1 = \frac{1}{\sqrt{2}}(W_1 + W_2)$, we obtain

$$\bar{L}_1 \;=\; \left(B_1^1 + B_2^1 \right) \left\{ L_1 \;-\; C_2\,\frac{dt}{d\bar{t}}\,\xi \right\}, \qquad (3.3.8)$$

where $C_2 = \frac{1}{\sqrt{2}}(c_2 - c_1)$ and $B_1^1 + B_2^1 \neq 0$. Since $\widetilde{W}_1$ and $\widetilde{W}_2$ are timelike and spacelike respectively, the first row $(P_1^1, \ldots, P_1^m)$ of the matrix $[\,P_\alpha^\beta(x)\,]$ is a timelike and the second row $(P_2^1, \ldots, P_2^m)$ of $[\,P_\alpha^\beta(x)\,]$ is a spacelike vector field of $\mathbf{R}_1^m$ respectively and both are perpendicular to each other. Thus we have

$$\left(P_1^1\right)^2 - \left(P_1^2\right)^2 - 1 \;=\; \left(P_2^1\right)^2 - \left(P_2^2\right)^2 + 1 \;=\; -P_1^1 P_2^1 + P_1^2 P_2^2.$$

From this relation we have the following relation

$$\det \begin{pmatrix} P_1^1 & P_2^1 \\ P_2^1 & P_2^2 \end{pmatrix} \;=\; \det \begin{pmatrix} 0 & P_1^1 + P_2^1 \\ -P_1^1 + P_1^2 & 0 \end{pmatrix} \;=\; 1.$$

Using (3.2.1), (3.3.1), (3.3.2) and (3.3.6) for $\alpha \in \{3, \ldots, m\}$ we have

$$W_\alpha^* \;=\; \frac{1}{\sqrt{2}} \left(A_\alpha^2 + A_\alpha^1 \right) L_1 + \frac{1}{\sqrt{2}} \left(A_\alpha^2 - A_\alpha^1 \right) L_2 + \sum_{\beta \geq 3}^{m} A_\alpha^\beta W_\beta,$$

$$\bar{W}_\alpha \;=\; \frac{1}{\sqrt{2}} \left(B_\alpha^2 + B_\alpha^1 \right) L_1 + \frac{1}{\sqrt{2}} \left(B_\alpha^2 - B_\alpha^1 \right) L_2 + \sum_{\beta \geq 3}^{m} B_\alpha^\beta W_\beta$$

$$-\; \sum_{i \geq 1}^{m} B_\alpha^i\, \epsilon_i\, c_i\, \frac{dt}{d\bar{t}}\,\xi.$$

The scalar product of $\widetilde{L}_1$ and $\widetilde{W}_\alpha$ provides

$$P_\alpha^1 \;=\; P_\alpha^2 \quad (3 \leq \alpha \leq m). \qquad (3.3.9)$$

Also, using (3.2.1), (3.3.1), (3.3.2) and (3.3.6) for $\alpha \in \{3, \ldots, m\}$ we have

$$L_2^* = \frac{1}{2}\left(A_2^1 - A_1^1 + A_2^2 - A_1^2\right)L_1 + \left(A_1^1 - A_1^2\right)L_2 - \sqrt{2}\sum_{\alpha \geq 3}^{m} A_1^{\alpha}W_{\alpha},$$

$$\bar{L}_2 = \frac{1}{2}\left(B_2^1 - B_1^1 + B_2^2 - B_1^2\right)L_1 + \left(B_1^1 - B_1^2\right)L_2 - \sqrt{2}\sum_{\alpha \geq 3}^{m} B_1^{\alpha}W_{\alpha}$$

$$- \frac{1}{\sqrt{2}}\sum_{i \geq 1}^{m}\left(B_2^i - B_1^i\right)\epsilon_i c_i \frac{dt}{dt}\xi.$$

From $g(\tilde{L}_2, \tilde{L}_2) = 0$, we have

$$\left(P_1^1 - P_1^2\right)\left(P_2^1 - P_1^1 + P_2^2 - P_1^2\right) + 2\sum_{\alpha \geq 3}^{m}\epsilon_{\alpha}\left(P_1^{\alpha}\right)^2 = 0.$$

Using this equation and the first equation of (3.3.6), we obtain

$$P_1^1 = P_2^2; \ P_1^2 = P_2^1 \ \Leftrightarrow \ P_2^1 - P_1^1 + P_2^2 - P_1^2 = 0 \ \Leftrightarrow \ \sum_{\alpha \geq 3}^{m}\epsilon_{\alpha}\left(P_1^{\alpha}\right)^2 = 0.$$

As a consequence of the following procedure for the fundamental transformation of matrices with respect to the rows and columns for the matrix $\left[P_{\beta}^{\alpha}\right]$

Step 1. Add 1 times the second row to the first row.

Step 2. Add -1 times the second column to the first column.

Step 3. Add $-\frac{P_2^2}{P_1^1 + P_2^1}$ times the first row to the second row.

Step 4. Add $-\frac{P_3^2}{P_1^1 + P_2^1}$ times the first row to the third row.

....

Step (m + 1). Add $-\frac{P_m^2}{P_1^1 + P_2^1}$ times the first row to the m-th row.

Step (m + 2). Add $-\frac{P_2^3}{P_2^2 - P_2^2}$ times the first column to the third column.

Step (m + 3). Add $-\frac{P_2^4}{P_2^1 - P_2^2}$ times the first column to the fourth column.

....

Step (2m - 1). Add $-\frac{P_2^3}{P_2^1 - P_2^2}$ times the first column to the m-th column.

and using the equations (3.3.6) and (3.3.9), the matrix $\left[\, P_\beta^\alpha \,\right]$ is transformed as

$$
T_1(P) = \begin{pmatrix}
0 & P_1^1 + P_2^1 & 0 & \cdots & 0 \\
-P_1^1 + P_1^2 & 0 & 0 & \cdots & 0 \\
0 & 0 & P_3^3 & \cdots & P_3^m \\
0 & 0 & P_4^3 & \cdots & P_4^m \\
\vdots & \vdots & \vdots & \cdots & \vdots \\
0 & 0 & P_m^3 & \cdots & P_m^m
\end{pmatrix}.
\tag{3.3.10}
$$

The transformations (3.3.10) are called the *canonical transformations*. Thus the screen vector bundle $S(TC^\perp)$ is a orthogonal direct sum of two invariant subspaces $Span\,\{L_1,\,L_2\} = Span\,\{W_1,\,W_2\}$ and $Span\,\{W_3,\,\ldots,\,W_m\}$ by the canonical transformations of the coordinate neighborhoods and the screen vector bundle of C.

Using the transformations of the screen vector bundle (3.3.10) that relate elements of F_2 and $\bar{F}_2$ on $\mathcal{U} \cap \bar{\mathcal{U}}$ such that $\tau_1 \neq 0$, we have (3.3.2) and

$$
\bar{N} = -\frac{1}{2}\frac{dt}{d\bar{t}}\sum_{\alpha \geq 1}^{m} \epsilon_\alpha (c_\alpha)^2 \xi + \frac{d\bar{t}}{dt}N + C_1 L_1 + C_2 L_2 + \sum_{\alpha \geq 3}^{m} c_\alpha W_\alpha,
$$

$$
\bar{L}_2 = (B_1^1 - B_1^2)\left\{ L_2 - C_1 \frac{dt}{d\bar{t}}\xi \right\}, \quad C_1 = \frac{1}{\sqrt{2}}(c_1 + c_2),
\tag{3.3.11}
$$

$$
\bar{W}_\alpha = \sum_{\beta \geq 3}^{m} B_\alpha^\beta \left\{ W_\beta - \epsilon_\beta c_\beta \frac{dt}{d\bar{t}}\xi \right\}, \quad \alpha,\,\beta \in \{3 \cdots,\,m\}.
$$

Theorem 3.3. *Let C be a null curve of a semi-Riemannian manifold M_2^{m+2} with a general Frenet frame F_2 of Type 2 with respect to the screen vector bundle $S(TC^\perp)$ such that $\tau_1 \neq 0$. Then there exists a screen vector bundle $\bar{S}(TC)$ which induces another Frenet frame $\bar{F}_2$ of Type 2 on $\bar{\mathcal{U}}$ such that $\bar{K}_4 = \bar{K}_7 = \bar{K}_8 = \bar{\ell} = \bar{\kappa}_4 = \bar{\tau}_4 = 0$ on $\mathcal{U} \cap \bar{\mathcal{U}}$.*

Proof. The transformations of the screen vector bundle relating the elements of F_2 and $\bar{F}_2$ on $\mathcal{U} \cap \bar{\mathcal{U}}$ are given by (3.3.8) and (3.3.11). From the fourth equation of the Frenet equations (3.2.6), we have

$$
\bar{K}_1 = (B_1^1 - B_1^2)\,K_1 \left(\frac{dt}{d\bar{t}}\right)^2 ;
$$

$$
\bar{K}_4 = \left\{ K_4 + K_1 C_2 \frac{dt}{d\bar{t}} - (B_1^1 - B_1^2)\frac{d(B_1^1 + B_2^1)}{dt} \right\}\frac{dt}{d\bar{t}},
$$

$$
\bar{K}_7 B_3^3 + \bar{K}_8 B_4^3 + \bar{\ell}B_5^3 = (B_1^1 - B_1^2)\left(K_7 + K_1 c_3 \frac{dt}{d\bar{t}} \right)\frac{dt}{d\bar{t}},
$$

$$
\bar{K}_7 B_3^4 + \bar{K}_8 B_4^4 + \bar{\ell}B_5^4 = (B_1^1 - B_1^2)\left(K_8 + K_1 c_4 \frac{dt}{d\bar{t}} \right)\frac{dt}{d\bar{t}},
$$

$$\bar{K}_7 B_3^5 + \bar{K}_8 B_4^5 + \bar{\ell} B_5^5 = (B_1^1 - B_1^2)\left(\ell + K_1 c_5 \frac{dt}{d\bar{t}}\right)\frac{dt}{d\bar{t}},$$

$$\bar{K}_7 B_3^\alpha + \bar{K}_8 B_4^\alpha + \bar{\ell} B_5^\alpha = K_1 c_\alpha (B_1^1 - B_1^2)\left(\frac{dt}{d\bar{t}}\right)^2.$$

Taking into account that

$$C_2 = -\frac{K_4}{K_1}\frac{d\bar{t}}{dt}; \quad c_3 = -\frac{K_7}{K_1}\frac{d\bar{t}}{dt}; \quad c_4 = -\frac{K_8}{K_1}\frac{d\bar{t}}{dt}; \quad c_5 = -\frac{\ell}{K_1}\frac{d\bar{t}}{dt};$$

$$c_\alpha = 0, \quad \alpha \in \{6, \cdots, m\}; \qquad B_1^1 - B_1^2 = c \text{ (constant)},$$

in the last equations, we have $\bar{K}_4 = \bar{K}_7 = \bar{K}_8 = \bar{\ell} = 0$. On the other hand, from (3.2.1), (3.2.4) and (3.2.6), we have $\bar{\kappa}_4 = -\bar{K}_4$; $\bar{\tau}_4 = \sqrt{2}\bar{\ell}$ and $\sqrt{2}\bar{K}_7 = \bar{\kappa}_6 + \bar{\kappa}_5$; $\sqrt{2}\bar{K}_8 = \bar{\kappa}_7 + \bar{\tau}_3$, we have $\bar{\kappa}_4 = \bar{\tau}_4 = 0$ and $\bar{\kappa}_6 = -\bar{\kappa}_5$; $\bar{\kappa}_7 = -\bar{\tau}_3$.

Remark. If we take $t = \bar{t}$ in Theorem 3.3, then

$$C_2 = -\frac{K_4}{K_1}; \quad c_3 = -\frac{K_7}{K_1}; \quad c_4 = -\frac{K_8}{K_1}; \quad c_5 = -\frac{\ell}{K_1}; \quad c_\alpha = 0, \ \alpha > 5;$$

$$\bar{N} = -\frac{1}{2}\left(\frac{K_7^2 + K_8^2 + \ell^2}{K_1^2}\right)\xi + N - \frac{K_4}{K_1}L_2 - \frac{K_7}{K_1}W_3 - \frac{K_8}{K_1}W_4 - \frac{\ell}{K_1}W_5,$$

$$\bar{L}_1 = L_1 + \frac{K_4}{K_1}\xi, \quad \bar{L}_2 = L_2, \quad \bar{W}_3 = W_3 + \frac{K_7}{K_1}\xi,$$

$$\bar{W}_4 = W_4 + \frac{K_8}{K_1}\xi, \quad \bar{W}_5 = W_5 + \frac{\ell}{K_1}\xi, \quad \bar{W}_i = W_i, \ i \in \{6, \cdots, m\}.$$

Relabel $N = \bar{N}$; $L_\alpha = \bar{L}_\alpha$, $\alpha \in \{1, 2\}$; $W_3 = \bar{W}_3$; $K_4 = \bar{K}_5$, $K_5 = \bar{K}_6$; $\kappa_5 = \bar{\kappa}_8$ and $S(TC^\perp) = \bar{S}(TC^\perp)$ and take only first five equations in (3.2.6) as follows:

$$\nabla_\xi \xi = h\xi + K_1 L_1,$$
$$\nabla_\xi N = -hN + K_2 L_1 + K_3 L_2 + \tau_2 W_3,$$
$$\nabla_\xi L_1 = -K_3 \xi + K_4 W_3 + K_5 W_4,$$
$$\nabla_\xi L_2 = -K_2 \xi - K_1 N,$$
$$\nabla_\xi W_3 = -\tau_2 \xi - K_4 L_2 + \kappa_5 W_4 + R_6,$$

where $R_6 \in \Gamma(S(TC^\perp))$ is a spacelike vector field perpendicular to L_1, L_2, W_3 and W_4. Define the new third torsion function τ_3 by $\tau_3 = \|R_6\|$ and set $W_5 = \frac{R_6}{\tau_3}$. Then W_5 is also a unit spacelike vector field along C. Thus we have

$$\nabla_\xi W_3 = -\tau_2 \xi - K_4 L_2 + \kappa_5 W_4 + \tau_3 W_5.$$

Using the same method and $\nabla_\xi W_4 = -K_5 L_2 - \kappa_5 W_3 + \kappa_6 W_5 + R_6$, we have

$$\nabla_\xi W_4 = -K_5 L_2 - \kappa_5 W_3 + \kappa_6 W_5 + \tau_4 W_6.$$

Repeating above process for all the 2 null and $(m-2)$ unit spacelike vectors of a semi-orthogonal basis $F_2 = \{L_1, L_2, W_3, \cdots, W_m\}$ of $\Gamma(S(TC^\perp))$ and simplifying, we obtain the following;

$$
\begin{aligned}
\nabla_\xi \xi &= h\xi + K_1 L_1, \\
\nabla_\xi N &= -hN + K_2 L_1 + K_3 L_2 + \tau_2 W_3, \\
\nabla_\xi L_1 &= -K_3 \xi + K_4 W_3 + K_5 W_4, \\
\nabla_\xi L_2 &= -K_2 \xi - K_1 N, \\
\nabla_\xi W_3 &= -\tau_2 \xi - K_4 L_2 + \kappa_5 W_4 + \tau_4 W_5, \\
\nabla_\xi W_4 &= -K_5 L_2 - \kappa_5 W_3 + \kappa_6 W_5 + \tau_5 W_6, \\
\nabla_\xi W_5 &= -\tau_4 W_3 - \kappa_6 W_4 + \kappa_7 W_6 + \tau_6 W_7,
\end{aligned}
\tag{3.3.12}
$$

$$
\begin{aligned}
\nabla_\xi W_i &= -\tau_{i-1} W_{i-2} - \kappa_{i+1} W_{i-1} + \kappa_{i+2} W_{i+1} + \tau_{i+1} W_{i+2}, \quad 5 \le i \le m-1, \\
\nabla_\xi W_m &= -\tau_{m-1} W_{m-2} - \kappa_{m+1} W_{m-1},
\end{aligned}
$$

or equivalently, there exists a frame $F_2 = \{\xi, N, W_1, \ldots, W_m\}$ satisfying

$$
\begin{aligned}
\nabla_\zeta \xi &= h\xi + \kappa_1 W_1 + \tau_1 W_2, \\
\nabla_\xi N &= -hN + \kappa_2 W_1 + \kappa_3 W_2 + \tau_2 W_3, \\
\nabla_\xi W_1 &= \kappa_2 \xi + \kappa_1 N - \kappa_4 W_3 - \tau_3 W_4, \\
\nabla_\xi W_2 &= -\kappa_3 \xi - \tau_1 N + \kappa_4 W_3 + \tau_3 W_4, \\
\nabla_\xi W_3 &= -\tau_2 \xi - \kappa_4 W_1 + \kappa_4 W_2 + \kappa_5 W_4 + \tau_4 W_5, \\
\nabla_\xi W_4 &= -\tau_3 W_1 + \tau_3 W_2 - \kappa_5 W_3 + \kappa_6 W_5 + \tau_5 W_6, \\
\nabla_\xi W_5 &= -\tau_4 W_3 - \kappa_6 W_4 + \kappa_7 W_6 + \tau_6 W_7,
\end{aligned}
\tag{3.3.13}
$$

$$
\begin{aligned}
\nabla_\xi W_i &= -\tau_{i-1} W_{i-2} - \kappa_{i+1} W_{i-1} + \kappa_{i+2} W_{i+1} + \tau_{i+1} W_{i+2}, \quad 5 \le i \le m-1, \\
\nabla_\xi W_m &= -\tau_{m-1} W_{m-2} - \kappa_{m+1} W_{m-1}.
\end{aligned}
$$

We call the frame F_2 a *Natural Frenet frame of Type 2* on M_2^{m+2} along C with respect to the given screen vector bundle $S(TC^\perp)$ and the equations (3.3.12) and (3.3.13) are called its *Natural Frenet equations of Type 2*. Finally, the functions $\{\kappa_1, \ldots, \kappa_{m+1}\}$ and $\{\tau_1, \ldots, \tau_{m-1}\}$ are called the *curvature* and *torsion* functions of C with respect to the Natural Frenet frame F_2.

Next, using the transformations (3.3.10) such that $\tau_1 = 0$, we have

Theorem 3.4. Let C be a null curve of a semi-Riemannian manifold M_2^{m+2} with a general Frenet frame F_2 of Type 2 with respect to the screen vector bundle $S(TC^\perp)$ such that $\kappa_1 \ne 0$ and $\tau_1 = 0$. Then there exists a screen vector bundle $\bar{S}(TC)$ which induces another Frenet frame $\bar{F}_2$ of Type 2 on $\bar{\mathcal{U}}$ such that $\bar{\kappa}_4 = \bar{\kappa}_5 = \bar{\tau}_3 = 0$.

Proof. By the method of Theorem 3.3, from the first equation of the general

Frenet equations (3.2.9) such that $\tau_1 = 0$, we have

$$\bar{\kappa}_1 B_1^1 = \kappa_1 \left(\frac{dt}{d\bar{t}}\right)^2 ; \qquad \bar{\kappa}_1 B_1^\alpha = 0, \quad \alpha \in \{2, \cdots, m\},$$

$B_1^\alpha = B_\alpha^1 = 0 \, (\alpha \neq 1)$ and $B_1^1 = B_1 = \pm 1$. Also, using (3.3.2) and the third equation of the general Frenet equations (3.2.9), we have

$$\bar{\kappa}_4 B_2^2 + \bar{\kappa}_5 B_3^2 + \bar{\tau}_3 B_4^2 = B_1 \left(\kappa_4 + \kappa_1 c_2 \frac{dt}{d\bar{t}}\right) \frac{dt}{d\bar{t}},$$

$$\bar{\kappa}_4 B_2^3 + \bar{\kappa}_5 B_3^3 + \bar{\tau}_3 B_4^3 = B_1 \left(\kappa_5 + \kappa_1 c_3 \frac{dt}{d\bar{t}}\right) \frac{dt}{d\bar{t}},$$

$$\bar{\kappa}_4 B_2^4 + \bar{\kappa}_5 B_3^4 + \bar{\tau}_3 B_4^4 = B_1 \left(\tau_3 + \kappa_1 c_4 \frac{dt}{d\bar{t}}\right) \frac{dt}{d\bar{t}},$$

$$\bar{\kappa}_4 B_2^\alpha + \bar{\kappa}_5 B_3^\alpha + \bar{\tau}_3 B_4^\alpha = B_1 \kappa_1 c_\alpha \left(\frac{dt}{d\bar{t}}\right)^2, \quad \alpha \in \{5, \cdots, m\}.$$

Taking into account that

$$c_2 = -\frac{\kappa_4}{\kappa_1}\frac{d\bar{t}}{dt}; \quad c_3 = -\frac{\kappa_5}{\kappa_1}\frac{d\bar{t}}{dt}; \quad c_4 = -\frac{\tau_3}{\kappa_1}\frac{d\bar{t}}{dt};$$

$$c_\alpha = 0, \quad \alpha \in \{5, \cdots, m\}$$

in the last equations, we have $\bar{\kappa}_4 = \bar{\kappa}_5 = \bar{\tau}_3 = 0$.

Remark. If we take $t = \bar{t}$ in Theorem 3.4, then $c_2 = -\frac{\kappa_4}{\kappa_1}$; $c_3 = -\frac{\kappa_5}{\kappa_1}$; $c_4 = -\frac{\tau_3}{\kappa_1}$; $c_\alpha = 0$, $\alpha > 4$ and

$$\bar{N} = -\frac{1}{2}\left(\frac{\kappa_4^2 + \kappa_5^2 + \tau_3^2}{\kappa_1^2}\right)\xi + N - \frac{\kappa_4}{\kappa_1}W_2 - \frac{\kappa_5}{\kappa_1}W_3 - \frac{\tau_3}{\kappa_1}W_4,$$

$$\bar{W}_2 = W_2 + \frac{\kappa_4}{\kappa_1}\xi, \quad \bar{W}_3 = W_3 + \frac{\kappa_5}{\kappa_1}\xi, \quad \bar{W}_4 = W_4 + \frac{\tau_3}{\kappa_1}\xi,$$

$$\bar{W}_i = W_i, \quad i \in \{1, 5, \cdots, m\}.$$

Relabel $N = \bar{N}$, $W_1 = \bar{W}_1$, $W_2 = \bar{W}_2$, $\kappa_i = \bar{\kappa}_i$, $i \in \{1, 2, 3\}$, $\tau_2 = \bar{\tau}_2$, $\kappa_4 = \bar{\kappa}_6$ and $S(TC^\perp) = \bar{S}(TC^\perp)$ in the process of the above theorem and take only the first four equations in (3.2.9) with $\tau_1 = 0$ as follows:

$$\nabla_\xi \xi = h\,\xi + \kappa_1 W_1,$$

$$\nabla_\xi N = -h\,N + \kappa_2 W_1 + \kappa_3 W_2 + \tau_2 W_3,$$

$$\epsilon_1 \nabla_\xi W_1 = -\kappa_2 \xi - \kappa_1 N,$$

$$\epsilon_2 \nabla_\xi W_2 = -\kappa_3 \xi + \kappa_4 W_3 + R_7,$$

where $R_7 \in \Gamma(S(TC^\perp))$. Thus R_7 is a vector field perpendicular to ξ, N, W_1 and W_2. Assume that R_7 is a non-null. Denote $\rho_7 = \|R_7\|$ and define the new torsion function τ_3 by $\tau_3 = \epsilon_4 \, \rho_7$, where ϵ_4 is the sign of R_7. If ρ_7 is also non-zero for any t, we set $\bar{W}_4 = \frac{R_7}{\rho_7}$, then $\bar{W}_4$ is also a non-null unit vector field with the same causality as R_7 along C. Thus we have

$$\nabla_\xi W_2 = -\kappa_3 \, \xi + \kappa_4 \, W_3 + \tau_3 \, \epsilon_4 \, \bar{W}_4.$$

Repeating above process for all the m unit vectors of an orthonormal basis $\{W_1, W_2, W_3, \bar{W}_4, \ldots, \bar{W}_m\}$, of $\Gamma(S(TC^\perp))$, and then setting

$$W_i = \epsilon_i \, \bar{W}_i, \qquad i \in \{3, \ldots, m\}$$

and after some simplification we obtain the following;

$$
\begin{aligned}
\nabla_\xi \xi &= h\xi + \kappa_1 W_1, \\
\nabla_\xi N &= -hN + \kappa_2 W_1 + \kappa_3 W_2 + \tau_2 W_3, \\
\epsilon_1 \nabla_\xi W_1 &= -\kappa_2 \xi - \kappa_1 N, \\
\epsilon_2 \nabla_\xi W_2 &= -\kappa_3 \xi + \kappa_4 W_3 + \tau_3 W_4, \\
\epsilon_3 \nabla_\xi W_3 &= -\tau_2 \xi - \kappa_4 W_2 + \kappa_5 W_4 + \tau_4 W_5, \\
\epsilon_4 \nabla_\xi W_4 &= -\tau_3 W_2 - \kappa_5 W_3 + \kappa_6 W_5 + \tau_5 W_6, \qquad\qquad (3.3.14)\\
\epsilon_5 \nabla_\xi W_5 &= -\tau_4 W_3 - \kappa_6 W_4 + \kappa_7 W_6 + \tau_6 W_7,
\end{aligned}
$$

$$\cdots\cdots\cdots\cdots\cdots\cdots\cdots\cdots\cdots\cdots$$

$$
\begin{aligned}
\epsilon_i \nabla_\xi W_i &= -\tau_{i-1} W_{i-2} - \kappa_{i+1} W_{i-1} + \kappa_{i+2} W_{i+1} + \tau_{i+1} W_{i+2}, \; 3 \le i \le m-1, \\
\epsilon_m \nabla_\xi W_m &= -\tau_{m-1} W_{m-2} - \kappa_{m+1} W_{m-1}.
\end{aligned}
$$

We call the frame $F_2 = \{\xi, N, W_1, \ldots, W_m\}$ a *Natural Frenet frame of Type 2* on M_2^{m+2} along C with respect to the given screen vector bundle $S(TC^\perp)$ and the equations (3.3.14) are called its *Natural Frenet equations of Type 2*. Finally, the functions $\{\kappa_1, \ldots, \kappa_{m+1}\}$ and $\{\tau_2, \cdots, \tau_{m-1}\}$ are called *curvature* and *torsion functions* of C with respect to the frame F_2.

Example 9. Consider a null curve C in $\mathbf{R}_2^6$ given by

$$C: \left(\frac{1}{3}t^3 - 2t, \; t^2, \; \frac{1}{3}t^3, \; \sqrt{2}t, \; \cos\sqrt{2}t, \; \sin\sqrt{2}t \right).$$

In example 5, we show that C have Frenet frames of Type 2. Now if we take the Natural Frenet frames $F_2 = \{\xi, N, L_1, L_2, W_3, W_4\}$ of Type 2 given by

$$
\begin{aligned}
\xi &= \left(t^2 - 2, \; 2t, \; t^2, \; \sqrt{2}, \; -\sqrt{2}\sin\sqrt{2}t, \; \sqrt{2}\cos\sqrt{2}t \right), \\
N &= -\frac{1}{16} \left(t^2 - 6, \; 2t, \; t^2 - 4, \; \sqrt{2}, \; 3\sqrt{2}\sin\sqrt{2}t, \; -3\sqrt{2}\cos\sqrt{2}t \right), \\
L_1 &= \left(-t, \; -1, \; -t, \; 0, \; \cos\sqrt{2}t, \; \sin\sqrt{2}t \right), \\
L_2 &= \frac{1}{2} \left(t, \; 1, \; t, \; 0, \; \cos\sqrt{2}t, \; \sin\sqrt{2}t \right),
\end{aligned}
$$

$$W_3 = \frac{1}{4}\left(t^2 - 2, \ 2t, \ t^2, \ 3\sqrt{2}, \ \sqrt{2}\sin\sqrt{2}t, \ -\sqrt{2}\cos\sqrt{2}t\right),$$

$$W_4 = \frac{1}{4}\left(t^2 + 2, \ 2t, \ t^2 + 4, \ -\sqrt{2}, \ \sqrt{2}\sin\sqrt{2}t, \ -\sqrt{2}\cos\sqrt{2}t\right).$$

The Natural Frenet equation (3.3.12) of Type 2 give

$$\nabla_\xi \xi = -2L_1, \qquad\qquad \nabla_\xi N = -\frac{1}{8}L_1 - \frac{4}{5}L_2,$$

$$\nabla_\xi L_1 = \frac{4}{5}\xi - W_3 - W_4, \quad \nabla_\xi L_2 = \frac{1}{8}\xi + 2N,$$

$$\nabla_\xi W_3 = \nabla_\xi W_4 = L_2.$$

Example 10. Let C be the null curve in $\mathbf{R}_2^6$ given by

$$C: \left(\cosh t, \ \frac{1}{12}t^3 + 2t, \ \frac{1}{2}t^2, \ \frac{1}{12}t^3, \ \sqrt{3}t, \ \sinh t\right), \quad t \in R.$$

In example 6, this curve have Frenet frames of Type 2. We choose the Natural Frenet frame $F_2 = \{\xi, N, L_1, L_2, W_3, W_4\}$ of Type 2 as follows;

$$\xi = \left(\sinh t, \ \frac{t^2}{4} + 2, \ t, \ \frac{t^2}{4}, \ \sqrt{3}, \ \cosh t\right),$$

$$N = -\frac{1}{8}\left(-3\sinh t, \ \frac{t^2}{4} + 4, \ t, \ \frac{t^2}{4} + 2, \ \sqrt{3}, \ -3\cosh t\right),$$

$$L_1 = \frac{1}{\sqrt{2}}\left(\cosh t, \ \frac{t}{2}, \ 1, \ \frac{t}{2}, \ 0, \ \sinh t\right),$$

$$L_2 = \frac{1}{\sqrt{2}}\left(-\cosh t, \ \frac{t}{2}, \ 1, \ \frac{t}{2}, \ 0, \ -\sinh t\right),$$

$$W_3 = -\frac{1}{2}\left(-\sinh t, \ \frac{t^2}{4} + 1, \ t, \ \frac{t^2}{4} - 1, \ \sqrt{3}, \ -\cosh t\right),$$

$$W_4 = \left(0, \ \frac{\sqrt{3}}{2}, \ 0, \ \frac{\sqrt{3}}{2}, \ 1, \ 0\right).$$

The Natural Frenet equation (3.3.12) of Type 2 give

$$\nabla_\xi \xi = \sqrt{2}L_1, \qquad\qquad \nabla_\xi N = \frac{1}{4\sqrt{2}}L_1 - \frac{1}{2\sqrt{2}}L_2,$$

$$\nabla_\xi L_1 = \frac{1}{2\sqrt{2}}\xi + \frac{1}{\sqrt{2}}W_3, \quad \nabla_\xi L_2 = -\frac{1}{4\sqrt{2}}\xi - \sqrt{2}N,$$

$$\nabla_\xi W_3 = -\frac{1}{\sqrt{2}}L_2, \qquad\qquad \nabla_\xi W_4 = 0.$$

3.4   Invariance of Frenet equations

In this section, we examine the dependence of the Frenet frames and implicitly the Frenet equations of Type 1 and Type 2 on both the transformations of the coordinate neighborhood and the screen vector bundle of C.

From proposition 3.1 and using the method of this proposition for the Natural Frenet frames, we have

Proposition 4.1. *Let C be a null curve of a semi-Riemannian manifold M_2^{m+2}. Then the type of Natural Frenet equations is invariant to the transformations of the coordinate neighborhood and the screen vector bundle of C.*

Following properties of Frenet equations (3.2.9), (3.3.13) and (3.3.14) hold:

(a) *The vanishing of the first curvature κ_1 on a neighborhood is independent of both the parameter transformations on C and the screen vector bundle transformations.*

(b) *It is possible to find a parameter on C such that $h = 0$ in Frenet equations of all possible types, using the same screen bundle.*

To prove **(a)** we let $\kappa_1 = 0$ on $\mathcal{U} \cap \bar{\mathcal{U}}$. Then $\bar{\kappa}_1 = 0$ on $\mathcal{U} \cap \bar{\mathcal{U}}$, otherwise there exists a point $x \in \mathcal{U} \cap \bar{\mathcal{U}}$ such that, for Type 1, $B_1^1(x) = \ldots = B_1^m(x) = 0$ and, for Type 2, $B_1^1(x) + B_2^1(x) = \ldots = B_1^m(x) + B_2^m(x) = 0$. This implies that the first and the second rows of the matrix $[B_\alpha^\beta(x)]$ are linearly dependent, which is not possible since this matrix belongs to $O(1, m-1)$. Hence it follows from the equation (3.3.4) that **(a)** holds.

To prove **(b)** we consider the following differential equation

$$\frac{d^2 t}{dt^{*2}} - h^* \frac{dt}{dt^*} = 0$$

whose general solution comes from

$$t = a \int_{t_0^*}^{t^*} exp \left(\int_{s_0}^{s} h^*(t^*) dt^* \right) ds + b, \quad a, \, b \in R.$$

It follows from the equation (3.3.3) that any of these solutions, with $a \neq 0$, might be taken as special parameter on C such that $h = 0$. Denote one such solution by $p = \frac{t-b}{a}$, where t is the general parameter as defined in above equation. We call p a *distinguished parameter* of C, in terms for which $h = 0$.

In case $\kappa_1 = 0$, then, since $\tau_1 = 0$ or $\tau_1 = \kappa_1$, the first equation of all Frenet equations takes the following familiar form

$$\frac{d^2 x^i}{dp^2} + \Gamma_{jk}^i \frac{dx^j}{dp} \frac{dx^k}{dp} = 0, \quad i \in \{0, \ldots, m+1\}$$

where Γ^i_{jk} are the Christoffel symbols of the second type induced by ∇. Hence C is a null geodesic of M. The converse follows easily. Thus we have

Theorem 4.1. *Let C be a null curve of M_2^{m+2}. Then C is a null geodesic of M_2^{m+2} if and only if the first curvature κ_1 vanishes identically on C.*

Theorem 4.2. *Let $C(p)$ be a Type 2 null curve of M_2^{m+2}, where p is a distinguished parameter on C. If C'' is null, then there exists a Natural Frenet frame $\{\xi, N, L_1, L_2, W_3, \ldots, W_m\}$ satisfying the following equations*

$$
\begin{aligned}
\nabla_\xi \xi &= K_1 L_1, \\
\nabla_\xi N &= K_2 L_1 + K_3 L_2 + \tau_2 W_3, \\
\nabla_\xi L_1 &= -K_3 \xi + K_4 W_3 + K_5 W_4, \\
\nabla_\xi L_2 &= -K_2 \xi - K_1 N, \\
\nabla_\xi W_3 &= -\tau_2 \xi - K_4 L_2 + \kappa_5 W_4 + \tau_4 W_5, \\
\nabla_\xi W_4 &= -K_5 L_2 - \kappa_5 W_3 + \kappa_6 W_5 + \tau_5 W_6, \\
\nabla_\xi W_5 &= -\tau_4 W_3 - \kappa_6 W_4 + \kappa_7 W_6 + \tau_6 W_7,
\end{aligned}
\tag{3.4.1}
$$

$$\ldots\ldots\ldots\ldots\ldots\ldots\ldots$$

$$
\begin{aligned}
\nabla_\xi W_i &= -\tau_{i-1} W_{i-2} - \kappa_{i+1} W_{i-1} + \kappa_{i+2} W_{i+1} + \tau_{i+1} W_{i+2}, \quad 5 \le i \le m-1, \\
\nabla_\xi W_m &= -\tau_{m-1} W_{m-2} - \kappa_{m+1} W_{m-1},
\end{aligned}
$$

If C'' is non-null, then there exists a Natural Frenet frame $\{\xi, N, W_1, \ldots, W_m\}$ satisfying the following equations

$$
\begin{aligned}
\nabla_\xi \xi &= \kappa_1 W_1, \\
\nabla_\xi N &= \kappa_2 W_1 + \kappa_3 W_2 + \tau_2 W_3, \\
\epsilon_1 \nabla_\xi W_1 &= -\kappa_2 \xi - \kappa_1 N, \\
\epsilon_2 \nabla_\xi W_2 &= -\kappa_3 \xi + \kappa_4 W_3 + \tau_3 W_4, \\
\epsilon_3 \nabla_\xi W_3 &= -\tau_2 \xi - \kappa_4 W_2 + \kappa_5 W_4 + \tau_4 W_5, \\
\epsilon_4 \nabla_\xi W_4 &= -\tau_3 W_2 - \kappa_5 W_3 + \kappa_6 W_5 + \tau_5 W_6, \\
\epsilon_5 \nabla_\xi W_5 &= -\tau_4 W_3 - \kappa_6 W_4 + \kappa_7 W_6 + \tau_6 W_7,
\end{aligned}
\tag{3.4.2}
$$

$$\ldots\ldots\ldots\ldots\ldots\ldots\ldots$$

$$
\begin{aligned}
\epsilon_i \nabla_\xi W_i &= -\tau_{i-1} W_{i-2} - \kappa_{i+1} W_{i-1} + \kappa_{i+2} W_{i+1} + \tau_{i+1} W_{i+2}, \quad 3 \le i \le m-1, \\
\epsilon_m \nabla_\xi W_m &= -\tau_{m-1} W_{m-2} - \kappa_{m+1} W_{m-1}.
\end{aligned}
$$

From (3.3.10), we have

Proposition 4.2. *Let C be a null curve of M_2^{m+2} and F and $\widetilde{F}$ be two general Frenet frames of Type 2 induced by the screen vector bundle $S(TC^\perp)$ such that $\tau_1 \ne 0$. Then the screen vector bundle $S(TC^\perp)$ is a orthogonal direct sum of two invariant subspaces $Span\{L_1, L_2\} = Span\{W_1, W_2\}$ and $Span\{W_3, \ldots, W_m\}$ by* the transformation of coordinate neighborhoods and $S(TC^\perp)$.

By exchanging the form of the general Frenet equations (of the same type) (in case $\tau_1 = \cdots = \tau_{i-1} = 0$ and $\tau_i \neq 0$) and let $\prod_{\alpha=1}^{i} \kappa_{2\alpha-1} \neq 0$, we find that the screen distribution $S(TC^\perp)$ is a orthogonal direct sum of two invariant subspaces

$$Span\{\, L_i,\, L_{i+1}\,\} = Span\{\, W_i,\, W_{i+1}\,\},\ 1 \le i \le (m-1);$$
$$Span\{\, W_1,\, \ldots,\, \widehat{W_i},\, \widehat{W_{i+1}},\, \ldots,\, W_m\,\}$$

by the transformation of coordinate neighborhoods of C, where overhat $\widehat{\ }$ denotes the deleted symbol for that term. Then the matrix $[\,A_\beta^\alpha(x)\,]$ is transformed as

$$\begin{pmatrix}
A_1 & & & & & & & & \\
 & \ddots & & & & & \mathbf{O} & & \\
 & & A_{i-1} & & & & & & \\
 & & & 0 & A_i^i + A_{i+1}^i & 0 & \cdots & 0 \\
 & & & -A_i^i + A_i^{i+1} & 0 & 0 & \cdots & 0 \\
 & & & 0 & 0 & A_{i+2}^{i+2} & \cdots & A_{i+2}^m \\
 & \mathbf{O} & & 0 & 0 & A_{i+3}^{i+2} & \cdots & A_{i+3}^m \\
 & & & \vdots & \vdots & \vdots & \cdots & \vdots \\
 & & & 0 & 0 & A_m^{i+2} & \cdots & A_m^m
\end{pmatrix},$$

where $A_\alpha = A_\alpha^\alpha = \pm 1$, $\alpha \in \{1, \cdots, i-1\}$.

Proposition 4.3. Let C be a null curve of M_2^{m+2} and F and F^* be two general Frenet frames of Type 2 induced by the screen vector bundle $S(TC^\perp)$ such that $\prod_{\alpha=1}^{i} \kappa_{2\alpha-1} \neq 0$ and $\tau_1 = \cdots = \tau_{i-1} = 0$; $\tau_i \neq 0$, $1 \le i \le m-1$. Then $S(TC^\perp)$ is an orthogonal direct sum of two invariant subspaces $Span\{L_i, L_{i+1}\} = Span\{W_i, W_{i+1}\}$ and $Span\{W_1, \ldots, \widehat{W_i}, \widehat{W_{i+1}}, \ldots, W_m\}$ by the transformation of coordinate neighborhoods of C.

Remark. From the above fact, we know that if $\prod_{\alpha=1}^{m} \kappa_{2\alpha-1} \neq 0$, the matrix $\left[A_\beta^\alpha(x)\right]$ corresponding to the general Frenet frames of Type 1 is transformed as

$$\begin{pmatrix}
A_1 & & & \mathbf{O} & \\
 & A_2 & & & \\
 & & \ddots & & \\
 & \mathbf{O} & & A_m
\end{pmatrix},$$

where $A_\alpha = A_\alpha^\alpha = \pm 1$, $\alpha \in \{1, \cdots, m\}$.

Using the Natural Frenet equations of Type 1 in (3.3.5) and the method of proposition 2.1 in chapter 2, we have

Proposition 4.4. *Let C be a null curve of M_2^{m+2} and F and F^* be two Natural Frenet frames of Type 1 on $\mathcal{U}$ and $\mathcal{U}^*$ respectively induced by the same screen vector bundle $S(TC^\perp)$. Suppose $\prod_{i=1}^{m+1} \kappa_i \neq 0$ on $\mathcal{U} \cap \mathcal{U}^* \neq \emptyset$. Then at any point of $\mathcal{U} \cap \mathcal{U}^*$ we have*

$$\kappa_1^* = \kappa_1 A_1 \left(\frac{dt}{dt^*} \right)^2,$$

$$\kappa_2^* = \kappa_2 A_1, \quad \kappa_3^* = \kappa_3 A_2, \tag{3.4.3}$$

$$\kappa_\alpha^* = \kappa_\alpha A_{\alpha-1} \frac{dt}{dt^*}, \quad 4 \leq \alpha \leq m+1, \text{ where } A_\alpha = \pm 1.$$

Hence, κ_2 and κ_3 are invariant functions up to a sign, with respect to the parameter transformations on C.

Proof. From the relations (3.3.3) we have $\tau_1 = \tau_1^* = 0$. Therefore, $k_1^* \neq 0$ on $\mathcal{U} \cap \mathcal{U}^*$ and $A_1^2 = \ldots = A_1^m = 0$. Since $[A_\alpha^\beta(x)]$ is a Lorentzian matrix, we infer that $A_1^1 = A_1 = \pm 1$ and $A_2^1 = \ldots = A_m^1 = 0$. Then from the second equation of the Natural Frenet equations (3.3.5) of Type 1 wit respect to F and F^* and taking into account that $\kappa_3 \neq 0$, we obtain $\kappa_3^* \neq 0$ on $\mathcal{U} \cap \mathcal{U}^*$ which implies $A_2^3 = A_3^2 = \ldots = A_2^m = A_m^2 = 0$ and $A_2^2 = A_2 = \pm 1$. Repeating this process for all other equations we obtain all the relations in (3.4.3), which completes the proof.

Using the general Frenet equations of Type 1 in (3.1.1) and the method of proposition 2.2 in chapter 2, we have

Proposition 4.5. *Let C be a null curve of M_2^{m+2} and F and F^* be two general Frenet frames of Type 1 on $\mathcal{U}$ and $\mathcal{U}^*$ respectively induced by the same screen vector bundle $S(TC^\perp)$. Suppose $\prod_{n=1}^{m} \kappa_{2n-1} \neq 0$ on $\mathcal{U} \cap \mathcal{U}^* \neq \emptyset$. Then at any point of $\mathcal{U} \cap \mathcal{U}^*$ we have*

$$\kappa_1^* = \kappa_1 A_1 \left(\frac{dt}{dt^*} \right)^2,$$

$$\kappa_2^* = \kappa_2 A_1, \quad \kappa_3^* = \kappa_3 A_2, \tag{3.4.4}$$

$$\kappa_\alpha^* = \kappa_\alpha A_{\alpha-1} \frac{dt}{dt^*}, \quad 4 \leq \alpha \leq 2m, \text{ where } A_\alpha = \pm 1.$$

Hence, κ_2 and κ_3 are invariant functions up to a sign, with respect to the parameter transformations on C.

Also, using the method of proposition 2.3 in chapter 2 for the different screen vector bundle with $\bar{F}$, we have

Proposition 4.6. *Let C be a null curve of M_2^{m+2} and F and $\bar{F}$ be two Natural Frenet frames of Type 1 on $\mathcal{U}$ and $\bar{\mathcal{U}}$ respectively. Suppose $\prod_{i=1}^{m+1} \kappa_i \neq 0$ on $\mathcal{U} \cap \bar{\mathcal{U}} \neq$*

$\emptyset$. Then, their curvature functions are related by

$$\bar{\kappa}_1 = \kappa_1 B_1 \left(\frac{dt}{d\bar{t}}\right)^2,$$

$$\bar{\kappa}_2 = \left\{\kappa_2 + \bar{h}\, c_1 + \frac{dc_1}{d\bar{t}} - \frac{1}{2}\kappa_1 c_1^2 \left(\frac{dt}{d\bar{t}}\right)^2\right\} B_1,$$

$$\bar{\kappa}_3 = \kappa_3 B_2,$$

$$\bar{\kappa}_\alpha = \kappa_\alpha B_{\alpha-1}\frac{dt}{d\bar{t}}, \quad \alpha \in \{4, \cdots, m\}, \text{ where } B_i = \pm 1,$$

and $c_2 = \cdots = c_m = 0$. Hence, κ_3 is invariant functions up to a sign, with respect to the transformations of the screen vector bundle of C.

Proof. From (3.3.2) and the first equation of (3.3.5), we have the first equation of this proposition and $\bar{\kappa}_1 B_1^\alpha = 0\,(\alpha \neq 1)$ on $\mathcal{U} \cap \bar{\mathcal{U}}$. Thus we have $B_1^\alpha = 0\,(\alpha \neq 1)$. Since $[\,B_\alpha^\beta(x)\,]$ is a Lorentzian matrix, we infer that $B_1^1 = B_1 = \pm 1$ and $B_\alpha^1 = 0\,(\alpha \neq 1)$. While, from (3.3.2) and the third equation of (3.3.5), we have the first and second equations of this proposition and $\bar{\kappa}_1 c_\alpha = 0\,(\alpha \neq 1)$ on $\mathcal{U} \cap \bar{\mathcal{U}}$. Thus we have $c_\alpha = 0\,(\alpha \neq 1)$. Also, from (3.3.2) and the second equation of (3.3.5), we have the second equation of proposition and $\bar{\kappa}_3 B_2^2 = \kappa_3$; $\bar{\kappa}_3 B_2^\alpha = 0\,(\alpha \geq 3)$. Thus we have $B_2 = B_2^2 = \pm 1$ and $B_2^\alpha = B_\alpha^2 = 0\,(\alpha \geq 3)$. Repeating this process for all other equations of (3.3.5) and set $B_\alpha = B_{\alpha-1}^{\alpha-1}B_\alpha^\alpha\,(\alpha \geq 3)$, we obtain all the relations in proposition, which completes the proof.

Let F and $\bar{F}$ be two Natural Frenet frames with respect to $(t, S(TC^\perp), \mathcal{U})$ and $(\bar{t}, \bar{S}(TC^\perp), \bar{\mathcal{U}})$ respectively such that $\tau_1 = 0$, then, by Proposition 4.6, we have $c_2 = \cdots = c_m = 0$. Thus the transformations of the screen vector bundle that relate elements of F and $\bar{F}$ on $\mathcal{U} \cap \bar{\mathcal{U}} \neq \emptyset$ are given by

$$\bar{\xi} = \frac{dt}{d\bar{t}}\xi,$$

$$\bar{N} = -\frac{1}{2}\frac{dt}{d\bar{t}}\epsilon_1 c_1^2 \xi + \frac{d\bar{t}}{dt}N + c_1 W_1,$$

$$\bar{W}_1 = B_1^1 \left(W_1 - \epsilon_1 \frac{dt}{d\bar{t}}c_1 \xi\right),$$

$$\bar{W}_\alpha = \sum_{\beta=2}^{m} B_\alpha^\beta W_\beta, \quad 2 \leq \alpha \leq m.$$

Using this and the method of proposition 4.3, we have

Theorem 4.3. Let C be a null curve of M_2^{m+2} and $F; \widetilde{F}$ be two Natural Frenet frames of Type 2 induced by the screen vector bundles $(t, S(TC^\perp), \mathcal{U})$ and $\left(\tilde{t}, \tilde{S}(TC^\perp), \tilde{\mathcal{U}}\right)$ respectively such that $\prod_{\alpha=1}^{i} \kappa_\alpha \neq 0$ and $\tau_1 = \cdots = \tau_{i-1} = 0$; $\tau_i \neq 0$, $1 \leq i \leq m-1$. Then $S(TC^\perp)$ is an orthogonal direct sum of two invariant subspaces $Span\{L_i, L_{i+1}\} = Span\{W_i, W_{i+1}\}$ and $Span\{W_1, \ldots, \widehat{W}_i, \widehat{W}_{i+1}, \ldots, W_m\}$ by the transformation of the coordinate neighborhoods and the screen vector bundle of C.

Using the parameter transformations (3.3.10) that relate elements of F_2 and F_2^* on $\mathcal{U} \cap \mathcal{U}^*$ such that $\tau_1 \neq 0$, we have (3.3.1), (3.3.7) and

$$L_2^* = \left(A_1^1 - A_1^2\right) L_2, \quad W_\alpha^* = \sum_{\beta \geq 3}^{m} A_\alpha^\beta W_\beta. \tag{3.4.5}$$

Proposition 4.7. *Let C be a null curve of M_2^{m+2} and F and F^* be two Natural Frenet frames of Type 2 on $\mathcal{U}$ and $\mathcal{U}^*$ respectively induced by the same screen vector bundle $S(TC^\perp)$. Suppose $\prod_{i=1}^{m-1} \tau_i \neq 0$ on $\mathcal{U} \cap \mathcal{U}^* \neq \emptyset$. Then at any point of $\mathcal{U} \cap \mathcal{U}^*$ we have*

$$K_1^* = K_1 \left(A_1^1 - A_1^2\right) \left(\frac{dt}{dt^*}\right)^2,$$

$$K_2^* = K_2 \left(A_1^1 - A_1^2\right), \quad K_3^* = K_3 \left(A_1^1 + A_2^1\right), \quad \tau_2^* = \tau_2 A_3,$$

$$K_4^* = K_4 A_3 \left(A_1^1 + A_2^1\right) \frac{dt}{dt^*}, \quad K_5^* = K_5 A_4 \left(A_1^1 + A_2^1\right) \frac{dt}{dt^*}, \tag{3.4.6}$$

$$\kappa_\alpha^* = \kappa_\alpha A_{\alpha-1} \frac{dt}{dt^*}, \quad 5 \leq \alpha \leq m+1, \text{ where } A_\alpha = \pm 1.$$

$$\tau_\alpha^* = \tau_\alpha A_{\alpha+1} \frac{dt}{dt^*}, \quad 4 \leq \alpha \leq m+1, \text{ where } A_\alpha = \pm 1.$$

Hence, K_2, K_3 and τ_2 are invariant functions up to a sign, with respect to the parameter transformations on C.

Proof. From the equation (3.3.7) and the first equations of (3.3.1) and (3.3.12) respectively, we have $K_1^* = K_1 (A_1^1 - A_1^2)(\frac{dt}{dt^*})^2$. Also, from (3.4.5) and the second equations of (3.3.1) and (3.3.12) respectively, we have $K_2^* = K_2 (A_1^1 - A_1^2)$, $K_3^* = K_3 (A_1^1 + A_2^1)$ and $\tau_2^* A_3^3 = \tau_2$; $\tau_2^* A_3^\alpha = 0$ for all $\alpha \in \{4, \cdots, m\}$. Since $\tau_2 \neq 0$, therefore, $\tau_2^* \neq 0$ on $\mathcal{U} \cap \mathcal{U}^*$ and $A_3^4 = \ldots = A_3^m = 0$. Since $[A_\alpha^\beta]$ is a Lorentzian matrix, we infer that $A_3^3 = A_3 = \pm 1$ and $A_4^3 = \ldots = A_m^3 = 0$. Then from the third equation of the Natural Frenet equations (3.3.12) of Type 2 with respect to F and F^*, we have $A_1^1 + A_2^1$ is a constant, $K_4^* = K_4 A_3(A_1^1 + A_2^1) \frac{dt}{dt^*}$, $K_5^* A_4^4 = K_5 (A_1^1 + A_2^1) \frac{dt}{dt^*}$ and $K_5^* A_4^\alpha = 0$ for all $\alpha \in \{5, \cdots, m\}$. Since $\tau_3 \neq 0$ and $K_5 = -\sqrt{2}\tau_3$, we have $K_5 \neq 0$, therefore $K_5^* \neq 0$ on $\mathcal{U} \cap \mathcal{U}^*$ and $A_4^5 = \ldots = A_4^m = 0$. We infer that $A_4^4 = A_4 = \pm 1$ and $A_5^4 = \ldots = A_m^4 = 0$. Repeating this process for all other equations of (3.3.12), we obtain all the relations in (3.4.6), which completes the proof.

Using the general Frenet equations (3.2.6) of Type 2 and the method of proposition 4.7, we have

Proposition 4.8. Let C be a null curve of M_2^{m+2} and F and F^* be two general Frenet frames of Type 2 on $\mathcal{U}$ and $\mathcal{U}^*$ respectively induced by the same screen vector bundle $S(TC^\perp)$. Suppose $\prod_{i=1}^{m-1} \tau_i \neq 0$ on $\mathcal{U} \cap \mathcal{U}^* \neq \emptyset$. Then at any point of $\mathcal{U} \cap \mathcal{U}^*$ we have

$$K_1^* = K_1 \left(A_1^1 - A_1^2\right) \left(\frac{dt}{dt^*}\right)^2,$$

$$K_2^* = K_2 \left(A_1^1 - A_1^2\right), \quad K_3^* = K_3 \left(A_1^1 + A_2^1\right), \quad \tau_2^* = \tau_2 A_3,$$

$$K_4^* = \left\{K_4 + \left(A_1^1 + A_2^1\right)\frac{d(A_1^1 - A_1^2)}{dt}\right\} \frac{dt}{dt^*},$$

$$K_5^* = K_5 A_3 \left(A_1^1 + A_2^1\right) \frac{dt}{dt^*}, \quad K_6^* = K_6 A_4 \left(A_1^1 + A_2^1\right) \frac{dt}{dt^*}, \qquad (3.4.7)$$

$$K_7^* = K_7 A_3 \left(A_1^1 - A_1^2\right) \frac{dt}{dt^*}, \quad K_8^* = K_8 A_4 \left(A_1^1 - A_1^2\right) \frac{dt}{dt^*},$$

$$\kappa_\alpha^* = \kappa_\alpha A_{2\alpha-3} \frac{dt}{dt^*}, \quad 8 \leq \alpha \leq m+1, \text{ where } A_\alpha = \pm 1.$$

$$\tau_\alpha^* = \tau_\alpha A_{2\alpha} \frac{dt}{dt^*}, \qquad 5 \leq \alpha \leq m+1, \text{ where } A_\alpha = \pm 1.$$

Hence, K_2, K_3 and τ_2 are invariant functions up to a sign, with respect to the parameter transformations on C.

Also, using the method of proposition 2.4 for the Natural Frenet equations (3.3.12) of Type 2, we have

Proposition 4.9. Let C be a null curve of M_2^{m+2} and F and $\bar{F}$ be two Natural Frenet frames of Type 2 on $\mathcal{U}$ and $\bar{\mathcal{U}}$ respectively. Suppose $\tau_1 \neq 0$ on $\mathcal{U} \cap \bar{\mathcal{U}} \neq \emptyset$. Then, their curvature functions are related by

$$\bar{K}_1 = K_1 \left(B_1^1 - B_1^2\right) \left(\frac{dt}{d\bar{t}}\right)^2,$$

$$\bar{K}_2 = \left\{K_2 + \bar{h}\, C_1 + \frac{dC_1}{d\bar{t}}\right\} \left(B_1^1 - B_1^2\right), \qquad (3.4.8)$$

$$\bar{K}_3 = K_3 \left(B_1^1 + B_2^1\right)$$

and $C_2 = c_3 = \cdots = c_m = 0$. Hence, K_3 is invariant functions up to a sign, with respect to the transformations of the screen vector bundle of C.

Proof. From (3.3.11) and the fourth equation of (3.3.12), we have the first equation of (3.4.8) and $\bar{K}_1 C_2 = 0$; $\bar{K}_1 c_\alpha = 0$ ($\alpha \geq 3$) on $\mathcal{U} \cap \bar{\mathcal{U}}$. Since $K_1 = \sqrt{2}\tau_1 \neq 0$, we have $C_2 = c_\alpha = 0$ ($\alpha \geq 3$). Thus (3.3.11) is reduced to

$$\bar{N} = \frac{d\bar{t}}{dt} N + C_1 L_1, \quad \bar{L}_1 = \left(B_1^1 + B_2^1\right) L_1,$$

$$\bar{L}_2 = \left(B_1^1 - B_1^2\right) \left\{L_2 - C_1 \frac{dt}{d\bar{t}} \xi\right\}, \quad C_1 = \frac{1}{\sqrt{2}}(c_1 + c_2), \qquad (3.4.9)$$

$$\bar{W}_\alpha = \sum_{\beta \geq 3}^{m} B_\alpha^\beta W_\beta, \quad \alpha, \beta \in \{3 \cdots, m\}.$$

From second equation of (3.3.12), we get second and third equations of (3.4.8).

3.5   Geometry of null curves in M_2^4

Let C be a null curve in a 4-dimensional semi-Riemannian manifold M_2^4. Suppose $F = \{ \frac{d}{dt}, N, W_1, W_2 \}$ and $\bar{F} = \{ \frac{d}{d\bar{t}}, \bar{N}, \bar{W}_1, \bar{W}_2 \}$ are two general Frenet frames of C with their respective screen vector bundles where W_1 is a timelike vector field and W_2 is a spacelike one. Then, we know from propositions 4.1 that they both are either Type 1 or Type 2.

Lemma 5.1. *Let C be a null curve of M_2^4 with two general Frenet frames F and $\bar{F}$ of Type 1 such that $\kappa_1 \neq 0$. Then, their curvature functions are related by*

$$\bar{\kappa}_1 = B_1 \kappa_1 \left(\frac{dt}{d\bar{t}} \right)^2,$$

$$\bar{\kappa}_2 = B_1 \left\{ \kappa_2 + c_1 \bar{h} + \frac{dc_1}{d\bar{t}} - c_2 \kappa_4 \frac{dt}{d\bar{t}} + \frac{\kappa_1}{2} \left((c_1)^2 - (c_2)^2 \right) \left(\frac{dt}{d\bar{t}} \right)^2 \right\},$$

$$\bar{\kappa}_3 = B_2 \left\{ \kappa_3 + \bar{h} c_2 + \frac{dc_2}{d\bar{t}} - c_1 \kappa_4 \frac{dt}{d\bar{t}} \right\}, \tag{3.5.1}$$

$$\bar{\kappa}_4 = B_1 B_2 \left\{ \kappa_4 + \kappa_1 \frac{dt}{d\bar{t}} c_2 \right\} \frac{dt}{d\bar{t}}.$$

Proof. For Type 1 it follows from (3.1.1) and the matrix $[B_\beta^\alpha(x)]$ is Lorentzian and $B_1^2 = B_2^1 = 0$; $B_1^1 = B_1 = \pm 1$; $B_2^2 = B_2 = \pm 1$. Therefore, by (3.3.2), the general transformations relating the elements of F and $\bar{F}$ on $\mathcal{U} \cap \bar{\mathcal{U}}$ are

$$\frac{d}{d\bar{t}} = \frac{dt}{d\bar{t}} \xi,$$

$$\bar{N} = \frac{1}{2} \frac{dt}{d\bar{t}} \left((c_1)^2 - (c_2)^2 \right) \xi + \frac{d\bar{t}}{dt} N + c_1 W_1 + c_2 W_2, \tag{3.5.2}$$

$$\bar{W}_1 = B_1 \left(W_1 + \frac{dt}{d\bar{t}} c_1 \xi \right),$$

$$\bar{W}_2 = B_2 \left(W_2 - \frac{dt}{d\bar{t}} c_2 \xi \right).$$

The relations (3.5.1) follow by straightforward calculations from the general Frenet equations of Type 1 and the use of (3.5.2).

Theorem 5.1. *Let C be a null curve of M_2^4 with a general Frenet frame F of Type 1 and a screen vector bundle $S(TC^\perp)$ such that $\kappa_1 \neq 0$. Then there exists a screen vector bundle $\bar{S}(TC^\perp)$ which induces another Frenet frame $\bar{F}$ of Type 1 on*

$\mathcal{U}$ such that $\bar{\kappa}_4 = 0$.

Proof. Define the following vector fields in terms of the elements of F on $\mathcal{U}$:

$$\bar{N} = -\frac{1}{2}\left(\frac{k_4}{\kappa_1}\right)^2 \xi + N - \frac{\kappa_4}{\kappa_1} W_2,$$

$$\bar{W}_1 = W_1; \quad \bar{W}_2 = W_2 + \frac{\kappa_4}{\kappa_1}\xi, \tag{3.5.3}$$

Let $\mathcal{U}^*$ be another coordinate neighborhood with parameter t^* on C such that $\mathcal{U} \cap \mathcal{U}^* \neq \emptyset$. By proposition 4.5 we have the following on $\mathcal{U} \cap \mathcal{U}^*$

$$\kappa_1^* = \kappa_1 A_1 \left(\frac{dt}{dt^*}\right)^2 ; \quad \kappa_4^* = \kappa_4 A_3 \frac{dt}{dt^*},$$

$$W_1^* = A_1 W_1, \qquad W_2^* = A_2 W_2. \tag{3.5.4}$$

Define $\{\bar{N}^*, \bar{W}_1^*, \bar{W}_2^*\}$ by (3.5.3) but on $\mathcal{U}^*$ with respect to F^* induced by the same $S(TC^\perp)$ on $\mathcal{U}^*$. Then by using (3.3.1), (3.3.2), (3.5.3) and (3.5.4) we obtain

$$\bar{N}^* = \frac{dt^*}{dt}\bar{N}, \quad \bar{W}_1^* = A_1 \bar{W}_1, \quad \bar{W}_2^* = A_2 \bar{W}_2.$$

Hence there exists a vector bundle $\bar{S}(TC^\perp)$ spanned on $\mathcal{U}$ by $\{\bar{W}_1, \bar{W}_2\}$ given by (3.5.3). Moreover, it is easy to check that $\bar{S}(TC^\perp)$ is complementary to TC in $TC^\perp$. The null transversal vector field with respect to $S(TC^\perp)$ is locally represented by $\bar{N}$ from (3.5.3). Finally taking $t = \bar{t}$ and $c_2 = -\frac{\kappa_4}{\kappa_1}$ in the fourth equation of (3.5.1), we obtain $\bar{\kappa}_4 = 0$ which completes the proof.

Example 11. Let C be the null curve of $\mathbf{R}_2^4$ given by

$$C: \left(\frac{1}{3}t^3 - 2t, \ t^2, \ \frac{1}{3}t^3, \ 2t\right).$$

Then

$$\xi = \left(t^2 - 2, \ 2t, \ t^2, \ 2\right).$$

If we take $V = \left(0, \ 0, \ 0, \ \frac{1}{2}\right)$, then $g(\xi, V) = 1$ and $g(V, V) = \frac{1}{4}$. Thus we have

$$N = -\frac{1}{8}\left(t^2 - 2, \ 2t, \ t^2, \ -2\right).$$

Then in order to write the first equation in (3.2.9) we have to know the causal character of the vector field $H(t) = \nabla_\xi \xi - h\xi$ along C. Since

$$\nabla_\xi \xi = 2\left(t, \ 1, \ t, \ 0\right)$$

and $h = g(\nabla_\xi \xi, N) = 0$, by direct calculations we obtain

$$g(H(t), H(t)) = -4.$$

Hence $H(t)$ is timelike vector field and this curve falls in the Type 1. Let

$$W_1 = (t, \ 1, \ t, \ 0),$$

then W_1 is a timelike vector field, $k_1 = 2$ and $\nabla_\xi \xi = 2 W_1$. From

$$\nabla_\xi N = -\frac{1}{4}(t, \ 1, \ t, \ 0)$$

and $k_2 = -g(\nabla_\xi N, W_1) = -\frac{1}{4}$, we have $\nabla_\xi N = -\frac{1}{4} W_1$ and $k_3 = 0$. Choose

$$W_2 = \frac{1}{2}(t^2, \ 2t, \ t^2 + 2, \ 0),$$

then $\nabla_\xi W_1 = -\frac{1}{4}\xi + 2N + W_2$; $\nabla_\xi W_2 = W_1$ and $k_4 = -1$.

From (3.4.3), since $k_1 = 2$ and $k_4 = -1$, let

$$\bar{N} = -\frac{1}{8}\xi + N + \frac{1}{2}W_2 = \frac{1}{2}(1, \ 0, \ 1, \ 0),$$

$$\bar{W}_1 = W_1; \quad \bar{W}_2 = W_2 - \frac{1}{2}\xi = (1, \ 0, \ 1, \ 1).$$

Then we have $\bar{k}_4 = -g(\nabla_\xi \bar{W}_1, \bar{W}_2) = 0$.

Also, using the method of proposition 4.9 for the general Frenet equations (3.2.6) of Type 2, we have

Lemma 5.2. *Let C be a null curve of M_2^4 with two general Frenet frames F and $\bar{F}$ of Type 2 such that $\tau_1 \neq 0$. Then their curvature functions are related by*

$$\bar{K}_1 = H K_1 \left(\frac{dt}{d\bar{t}}\right)^2,$$

$$\bar{K}_2 = H \left\{ K_2 + \bar{h} \, C_1 + \frac{dC_1}{d\bar{t}} - C_1 K_4 \frac{dt}{d\bar{t}} - C_1 C_2 \left(\frac{dt}{d\bar{t}}\right)^2 \right\}$$

$$\bar{K}_3 = G \left\{ K_3 + \bar{h} \, C_2 + \frac{dC_2}{d\bar{t}} + C_2 K_4 \frac{dt}{d\bar{t}} \right\}, \qquad (3.5.5)$$

$$\bar{K}_4 = \left\{ K_4 + K_1 C_2 \frac{dt}{d\bar{t}} - H \frac{dG}{dt} \right\} \frac{dt}{d\bar{t}},$$

where $H = B_1^1 - B_1^2$, $G = B_1^1 + B_2^1$, $C_1 = \frac{1}{\sqrt{2}}(c_1 + c_2)$ and $C_2 = \frac{1}{\sqrt{2}}(c_2 - c_1)$.

Proof. For $m = 2$, the matrix $[B_j^i(x)]$ is made up of a 2×2 Lorentz matrix. Therefore, using (3.3.8) and (3.3.11), the general transformations are given by

$$\frac{d}{d\bar{t}} = \frac{dt}{d\bar{t}} \xi,$$

$$\bar{N} = -C_1 C_2 \frac{dt}{d\bar{t}} \xi + \frac{d\bar{t}}{dt} N + C_1 L_1 + C_2 L_2,$$

$$\bar{L}_1 = G \left\{ L_1 - \frac{dt}{d\bar{t}} C_2 \xi \right\}, \tag{3.5.6}$$

$$\bar{L}_2 = H \left\{ L_2 - \frac{dt}{d\bar{t}} C_1 \xi \right\},$$

Straightforward calculations from above relations and the use of (3.2.6) implies (3.5.5), which proves this lemma.

By a similar procedure, one can prove the following:

Theorem 5.2. *Let C be a null curve of M_2^4 with screen vector bundle $S(TC^\perp)$ and a general Frenet frame F of Type 2 such that $\tau_1 \neq 0$. Then there exists a screen $\bar{S}(TC^\perp)$ which induces another frame $\bar{F}$ such that $\bar{K}_4 = 0$.*

Example 12. Let C be the null curve of $\mathbf{R}_2^4$ given by

$$C : \left(\frac{1}{3}t^3 - t, \ \frac{1}{3}t^3 + t, \ \frac{\sqrt{2}}{3}t^3, \ \sqrt{2}t \right), \qquad t \neq 0.$$

Then

$$\xi = \left(t^2 - 1, \ t^2 + 1, \ \sqrt{2}t^2, \ \sqrt{2} \right).$$

If we take $V = \left(0, \ 0, \ 0, \ \frac{1}{\sqrt{2}} \right)$, then $g(\xi, V) = 1$ and $g(V, V) = \frac{1}{2}$. Thus we have

$$N = -\frac{1}{4} \left(t^2 - 1, \ t^2 + 1, \ \sqrt{2}t^2, \ -\sqrt{2} \right).$$

Then in order to write the first equation in (3.2.9) we have to know the causal character of the vector field $H(t) = \nabla_\xi \xi - h\xi$ along C. Since

$$\nabla_\xi \xi = 2 \left(t, \ t, \ \sqrt{2}t, \ 0 \right).$$

and $h = g(\nabla_\xi \xi, N) = 0$, by direct calculations we obtain

$$g(H(t), H(t)) = 0.$$

Hence $H(t)$ is null vector field and this curve falls in the Type 2. Let

$$L_1 = \left(t, \ t, \ \sqrt{2}t, \ 0 \right),$$

then L_1 is a timelike vector field, $K_1 = 2$ and $\nabla_\xi \xi = 2 L_1$. Take

$$L_2 = -\frac{1}{4t^2} \left(t, \ t, \ -\sqrt{2}t, \ 0 \right),$$

then L_2 is a null vector field such that $g(L_1, L_2) = 1$. Since

$$\nabla_\xi N = -\frac{1}{2}\left(t,\ t,\ \sqrt{2}t,\ 0 \right),$$

$K_2 = g(\nabla_\xi N, L_2) = -\frac{1}{2}$ and $K_3 = g(\nabla_\xi N, L_1) = 0$, we have $\nabla_\xi N = -\frac{1}{2}L_1$. Also from $\nabla_\xi L_1 = (1,\ 1,\ \sqrt{2},\ 0)$ and $K_4 = g(\nabla_\xi L_1, L_2) = \frac{1}{t}$, we have

$$\nabla_\xi L_1 = \frac{1}{t}L_1; \quad \nabla_\xi L_2 = \frac{1}{2}\xi - 2N + \frac{1}{t}L_2.$$

Since $K_1 = 2$ and $K_4 = \frac{1}{t}$, let

$$\bar{N} = N + \frac{1}{2t}L_2; \quad \bar{L}_1 = L_1 - \frac{1}{2t}\xi; \quad \bar{L}_2 = L_2,$$

then we have $\bar{K}_4 = g(\nabla_\xi \bar{L}_1, \bar{L}_2) = 0$.

Example 13. Let C be the null curve of $\mathbf{R}_2^4$ given by

$$C: \left(\frac{1}{3}(2t-1)^{\frac{3}{2}},\ \frac{1}{2}t^2 - t,\ t\sin t + \cos t,\ \sin t - t\cos t \right), \quad t > \frac{1}{2}.$$

Then

$$\xi = \left(\sqrt{2t-1},\ t-1,\ t\cos t,\ t\sin t \right).$$

If we take $V = (0,\ 0,\ \cos t,\ \sin t)$, then $g(\xi, V) = t$ and $g(V, V) = 1$. Thus we have

$$N = -\frac{1}{2t^2}\left(\sqrt{2t-1},\ t-1,\ -t\cos t,\ -t\sin t \right).$$

Since

$$\nabla_\xi \xi = \left(\frac{1}{\sqrt{2t-1}},\ 1,\ \cos t - t\sin t,\ \sin t + t\cos t \right),$$

and $h = g(\nabla_\xi \xi, N) = \frac{1}{t}$, we have

$$H(t) = \nabla_\xi \xi - \frac{1}{t}\xi = \left(\frac{-t+1}{t\sqrt{2t-1}},\ \frac{1}{t},\ -t\sin t,\ t\cos t \right).$$

Thus

$$g(H(t), H(t)) = \frac{(t-1)(2t^2 + t + 1)}{2t - 1}.$$

This imply that $H(t)$ is spacelike, timelike or null according as $t > 1$, $\frac{1}{2} < t < 1$ or $t = 1$, respectively. If $t \neq 1$. Let

$$W_1 = \frac{\sqrt{2t-1}}{\sqrt{2t^3 - t^2 - 1}}\left(\frac{-t+1}{t\sqrt{2t-1}},\ \frac{1}{t},\ -t\sin t,\ t\cos t \right),$$

we have

$$\nabla_\xi \xi = \frac{1}{t}\xi + \frac{\sqrt{2t^3 - t^2 - 1}}{\sqrt{2t-1}}W_1.$$

From

$$\nabla_\xi N = -\frac{1}{2t^2}\left(\frac{-3t+2}{t\sqrt{2t-1}}, \; \frac{-t+2}{t}, \; \cos t + t\sin t, \; \sin t - t\cos t\right)$$

and $k_2 = g(\nabla_\xi N, W_1) = \frac{2t^3-t^2+1}{2t^2\sqrt{2t^3-t^2-1}\sqrt{2t-1}}$, we have

$$\nabla_\xi N + \frac{1}{t}N - \frac{2t^3-t^2+1}{2t^2\sqrt{2t^3-t^2-1}\sqrt{2t-1}}W_1$$

$$= -\frac{1}{t(2t^3-t^2-1)}\left((-t+1)\sqrt{2t-1}, \; 2t-1, \; -\sin t, \; \cos t\right).$$

Choose

$$W_2 = -\frac{1}{\sqrt{2t^3-t^2-1}}\left((-t+1)\sqrt{2t-1}, \; 2t-1, \; -\sin t, \; \cos t\right),$$

then $\nabla_\xi N = -\frac{1}{t}N + \frac{2t^3-t^2+1}{2t^2\sqrt{2t^3-t^2-1}\sqrt{2t-1}}W_1 + \frac{1}{t\sqrt{2t^3-t^2-1}}W_2$.

By straightforward calculations we have

$$\nabla_\xi W_1 = -\frac{2t^3-t^2+1}{2t^2\sqrt{2t^3-t^2-1}\sqrt{2t-1}}\xi - \frac{\sqrt{2t^3-t^2-1}}{\sqrt{2t-1}}N$$
$$- \frac{6t^4-5t^3+t^2-3t+1}{(2t^3-t^2-1)^2\sqrt{2t-1}}W_2,$$

$$\nabla_\xi W_2 - -\frac{1}{t\sqrt{2t^3-t^2-1}}\xi + \frac{6t^4-5t^3+t^2-3t+1}{(2t^3-t^2-1)^2\sqrt{2t-1}}W_2.$$

Since $k_1 = \frac{\sqrt{2t^3-t^2-1}}{\sqrt{2t-1}}$ and $k_4 = -\frac{6t^4-5t^3+t^2-3t+1}{(2t^3-t^2-1)^2\sqrt{2t-1}}$, let

$$\bar{W}_1 = W_1; \quad \bar{W}_2 = W_2 - \frac{6t^4-5t^3+t^2-3t+1}{(2t^3-t^2-1)^{\frac{5}{2}}}\xi,$$

then $\bar{k}_4 = g(\nabla_\xi \bar{W}_1, \bar{W}_2) = 0$.

3.6   Geometry of null curves in M_2^6

Let C be a null curve in a 6-dimensional semi-Riemannian manifold M_2^6. Suppose $F = \{\xi, N, W_1, W_2, W_3, W_4\}$ and $\bar{F} = \{\frac{d}{dt}, \bar{N}, \bar{W}_1, \bar{W}_2, \bar{W}_3, \bar{W}_4\}$ are two general Frenet frames of C with their respective screen spaces. Then, we know from proposition 4.1 that they both are either of Type 1 or of Type 2.

Lemma 6.1. Let C be a null curve of M_2^6 with two general Frenet frames F and $\bar{F}$ of Type 1 such that $\kappa_1 \neq 0$. Then, their curvature functions are related by

$$\bar{\kappa}_1 = B_1\kappa_1\left(\frac{dt}{d\bar{t}}\right)^2,$$

$$\bar{\kappa}_2 = B_1 \left\{ \kappa_2 + c_1 \bar{h} + \frac{dc_1}{d\bar{t}} - (c_2\kappa_4 + c_3\kappa_5)\frac{dt}{d\bar{t}} - \frac{\kappa_1}{2} \sum_{i=1}^{4} \epsilon_i (c_i)^2 \left(\frac{dt}{d\bar{t}}\right)^2 \right\},$$

$$\bar{\kappa}_3 = \kappa_3 B_2^2 + \sum_{\alpha=2}^{4} B_2^\alpha \left(\bar{h}c_\alpha + \frac{dc_2}{d\bar{t}}\right) - \{B_2^2(c_1\kappa_4 + c_3\kappa_6 + c_4\kappa_7)$$

$$+ \quad B_2^3(c_1\kappa_5 - c_2\kappa_6 + c_4\kappa_8) - B_2^4(c_2\kappa_7 + c_3\kappa_8)\}\frac{dt}{d\bar{t}}, \qquad (3.6.1)$$

$$\bar{\kappa}_4 = B_1 \left\{ B_2^2 \left(\kappa_4 + \kappa_1\frac{dt}{d\bar{t}}c_2\right) + B_2^3 \left(\kappa_5 + \kappa_1\frac{dt}{d\bar{t}}c_3\right) + B_2^4\kappa_1\frac{dt}{d\bar{t}}c_4 \right\}\frac{dt}{d\bar{t}},$$

$$\bar{\kappa}_5 = B_1 \left\{ B_3^2 \left(\kappa_4 + \kappa_1\frac{dt}{d\bar{t}}c_2\right) + B_3^3 \left(\kappa_5 + \kappa_1\frac{dt}{d\bar{t}}c_3\right) + B_3^4\kappa_1\frac{dt}{d\bar{t}}c_4 \right\}\frac{dt}{d\bar{t}},$$

$$\bar{\kappa}_6 = \left\{ \kappa_6 \left(B_2^2 B_3^3 - B_2^3 B_3^2\right) + \kappa_7 \left(B_2^2 B_3^4 - B_2^4 B_3^2\right) \right.$$

$$+ \quad \kappa_8 \left(B_2^3 B_3^4 - B_2^4 B_3^3\right) + \sum_{\alpha=2}^{4} B_3^\alpha \frac{d B_2^\alpha}{dt} \right\}\frac{dt}{d\bar{t}},$$

$$\bar{\kappa}_7 = \left\{ \kappa_6 \left(B_2^2 B_4^3 - B_2^3 B_4^2\right) + \kappa_7 \left(B_2^2 B_4^4 - B_2^4 B_4^2\right) \right.$$

$$+ \quad \kappa_8 \left(B_2^3 B_4^4 - B_2^4 B_4^3\right) + \sum_{\alpha=2}^{4} B_4^\alpha \frac{d B_2^\alpha}{dt} \right\}\frac{dt}{d\bar{t}},$$

$$\bar{\kappa}_8 = \left\{ \kappa_6 \left(B_3^2 B_4^3 - B_3^3 B_4^2\right) + \kappa_7 \left(B_3^2 B_4^4 - B_3^4 B_4^2\right) \right.$$

$$+ \quad \kappa_8 \left(B_3^3 B_4^4 - B_3^4 B_4^3\right) + \sum_{\alpha=2}^{4} B_4^\alpha \frac{d B_3^\alpha}{dt} \right\}\frac{dt}{d\bar{t}},$$

$$0 = B_4^2 \left(\kappa_4 + \kappa_1\frac{dt}{d\bar{t}}c_2\right) + B_4^3 \left(\kappa_5 + \kappa_1\frac{dt}{d\bar{t}}c_3\right) + B_4^4\kappa_1\frac{dt}{d\bar{t}}c_4.$$

Proof. For the Type 1, (3.3.4) implies $B_1^i = B_i^1 = 0 \,(i \neq 1)$; $B_1^1 = B_1 = \pm1$. Therefore, the elements of F and $\bar{F}$ on $\mathcal{U} \cap \bar{\mathcal{U}}$ are related by

$$\frac{d}{d\bar{t}} = \frac{dt}{d\bar{t}}\xi,$$

$$\bar{N} = -\frac{1}{2}\frac{dt}{d\bar{t}} \sum_{i=1}^{4} \epsilon_i (c_i)^2 \xi + \frac{d\bar{t}}{d\bar{t}} N + \sum_{i=1}^{4} c_i W_i, \qquad (3.6.2)$$

$$\bar{W}_1 = B_1 \left(W_1 + \frac{dt}{d\bar{t}}c_1 \xi\right),$$

$$\bar{W}_i = \sum_{j=2}^{4} B_i^j \left(W_j - \frac{dt}{d\bar{t}}c_j \xi\right), \qquad i \in \{2, \ldots, 4\}.$$

The relations (3.6.1) follow by straightforward calculations from the general Frenet equations of Type 1 and the use of (3.6.2).

Theorem 6.1. Let C be a null curve of M_2^6 with a general Frenet frame F of Type 1 and a screen vector bundle $S(TC^\perp)$ on $\mathcal{U} \subset C$ such that $\kappa_1 \neq 0$ on $\mathcal{U}$. Then there exists a screen vector bundle $\bar{S}(TC)$ which induces another Frenet frame $\bar{F}$ of Type 1 on $\mathcal{U}$ such that $\bar{\kappa}_4 = \bar{\kappa}_5 = 0$.

Proof. Define the following vector fields in terms of the elements of F on $\mathcal{U}$:

$$\bar{N} = -\frac{1}{2}\left(\frac{k_4^2 + \kappa_5^2}{\kappa_1^2}\right)\xi + N - \frac{\kappa_4}{\kappa_1}W_2 - \frac{\kappa_5}{\kappa_1}W_3,$$

$$\bar{W}_2 = W_2 + \frac{\kappa_4}{\kappa_1}\xi,$$

$$\bar{W}_3 = W_3 + \frac{\kappa_5}{\kappa_1}\xi, \tag{3.6.3}$$

$$\bar{W}_i = W_i, \qquad i \in \{1,\, 4\}.$$

Let $\mathcal{U}^*$ be another coordinate neighborhood with parameter t^* on C such that $\mathcal{U} \cap \mathcal{U}^* \neq \emptyset$. By proposition 4.5 we have the following on $\mathcal{U} \cap \mathcal{U}^*$

$$\kappa_1^* = \kappa_1 A_1 \left(\frac{dt}{dt^*}\right)^2; \quad \kappa_4^* = \kappa_4 A_3 \frac{dt}{dt^*},$$

$$\kappa_5^* = \kappa_5 A_4 \frac{dt}{dt^*}, \quad W_i^* = A_i W_i, \quad i \in \{1,\, 2,\, 3,\, 4\}. \tag{3.6.4}$$

Define $\{\bar{N}^*, \bar{W}_1^*, \ldots, \bar{W}_4^*\}$ by (3.6.4) but on $\mathcal{U}^*$ with respect to F^*, induced by the same $S(TC^\perp)$ on $\mathcal{U}^*$. Then by using (3.3.1), (3.3.2), (3.6.3) and (3.6.4) we obtain

$$\bar{N}^* = \frac{dt^*}{dt}\bar{N}, \qquad \bar{W}_i^* = A_i \bar{W}_i, \quad i \in \{1,\, 2,\, 3,\, 4\}.$$

Hence there exists a vector bundle $\bar{S}(TC^\perp)$ spanned on $\mathcal{U}$ by $\{\bar{W}_1, \ldots, \bar{W}_4\}$ given by (3.6.3). Moreover, it is easy to check that this vector bundle is complementary to TC in $TC^\perp$. The null transversal vector field, with respect to $S(TC^\perp)$, is locally represented by $\bar{N}$ from (3.6.3). Finally taking into account that $t = \bar{t}$ and $c_2 = -\frac{\kappa_4}{\kappa_1}$; $c_3 = -\frac{\kappa_5}{\kappa_1}$ in the fourth and the fifth equations of (3.6.1), we obtain $\bar{\kappa}_4 = \bar{\kappa}_5 = 0$ which completes the proof.

Example 14. Consider a null curve in $\mathbf{R}_2^6$ given by

$$C: \left(\frac{4}{3}t^3 - 2t,\; 2t^2,\; \frac{4}{3}t^3,\; \sqrt{2}t,\; \cos\sqrt{2}t,\; \sin\sqrt{2}t\right).$$

This curve falls in the Type 1 and TM have a Frenet frame

$$F = \{\xi,\, N,\, W_1,\, W_2,\, W_3,\, W_4\}$$

as in Example 5. Since $\kappa_1 = -2\sqrt{3}$, $\kappa_4 = 0$ and $\kappa_5 = \sqrt{\frac{17}{3}}$, let

$$\bar{N} = -\frac{17}{72}\xi + N + \frac{\sqrt{17}}{6}W_3,$$

$$\bar{W}_3 = W_3 - \frac{\sqrt{17}}{6}\xi, \quad \bar{W}_i = W_i, \qquad i \in \{1,\, 2,\, 4\},$$

then we have $\bar\kappa_4 = g(\nabla_\xi \bar W_1, \bar W_2) = 0$ and $\bar\kappa_5 = g(\nabla_\xi \bar W_1, \bar W_3) = 0$.

At this point we assume that the transformations (3.3.2) are *diagonal transformations*, that is, they satisfy $B_i^j = B_j^i = 0 \, (i \neq j)$. For this case, it follows from the last equation of (3.6.1) that $c_4 = 0$. Using this we obtain

Theorem 6.2. *Let C be a null curve of M_2^6 with general Frenet frame of Type 1 such that $\kappa_1 \neq 0$. Then, there exist a null 2-surface which is invariant with respect to both the parameter transformations on C and the diagonal screen vector bundle transformations.*

Proof. Let C^* be an integral curve of the vector field W_4. Since, by lemma 6.1, $c_4 = 0$ for a diagonal screen vector bundle transformation, the 2-surface $S = C \times C^*$ is always invariant with respect to this class of screen vector bundle transformations. S can neither be Lorentz nor definite as its two base vectors $\{\tau, W_4\}$ contain a single null vector τ. Thus, S must be null which completes the proof.

Now we consider the case when F and $\bar F$ are both general Frenet frames of Type 2. Using the equations (3.2.6) and the method of lemma 6.1, we have

Lemma 6.2. *Let C be a null curve of M_2^6, $\kappa_1 \neq 0$, and two general Frenet frames F and $\bar F$ of Type 2. Then their curvature functions are related by*

$$\bar K_1 = H K_1 \left(\frac{dt}{d\bar t}\right)^2,$$

$$\bar K_2 = H \left\{ K_2 + \bar h C_1 + \frac{dC_1}{d\bar t} - \frac{1}{2}\left(\frac{dt}{d\bar t}\right)^2 \sum \epsilon_i \, (c_i)^2 \right.$$

$$\left. - (C_1 \kappa_4 + c_3 K_7 + c_4 K_8)\frac{dt}{d\bar t} \right\},$$

$$\bar K_3 = G \left\{ K_3 + \bar h \, C_2 + \frac{dC_2}{d\bar t} + (C_2 \kappa_4 - c_3 K_5 - c_4 K_6)\frac{dt}{d\bar t} \right\},$$

$$\bar K_4 = \left\{ \kappa_4 + K_1 C_2 \frac{dt}{d\bar t} - H \frac{dG}{dt} \right\} \frac{dt}{d\bar t},$$

$$\bar K_5 = G \left\{ K_5 B_3^3 + K_6 B_3^4 \right\} \frac{dt}{d\bar t}, \qquad\qquad (3.6.5)$$

$$\bar K_6 = G \left\{ K_5 B_4^3 + K_6 B_4^4 \right\} \frac{dt}{d\bar t},$$

$$\bar K_7 = H \left\{ \left(K_7 + K_1\frac{dt}{d\bar t}c_3\right) B_3^3 + \left(K_8 + K_1\frac{dt}{d\bar t}c_4\right) B_3^4 \right\} \frac{dt}{d\bar t},$$

$$\bar K_8 = H \left\{ \left(K_7 + K_1\frac{dt}{d\bar t}c_3\right) B_4^3 + \left(K_8 + K_1\frac{dt}{d\bar t}c_4\right) B_4^4 \right\} \frac{dt}{d\bar t},$$

where $H = B_1^1 - B_1^2$, $G = B_1^1 + B_2^1$, $C_1 = \frac{1}{\sqrt 2}(c_1 + c_2)$ *and* $C_2 = \frac{1}{\sqrt 2}(c_2 - c_1)$.

Proof. For $m = 4$, the matrix $[B_j^i(x)]$, in the general relations (3.3.2), is made up of two 2×2 Lorentz matrices. Therefore, using (3.2.1), the general transformations are given by

$$\frac{d}{d\bar{t}} = \frac{dt}{d\bar{t}} \xi,$$

$$\bar{N} = -\frac{1}{2} \frac{dt}{d\bar{t}} \sum_{i=1}^{4} \epsilon_i (c_i)^2 \xi + \frac{d\bar{t}}{dt} N + C_1 L_1 + C_2 L_2 + c_3 W_3 + c_4 W_4,$$

$$\bar{L}_1 = G \left\{ L_1 - \frac{dt}{d\bar{t}} C_2 \xi \right\},$$

$$\bar{L}_2 = H \left\{ L_2 - \frac{dt}{d\bar{t}} C_1 \xi \right\}, \tag{3.6.6}$$

$$\bar{W}_\alpha = \sum_{\beta=3}^{4} B_\alpha^\beta \left(W_\beta - \frac{dt}{d\bar{t}} c_\beta \xi \right).$$

In this case we have $B_1^3 = B_3^1 = B_1^4 = B_4^1 = B_2^3 = B_3^2 = B_2^4 = B_4^2 = 0$. Straightforward calculations from above relations and the use of (3.2.6) implies (3.6.5), which proves this lemma.

By a procedure, one can prove the following:

Theorem 6.3. *Let C be a null curve of M_2^6 with screen vector bundle $S(TC^\perp)$ and a general Frenet frame F of Type 2 such that $\kappa_1 \neq 0$ on $\mathcal{U}$. Then there exists a screen vector bundle $\bar{S}(TC^\perp)$ which induces another Frenet frame $\bar{F}$ on $\mathcal{U}$ such that $\bar{K}_7 = \bar{K}_8 = 0$ on $\mathcal{U}$.*

Example 15. Consider a null curve in $\mathbf{R}_2^6$ given by

$$C: \left(\frac{1}{3}t^3 - 2t, \ t^2, \ \frac{1}{3}t^3, \ \sqrt{2}t, \ \cos\sqrt{2}t, \ \sin\sqrt{2}t \right).$$

This curve falls in the Type 2 and TM have a Frenet frame

$$F = \{ \xi, \ N, \ L_1, \ L_2, \ W_3, \ W_4 \}$$

as in Example 5. Since $K_1 = -2$, $K_7 = -\frac{1}{2}$ and $K_5 = \frac{1}{2}$, let

$$\bar{N} = -\frac{1}{16}\xi + N + \frac{1}{4}W_3 - \frac{1}{4}W_4,$$

$$\bar{W}_3 = W_3 - \frac{1}{4}\xi, \quad \bar{W}_4 = W_4 + \frac{1}{4}\xi,$$

$$\bar{L}_i = L_i, \quad i \in \{1, 2\},$$

then we have $\bar{K}_7 = g(\nabla_\xi \bar{L}_1, \bar{W}_3) = 0$ and $\bar{K}_8 = g(\nabla_\xi \bar{L}_1, \bar{W}_4) = 0$.

3.7 Fundamental theorem of null curves in $\mathbf{R}_2^{m+2}$

Let $\mathbf{R}_2^{m+2}$ be the $(m+2)$-dimensional semi-Euclidean space of index 2 with the semi-Euclidean metric

$$g(x,\, y) = -x^0\, y^0 - x^1\, y^1 + \sum_{a=2}^{m+1} x^a\, y^a.$$

Suppose C is a null curve in $\mathbf{R}_2^{m+2}$ locally given by the equations

$$x^i = x^i\,(t), \qquad t \in I \subset R, \qquad i \in \{0, 1, \ldots, m+1\}.$$

First, we define in $\mathbf{R}_2^{m+2}$ the natural orthonormal basis

$$N_0^o = \left(\frac{1}{\sqrt{2}},\, 0,\, \frac{1}{\sqrt{2}},\, 0,\, \ldots,\, 0\right), \quad N_1^o = \left(-\frac{1}{\sqrt{2}},\, 0,\, \frac{1}{\sqrt{2}},\, 0,\, \ldots,\, 0\right),$$

$$W_1^o = (0,\, 1,\, 0,\, 0,\, \ldots,\, 0)\,, \qquad W_2^o = (0,\, 0,\, 0,\, 1,\, 0,\, \ldots,\, 0) \qquad (3.7.1)$$

$$\cdots\cdots\cdots\cdots\cdots \qquad\qquad \cdots\cdots\cdots\cdots\cdots$$

$$W_{m-1}^o = (0,\, 0,\, \ldots,\, 0,\, 1,\, 0), \qquad W_m^o = (0,\, 0,\, \ldots,\, 0,\, 0,\, 1)$$

where $\{N_0^o,\, N_1^o\}$ are null vectors such that $g\,(N_0^o,\, N_1^o) = 1$, W_1^o is a timelike vector and $\{W_2^o,\, \ldots,\, W_m^o\}$ are orthonormal spacelike vectors. It is easy to see that

$$N_0^{oi}\, N_1^{oj} + N_0^{oj}\, N_1^{oi} - W_1^{oi}\, W_1^{oj} + \sum_{\alpha=2}^{m} W_\alpha^{oi}\, W_\alpha^{oj} = h^{ij}, \qquad (3.7.2)$$

for any $i,\, j \in \{0,\, \ldots,\, (m+1)\}$, where we put

$$h^{ij} = \begin{cases} -1, & i = j = 0,\, 1 \\ 1, & i = j \neq 0,\, 1 \\ 0, & i \neq j. \end{cases}$$

We are now in a position to state the fundamental existence and uniqueness theorem for null curves of semi-Euclidean space $\mathbf{R}_2^{m+2}$.

Theorem 7.1. *Let* $\kappa_1,\, \ldots,\, k_{2m}\ :\ [-\varepsilon,\, \varepsilon]\ \to\ \mathbf{R}$ *be everywhere continuous functions,* $x_o = (x_o^i)$ *be a fixed point of* $\mathbf{R}_2^{m+2}$ *and let* $N_0^o,\, N_1^o,\, W_1^o,\, \ldots,\, W_m^o$ *be an orthonormal basis. Then there exists a unique null curve* $C\ :\ [-\varepsilon,\, \varepsilon]\ \to\ \mathbf{R}_2^{m+2}$ *given by the equations* $x^i = x^i(p)$, *where p is a distinguished parameter on C, such that* $C(0) = x_o$ *whose curvature functions are* $\{\kappa_1,\, \ldots,\, k_{2m}\}$ *and whose Frenet frames of Type 1*

$$\left\{\frac{d}{dp},\, N,\, W_1,\, \ldots,\, W_m\right\}$$

satisfies $\frac{d}{dp}(0) = N_0^o$, $N(0) = N_1^o$ and $W_i(0) = W_i^o$ for $1 \leq i \leq m$.

Proof. Using the distinguished parameter p and the equations (3.1.1) with $\epsilon_1 = -1$ and $\epsilon_\alpha = 1$, $\alpha \in \{2, \ldots, m\}$ and note that $\nabla_{\frac{d}{dp}} X$ is just X' for any vector field X defined on $\mathcal{U}$ we consider the system of differential equation

$$
\begin{aligned}
N_0' &= \kappa_1 W_1 \\
N_1' &= \kappa_2 W_1 + \kappa_3 W_2 \\
W_1' &= \kappa_2 N_0 + \kappa_1 N_1 - \kappa_4 W_2 - \kappa_5 W_3 \\
W_2' &= -\kappa_3 N_0 - \kappa_4 W_4 + \kappa_6 W_3 + \kappa_7 W_4
\end{aligned}
\tag{3.7.3}
$$

$$
\cdots\cdots\cdots\cdots\cdots\cdots\cdots
$$

$$
\begin{aligned}
W_{m-1}' &= -\kappa_{2m-3} W_{m-3} - \kappa_{2m-2} W_{m-2} + \kappa_{2m} W_m \\
W_m' &= -\kappa_{2m-1} W_{m-2} - \kappa_{2m} W_{m-1}.
\end{aligned}
$$

Then there exists a unique solution $\{N_0, N_1, W_1, \ldots, W_m\}$ satisfying the initial conditions $N_i(0) = N_i^o$, $W_\alpha(0) = W_\alpha^o$, $i \in \{1, 2\}$, $\alpha \in \{1, \ldots, m\}$. Now we claim that $\{N_i(p), W_\alpha(p)\}$ is a pseudo-orthonormal basis such that $\{N_0, N_1\}$, $\{W_1\}$ and $\{W_2, \ldots, W_m\}$ are null, timelike and spacelike vectors respectively, for $p \in [-\varepsilon, \varepsilon]$. To this end, by direct calculations using (3.7.3), we obtain

$$
\frac{d}{dp} \left\{ N_0^i N_1^j + N_0^j N_1^i - W_1^i W_1^j + \sum_{\alpha=2}^{m} W_\alpha^i W_\alpha^j \right\} = 0.
\tag{3.7.4}
$$

As for $p = 0$ we have (3.7.2), from (3.7.4) it follows that

$$
N_0^i N_1^j + N_0^j N_1^i - W_1^i W_1^j + \sum_{\alpha=2}^{m} W_\alpha^i W_\alpha^j = h^{ij}.
\tag{3.7.5}
$$

Further on, construct the field of frames

$$
W_{m+1} = \frac{1}{\sqrt{2}}(N_0 - N_1), \qquad W_{m+2} = \frac{1}{\sqrt{2}}(N_0 + N_1).
\tag{3.7.6}
$$

Then (3.7.5) becomes

$$
-W_1^i W_1^j - W_{m+1}^i W_{m+1}^j + W_{m+2}^i W_{m+2}^j + \sum_{\alpha=2}^{m} W_\alpha^i W_\alpha^j = h^{ij}.
\tag{3.7.7}
$$

We define for each $p \in [-\varepsilon, \varepsilon]$ the matrix $D(p) = [d^{ij}(p)]$ such that

$$
d^{ab} = W_b^a; \ d^{aB} = \sqrt{-1} W_B^a; \ d^{Ab} = -\sqrt{-1} W_b^A; \ d^{AB} = W_B^A,
$$

for any $a, b \in \{1; m+1\}$; $A, B \in \{2, \ldots, m; m+2\}$. By using (3.7.7) it is easy to check that $D(p)D(p)^t = I_{m+2}$, which implies that $\{W_1, \ldots, W_{m+2}\}$ is an orthonormal basis for any $p \in [-\varepsilon, \varepsilon]$. Then from (3.7.6) we conclude that

$\{N_0, N_1, W_1 W_2, \ldots, W_m\}$ is a pseudo-orthonormal basis for any $p \in [-\varepsilon, \varepsilon]$. The null curve is obtained by integrating the system

$$\frac{dx^i}{dp} = N_0^i(t), \qquad x^i(0) = x_o^i.$$

Taking into account of (3.7.3) we see

$$F = \left\{ \frac{d}{dp} = W_0, W_1, \ldots, W_m \right\}$$

is a general Frenet frame of Type 1 with curvature functions $\{\kappa_1, \ldots, \kappa_{2m}\}$ for C. This completes the proof of theorem.

Next, we define in $\mathbf{R}_2^{m+2}$ the pseudo-orthonormal basis

$$N_0^o = \left(\frac{1}{\sqrt{2}}, 0, \frac{1}{\sqrt{2}}, 0, 0, \ldots, 0 \right), \quad N_1^o = \left(-\frac{1}{\sqrt{2}}, 0, \frac{1}{\sqrt{2}}, 0, 0, \ldots, 0 \right),$$

$$N_2^o = \left(0, \frac{1}{\sqrt{2}}, 0, \frac{1}{\sqrt{2}}, 0, \ldots, 0 \right), \quad N_3^o = \left(0, -\frac{1}{\sqrt{2}}, 0, \frac{1}{\sqrt{2}}, 0, \ldots, 0 \right),$$

$$W_3^o = (0, 0, 0, 0, 1, \ldots, 0), \qquad W_4^o = (0, 0, 0, 0, 0, 1 \ldots, 0) \qquad (3.7.8)$$

$$\cdots\cdots\cdots\cdots\cdots\cdots\cdots\qquad\cdots\cdots\cdots\cdots\cdots\cdots\cdots$$

$$W_{m-1}^o = (0, 0, \ldots, 0, 1, 0), \qquad W_m^o = (0, 0, 0, \ldots, 0, 0, 1)$$

where $\{N_0^o, N_1^o, N_2^o, N_3^o\}$ are null vectors such that $g(N_0^o, N_1^o) = 1$ and $g(N_2^o, N_3^o) = 1$ and $\{W_3^o, \ldots, W_m^o\}$ are orthonormal spacelike vectors. In this case also we find

$$N_0^{oi} N_1^{oj} + N_0^{oj} N_1^{oi} + N_2^{oi} N_3^{oj} + N_2^{oj} N_3^{oi} + \sum_{\alpha=3}^{m} W_\alpha^{oi} W_\alpha^{oj} = h^{ij}, \qquad (3.7.9)$$

for any $i, j \in \{0, \ldots, m+1\}$.

Theorem 7.2. *Let* $\kappa_1, \ldots, k_{2m}; \tau_1, \ldots, \tau_{m-3} : [-\varepsilon, \varepsilon] \to \mathbf{R}$ *be everywhere continuous functions,* $x_o = (x_o^i)$ *be a fixed point of* $\mathbf{R}_2^{m+2}$ *and let* $\{N_i^o, W_\alpha^o\}, 0 \le i \le 3; 3 \le \alpha \le m$ *be a pseudo-orthonormal basis in (3.7.8). Then there exists a unique null curve* $C : [-\varepsilon, \varepsilon] \to \mathbf{R}_2^{m+2}$ *such that* $x^i = x^i(p), C(0) = x_o$ *and* $\{\kappa_1, \ldots, k_{2m}\}$ *are the curvature functions and* $\{\tau_1, \ldots, \tau_{m-3}\}$ *the torsion functions with respect to a general Frenet frame of Type 2*

$$F = \left\{ \frac{d}{dp}, N, L_1, L_2, W_3, \ldots, W_m \right\}$$

that satisfies

$$\frac{d}{dp}(0) = N_0^o, \ N(0) = N_1^o, \ L_1(0) = N_2^o, \ L_2(0) = N_3^o,$$

$$W_\alpha(0) = W_\alpha^o, \quad \alpha \in \{3, \ldots, m\}.$$

Proof. Using the distinguished parameter p and the equations (3.2.6) (or, equivalently (3.2.4)) we consider the system of differential equations

$$
\begin{aligned}
N_0' &= K_1\, N_3 \\
N_1' &= K_2\, N_2 + K_3\, N_3 + \tau_2\, W_3 \\
N_2' &= -K_3\, N_0 + \kappa_4\, N_2 + K_5\, W_3 + K_6\, W_4 \\
N_3' &= -K_2\, N_0 - K_1\, N_1 - \kappa_4\, N_3 + K_7\, W_3 + K_8\, W_4 \qquad (3.7.10) \\
W_3' &= -\tau_2\, N_0 - K_7\, N_2 - K_5\, N_3 + \kappa_8\, W_4 + \kappa_9\, W_5
\end{aligned}
$$

$$\cdots\cdots\cdots\cdots\cdots\cdots\cdots\cdots\cdots$$

$$
\begin{aligned}
W_{m-1}' &= -\kappa_{2m-3}\, W_{m-3} - \kappa_{2m-2}\, W_{m-2} + \kappa_{2m}\, W_m \\
W_m' &= -\kappa_{2m-1}\, W_{m-2} - \kappa_{2m}\, W_{m-1}.
\end{aligned}
$$

Then, there exists a unique solution $\{N_i(p),\, W_\alpha(p)\}$ satisfying the initial conditions

$$N_i(0) = N_i^o, \ W_\alpha(0) = W_\alpha^o, \ i \in \{0, 1, 2, 3\}, \ \alpha \in \{3, 4, \ldots, m\}$$

such that

$$N_0^i\, N_1^j + N_0^j\, N_1^i + N_2^i\, N_3^j + N_2^j\, N_3^i + \sum_{\alpha=3}^{m} W_\alpha^i\, W_\alpha^j = h^{ij}, \qquad (3.7.11)$$

Now we claim that $\{N_0(p), N_1(p), N_2(p), N_3(p)\}$ are null vectors such that $g\left(N_0(p), N_1(p)\right) = 1$ and $g\left(N_2(p), N_3(p)\right) = 1$ and $\{W_\alpha(p)\}$ are spacelike vectors, for $p \in [-\varepsilon, \varepsilon]$. To this end, construct the field of frames

$$
\begin{aligned}
W_0 &= \frac{1}{\sqrt{2}}\,(N_0 - N_1), & W_1 &= \frac{1}{\sqrt{2}}\,(N_0 + N_1), \qquad (3.7.12) \\[2mm]
W_{m+1} &= \frac{1}{\sqrt{2}}\,(N_2 - N_3), & W_{m+2} &= \frac{1}{\sqrt{2}}\,(N_2 + N_3).
\end{aligned}
$$

Then (3.7.11) becomes

$$-W_0^i\, W_0^j - W_{m+1}^i\, W_{m+1}^j + W_1^i\, W_1^j - W_{m+2}^i\, W_{m+2}^j + \sum_{\alpha=3}^{m} W_\alpha^i\, W_\alpha^j = h^{ij}$$

and it is easy to check that the matrix $D(p) = [d^{ij}(p)]$ satisfies $D(p)D(p)^t = I_{m+2}$, which implies also that $\{W_0, W_1, \ldots, W_{m+2}\}$ is an orthonormal basis for any $p \in$

$[-\varepsilon, \varepsilon]$. Then from (3.7.12) we conclude that $\{N_i, W_\alpha\}$ is a pseudo-orthonormal basis for any $p \in [-\varepsilon, \varepsilon]$. Thus there is a null C such that $C(0) = x_0$ and

$$F = \left\{ \frac{d}{dp} = N_0, \, N = N_1, \, L_1 = N_2, \, L_2 = N_3, \, W_3, \, \ldots, \, W_m \right\},$$

is a general Frenet frame of Type 2 with curvature functions $\{\kappa_1, \ldots, k_{2m}\}$ and torsion functions $\{\tau_1, \ldots, \tau_{m-1}\}$ for C.

3.8 Brief notes and exercises

(a) Semi-Euclidean spaces of degenerate degree r. Let (V, g) be an n-dimensional semi-Euclidean space with a constant index $q \geq 1$ and an orthonormal basis $B = \{e_1, \ldots, e_n\}$. Then, the *Radical* of (V, g) is its subspace defined by

$$Rad\,V = \{\xi \in V; \, g(\xi, v) = 0, \forall v \in V\}.$$

Let r_i be the dimension of the radical of the span $\{e_1, \ldots, e_i\}$ for all i. The sequence $\{r_i; \, 0 \leq i \leq n\}$ will be called the *nullity degree sequence* of the basis B. It is easy to see that $|r_i - r_{i-1}|$ is either 0 or 1 for all $i = 1, \ldots, n$ as well as $r_n = 0$. The positive number

$$r = \frac{1}{2} \sum_{i=1}^{n} |r_i - r_{i-1}|$$

is said to be the *degeneration degree* of the basis B. A basis

$$B = \{\xi_1, N_1, \ldots, \xi_r, N_r, W_1, \ldots, W_m\}, \quad \text{where} \quad 2r \leq 2q \leq n = m - 2r$$

of a semi-Euclidean space $\bar{\mathbf{R}}_q^n$ is said to be quasi-orthonormal if it satisfies

$$
\begin{aligned}
g(\xi_i, \xi_i) = g(N_i, N_i) &= 0, \quad g(\xi_i, N_j) = \eta_i\,\delta_{ij}, \\
g(\xi_i, W_a) = g(N_i, W_a) &= 0, \quad g(W_a, W_b) = \epsilon_a\,\delta_{ab},
\end{aligned}
$$

where $i, j \in \{1, \ldots, r\}$; $\eta_i = g(\xi_i, N_i) = \pm 1$; $a, b \in \{1, \ldots, m\}$; $\epsilon_a = -1$ if $1 \leq a \leq q - r$ and $\epsilon_a = 1$ if $q - r + 1 \leq a \leq m$.

Proposition. *Let* $B = \{e_1, \ldots, e_r\}$ *be an ordered basis of a semi-Euclidean space of constant index* q *and of degeneration degree* r. *Then,* $r \leq q$ *is well-defined, that is, it is an integer.*

(b) Degenerate curves in $\mathbf{R}_2^n$. Ferrández et al. [36] studied degenerate curves in semi-Euclidean spaces $\mathbf{R}_2^n$ of index 2 with degeneration degrees one and two. They developed Cartan reference frames along these degenerate curves and obtained one family of null curves for the degeneration degree one and two different families of null curves for the degeneration degree two. They also presented existence, uniqueness and congruence theorems with respect to those three families of null curves. Since the proofs of their theorems also involves discussion on non-null curves (out of course for this book) we leave it as readers choice to see their paper [36] for details.

Exercises

Find the Frenet frames and the curvature functions of

(1) A null curve C in $\mathbf{R}_2^4$ given by

$$x^0 = \log(\sec t + \tan t), \ x^1 = \cosh t, \ x^2 = \sinh t, \ x^3 = \log \sec t, \ \ t \in \mathbf{R}.$$

(2) A null curve C in $\mathbf{R}_2^4$ given by

$$x^0 = t, \quad x^1 = \frac{1}{2}t^2, \quad x^2 = t \sin t, \quad x^3 = t \cos t, \ \ t \in \mathbf{R}.$$

(3) A null curve C in $\mathbf{R}_2^5$ given by

$$x^0 = \frac{1}{\sqrt{2}}t^2, \ x^1 = t \cosh t, \ x^2 = t \sinh t, \ x^3 = t \cos t, \ x^4 = t \sin t, \ t \in \mathbf{R}.$$

(4) A null curve C in $\mathbf{R}_2^5$ given by

$$x^0 = t \sin t, \ x^1 = t \cos t, \ x^2 = \sin t, \ x^3 = \cos t, \ x^4 = \frac{1}{2}t^2, \ \ t \in \mathbf{R}.$$

(5) A null curve C in $\mathbf{R}_2^6$ given by

$$x^0 = \frac{4}{3}t^3 + t, \quad x^1 = \sinh t, \quad x^2 = 2t^2,$$

$$x^3 = \frac{4}{3}t^3 - t, \quad x^4 = t, \quad x^5 = t \cosh t, \quad t \in \mathbf{R}.$$

(6) A null curve C in $\mathbf{R}_2^6$ given by

$$x^0 = \frac{1}{3}t^3 + t, \quad x^1 = \frac{3}{\sqrt{2}}t \cosh t, \quad x^2 = \frac{3}{\sqrt{2}}t \sinh t,$$

$$x^3 = \sqrt{\frac{3}{2}} t \cos t, \quad x^4 = \sqrt{\frac{3}{2}} t \sin t, \quad x^5 = \frac{1}{3}t^3 - 2t, \quad t \in \mathbf{R}.$$

(7) A null curve C in $\mathbf{R}_2^7$ given by

$$x^0 = \frac{4}{3}t^3 + t, \quad x^1 = \sinh t, \quad x^2 = \cosh t, \quad x^3 = \sin t,$$

$$x^4 = \cos t, \quad x^5 = 2t^2, \quad x^6 = \frac{4}{3}t^3 - t, \quad t \in \mathbf{R}.$$

Chapter 4

Null curves in M_q^{m+2}

In this chapter, we present general theory of a null curve C of an $(m + 2)$-dimensional semi-Riemannian manifold (M, g) of index $q \, (\geq 3)$, denoted by M_q^{m+2}. We know from chapter 1 that any screen vector bundle $S(TC^\perp)$ of C will be semi-Riemannian of index $(q - 1)$. We first introduce two classes of null curves C, called Type 1 and Type 2, such that the Frenet frame of Type 1 is made up of two null vector fields, $(q - 1)$-timelike and $(m - q + 1)$-spacelike vector fields and the Frenet frame of Type 2 is made up of two null vector fields and additional two null vector fields L_i and L_{i+1} such that $g(L_i, L_{i+1}) = 1$, $(q - 2)$-timelike and $(m - q)$-spacelike vector fields. Next, we introduce another classes of null curves C, called Type $r \, (3 \leq r \leq q)$, such that the Frenet frame is made up of two null vector fields, additional $2(r-1)$ null vector fields $L_{\sigma(k)}$ and $L_{\sigma(k)+1}$ such that $g(L_{\sigma(k)}, L_{\sigma(k)+1}) = 1$, $(q - r)$-timelike and $(m + 2 - q - r)$-spacelike vector fields. We also introduce another new Frenet equations, called *Natural Frenet equations*. We study their invariant properties and support the results by examples.

4.1 Frenet equations of Type 1 and Type 2

Let C be a null curve in an $(m + 2)$-dimensional semi-Riemannian manifold M_q^{m+2} of index $q \, (3 \leq q \leq n = \frac{m+2}{2})$ and $C = span \, \{\xi\}$. For the Type 1, by using the process of the section 3.1 of chapter 3, we obtain the following equations

$$
\begin{aligned}
\nabla_\xi \xi &= h\,\xi + \kappa_1\,W_1 \\
\nabla_\xi N &= -h\,N + \kappa_2\,W_1 + \kappa_3\,W_2 \\
\epsilon_1 \nabla_\xi W_1 &= -k_2\,\xi - \kappa_1\,N + \kappa_4\,W_2 + \kappa_5\,W_3 \\
\epsilon_2 \nabla_\xi W_2 &= -\kappa_3\,\xi - \kappa_4\,W_1 + \kappa_6\,W_3 + \kappa_7\,W_4 \\
\epsilon_3 \nabla_\xi W_3 &= -\kappa_5\,W_1 - \kappa_6\,W_2 + \kappa_8\,W_4 + \kappa_9\,W_5
\end{aligned}
\qquad (4.1.1)
$$

$$
\cdots\cdots\cdots\cdots\cdots
$$

$$
\epsilon_{m-1} \nabla_\xi W_{m-1} = -\kappa_{2m-3}\,W_{m-3} - \kappa_{2m-2}\,W_{m-2} + \kappa_{2m}\,W_m
$$

$$\epsilon_m \nabla_\xi W_m = -\kappa_{2m-1} W_{m-2} - \kappa_{2m} W_{m-1},$$

where h and $\{\kappa_1, \ldots, \kappa_{2m}\}$ are smooth functions on $\mathcal{U}$, $\{W_1, \ldots, W_m\}$ is an orthonormal basis of $\Gamma(S(TC^\perp)|_\mathcal{U})$ and $(\epsilon_1, \ldots, \epsilon_m)$ is the signature of the manifold M_q^{m+2} such that $\epsilon_i \delta_{ij} = g(W_i, W_j)$. We call

$$F_1 = \{\xi, N, W_1, \ldots, W_m\}$$

a *general Frenet frame of Type 1* on M_q^{m+2} along C with respect to the given screen vector bundle $S(TC^\perp)$, the equations (4.1.1) the *general Frenet equations of Type 1* and the functions $\{\kappa_1, \ldots, \kappa_{2m}\}$ the *curvature functions* of C.

For the Type 2, by using the process of the section 3.2 of chapter 3 we obtain the following equations

$$
\begin{aligned}
\nabla_\xi \xi &= h\xi + \kappa_1 W_1 + \tau_1 W_2 \\
\nabla_\xi N &= -hN + \kappa_2 W_1 + \kappa_3 W_2 + \tau_2 W_3 \\
\epsilon_1 \nabla_\xi W_1 &= -\kappa_2 \xi - \kappa_1 N + \kappa_4 W_2 + \kappa_5 W_3 + \tau_3 W_4 \\
\epsilon_2 \nabla_\xi W_2 &= -\kappa_3 \xi - \tau_1 N - \kappa_4 W_1 + \kappa_6 W_3 + \kappa_7 W_4 + \tau_4 W_5 \\
\epsilon_3 \nabla_\xi W_3 &= -\tau_2 \xi - k_5 W_1 - \kappa_6 W_2 + \kappa_8 W_4 + \kappa_9 W_5 + \tau_5 W_6 \\
\epsilon_4 \nabla_\xi W_4 &= -\tau_3 W_1 - \kappa_7 W_2 - \kappa_8 W_3 + \kappa_{10} W_5 + \kappa_{11} W_6 + \tau_6 W_7
\end{aligned}
\tag{4.1.2}
$$

$$\cdots\cdots\cdots\cdots\cdots\cdots$$

$$\epsilon_{m-1} \nabla_\xi W_{m-1} = -\tau_{m-2} W_{m-5} - \kappa_{2m-3} W_{m-3} - \kappa_{2m-2} W_{m-2} + \kappa_{2m} W_m$$
$$\epsilon_m \nabla_\xi W_m = -\tau_{m-1} W_{m-3} - \kappa_{2m-1} W_{m-2} - \kappa_{2m} W_{m-1}.$$

In the above case, we call

$$F_2 = \{\xi, N, W_1, \cdots, W_m\}$$

a *general Frenet frame of Type 2* on M_q^{m+2} along C with respect to a given screen vector bundle $S(TC^\perp)$, the equations (4.1.2) its *general Frenet equations of Type 2* and the functions $\{\kappa_1, \cdots, \kappa_{2m}\}$ and $\{\tau_1, \cdots, \tau_{m-1}\}$ are called the *curvature* and *torsion functions* of C.

Remark. Let the vector fields

$$\nabla_\xi \xi - h\xi,$$
$$\nabla_\xi N + hN - k_2 W_1,$$
$$\epsilon_1 \nabla_\xi W_1 + k_2 \xi + k_1 N - k_4 W_2,$$

$$\cdots\cdots\cdots\cdots\cdots$$

$$\epsilon_{m-4} \nabla_\xi W_{m-4} + \kappa_{2m-9} W_{m-6} + \kappa_{2m-8} W_{m-5} - \kappa_{2m-4} W_{m-3}, \quad \text{or}$$
$$\epsilon_{m-3} \nabla_\xi W_{m-3} + \kappa_{2m-7} W_{m-5} + \kappa_{2m-6} W_{m-4} - \kappa_{2m-4} W_{m-2}$$

in $\Gamma(S(TC^\perp))$ be null in turn. Then, we obtain the Frenet equation (4.1.2) which have the torsion functions

$$\{\tau_1 = \kappa_1, \tau_2, \cdots\cdots, \tau_{m-1}\},$$

$$\{\tau_1 = 0,\ \tau_2 = \kappa_3,\ \tau_3,\ \cdots\cdots,\ \tau_{m-1}\},$$

$$\dotfill$$

$$\{\tau_1 = 0,\ \cdots,\ \tau_{m-3} = 0,\ \tau_{m-2} = \kappa_{2m-4},\ \tau_{m-1}\}\ \text{ or}$$
$$\{\tau_1 = 0,\ \tau_2 = 0,\ \cdots,\ \tau_{m-2} = 0,\ \tau_{m-1} = \kappa_{2m-3}\}$$

respectively. Also, if the vector field $\nabla_\xi \xi - h\xi$ is null, then the Frenet equations (4.1.2) reduced to the following equations

$$
\begin{aligned}
\nabla_\xi \xi &= h\xi + K_1 L_1,\\
\nabla_\xi N &= -h N + K_2 L_1 + K_3 L_2 + \tau_2 W_3,\\
\nabla_\xi L_1 &= -K_3 \xi + K_4 L_1 + K_5 W_3 + K_6 W_4 + \ell W_5,\\
\nabla_\xi L_2 &= -K_2 \xi - K_1 N - K_4 L_2 + K_7 W_3 + K_8 W_4 + \ell W_5,\\
\epsilon_3 \nabla_\xi W_3 &= -\tau_2 \xi - K_7 L_1 - K_5 L_2 + \kappa_8 W_4 + \kappa_9 W_5 + \tau_5 W_6,\\
\epsilon_4 \nabla_\xi W_4 &= -K_8 L_1 - K_6 L_2 - \kappa_8 W_3 + \kappa_{10} W_5 + \kappa_{11} W_6 + \tau_6 W_7,
\end{aligned}
\tag{4.1.3}
$$

$$\dotfill$$

$$
\begin{aligned}
\epsilon_{m-1} \nabla_\xi W_{m-1} &= -\tau_{m-2} W_{m-4} - \kappa_{2m-3} W_{m-3} - \kappa_{2m-2} W_{m-2} + \kappa_{2m} W_m,\\
\epsilon_m \nabla_\xi W_m &= -\tau_{m-1} W_{m-3} - \kappa_{2m-1} W_{m-2} - \kappa_{2m} W_{m-1}.
\end{aligned}
$$

If the vector field

$$\epsilon_{i-2} \nabla_\xi W_{i-2} + \kappa_{2i-7} W_{i-5} + \kappa_{2i-6} W_{i-4} - \kappa_{2i-4} W_{i-2} - \kappa_{2i-3} W_{i-1},$$

is null then the Frenet equations (4.1.2) reduce to

$$
\begin{aligned}
\nabla_\xi \xi &= h\xi + \kappa_1 W_1,\\
\nabla_\xi N &= -h N + \kappa_2 W_1 + \kappa_3 W_2,\\
\epsilon_1 \nabla_\xi W_1 &= -\kappa_2 \xi - \kappa_1 N + \kappa_4 W_2 + \kappa_5 W_3,
\end{aligned}
\tag{4.1.4}
$$

$$\dotfill$$

$$
\begin{aligned}
\nabla_\xi L_i &= -K_{2i+1} W_{i-1} + K_{2i+3} L_i + K_{2i+4} W_{i+2} + K_{2i+5} W_{i+3} + \ell W_{i+2},\\
\nabla_\xi L_{i+1} &= -K_{2i-1} W_{i-2} - K_{2i} W_{i-1} - K_{2i+3} L_{i+1} + K_{2i+6} W_{i+2}\\
&\quad + K_{2i+7} W_{i+3} + \ell W_{i+2},
\end{aligned}
$$

$$\dotfill$$

$$
\begin{aligned}
\epsilon_{m-1} \nabla_\xi W_{m-1} &= -\tau_{m-2} W_{m-4} - \kappa_{2m-3} W_{m-3} - \kappa_{2m-2} W_{m-2} + \kappa_{2m} W_m,\\
\epsilon_m \nabla_\xi W_m &= -\tau_{m-1} W_{m-3} - \kappa_{2m-1} W_{m-2} - \kappa_{2m} W_{m-1}.
\end{aligned}
$$

For these cases, we also call

$$F_2 = \{\ \xi,\ N,\ W_1,\ \ldots,\ W_{i-1},\ L_i,\ L_{i+1},\ W_{i+2},\ \ldots,\ W_m\ \},$$

where $1 \le i \le (m-1)$, a *general Frenet frame of Type 2* on M_q^{m+2} along C, with respect to a given screen vector bundle $S(TC^\perp)$ and the both the equations (4.1.3) and (4.1.4) its *general Frenet equations of Type 2*.

4.2 Frenet equations of Type $q\,(\geq 3)$

First of all, we study a class of null curves C in a semi-Riemannian manifold M_q^{m+2} whose Frenet frame is generated by a pseudo-orthonormal basis consisting of two null vector fields ξ and N and additional four null vector fields L_i, L_{i+1} and L_j, L_{j+1} such that $g(L_\alpha, L_{\alpha+1}) = 1$, $\alpha = i, j$. We denote the Frenet frame and the Frenet equations of this particular class of C by *Type 3*. Since the screen distribution $S(TC^\perp)$ is a semi-Riemannian vector bundle of index $(q-1)$, we let

$$\epsilon_i \, \bar{W}_i = \frac{L_i - L_{i+1}}{\sqrt{2}}, \qquad \epsilon_{i+1} \, \bar{W}_{i+1} = \frac{L_i + L_{i+1}}{\sqrt{2}}, \tag{4.2.1}$$

where $\epsilon_\alpha = g(\bar{W}_\alpha, \bar{W}_\alpha)$ is the sign of $\bar{W}_\alpha$. Then, by the method of the section 2.2 of Chapter 3 for Type 2, we have two null vector fields $\{L_1, L_2\}$ or equivalently, using the equation (4.2.1), one timelike and one spacelike vector fields $\{\bar{W}_1, \bar{W}_2\}$ which satisfy the following two equations

$$\begin{aligned}
\nabla_\xi \xi &= h\xi + K_1 L_1 \\
&= h\xi + \kappa_1 \, \epsilon_1 \, \bar{W}_1 + \tau_1 \, \epsilon_2 \, \bar{W}_2, \\
\nabla_\xi N &= -hN + K_2 \, L_1 + K_3 \, L_2 + S_1 \\
&= -hN + \kappa_2 \, \epsilon_1 \, \bar{W}_1 + \kappa_3 \, \epsilon_2 \, \bar{W}_2 + S_1.
\end{aligned}$$

Now we let the vector field S_1 be null and define the curvature function T_4 by $S_1 = T_4 \, L_3$. Then we have

$$\begin{aligned}
\nabla_\xi N &= -hN + K_2 \, L_1 + K_3 \, L_2 + T_4 \, L_3 \\
&= -hN + \kappa_2 \, \epsilon_1 \, \bar{W}_1 + \kappa_3 \, \epsilon_2 \, \bar{W}_2 + T_4 \, L_3,
\end{aligned}$$

where L_3 is a null vector field along C perpendicular to ξ, N, L_1, L_2, $\bar{W}_1$ and $\bar{W}_2$. For this case, there exists another unique null vector field L_4 along C such that $g(L_3, L_4) = 1$ and L_4 is also everywhere perpendicular to ξ, N, L_1, L_2, $\bar{W}_1$ and $\bar{W}_2$. Set this case so that the equations (4.2.1) hold for $i = 3$. Therefore $\bar{W}_3$ and $\bar{W}_4$ are perpendicular to ξ, N, L_1, L_2, $\bar{W}_1$ and $\bar{W}_2$ and we have

$$\nabla_\xi N = -hN + \kappa_2 \, \epsilon_1 \, \bar{W}_1 + \kappa_3 \, \epsilon_2 \, \bar{W}_2 + \tau_2 \, \epsilon_3 \, \bar{W}_3 + \mu_1 \, \epsilon_4 \, \bar{W}_4$$

where $\mu_1 = \tau_2 = \frac{T_4}{\sqrt{2}}$. Also, we have the following results

$$g(\nabla_\xi \bar{W}_1, \xi) = -\kappa_1, \quad g(\nabla_\xi \bar{W}_1, N) = -\kappa_2, \quad g(\nabla_\xi \bar{W}_1, \bar{W}_1) = 0,$$

$$g(\nabla_\xi \bar{W}_1, \bar{W}_2) = \kappa_4, \quad g(\nabla_\xi \bar{W}_1, \bar{W}_3) = \kappa_5, \quad g(\nabla_\xi \bar{W}_1, \bar{W}_4) = \tau_3.$$

Thus we obtain

$$\nabla_\xi \bar{W}_1 = -\kappa_2 \, \xi - \kappa_1 \, N + \kappa_4 \, \epsilon_2 \, \bar{W}_2 + \kappa_5 \, \epsilon_3 \, \bar{W}_3 + \tau_3 \, \epsilon_4 \, \bar{W}_4 + S_2,$$

where $S_2 \in \Gamma(S(TC^\perp))$ is a spacelike vector field perpendicular to $\xi, N, W_1, \cdots, W_4$. If we define a function μ_2 by $\mu_2 = \|S_2\|$ and set $\bar{W}_5 = \frac{S_2}{\mu_2}$, then we have

$$\nabla_\xi \bar{W}_1 = -\kappa_2 \, \xi - \kappa_1 \, N + \kappa_4 \, \epsilon_2 \, \bar{W}_2 + \kappa_5 \, \epsilon_3 \, \bar{W}_3 + \tau_3 \, \epsilon_4 \, \bar{W}_4 + \mu_2 \, \epsilon_5 \, \bar{W}_5.$$

In a similar way we get

$$\nabla_\xi \bar{W}_2 = - \kappa_3 \, \xi - \tau_1 \, N - \kappa_4 \, \epsilon_1 \, \bar{W}_1 + \kappa_6 \, \epsilon_3 \, \bar{W}_3$$
$$+ \kappa_7 \, \epsilon_4 \, \bar{W}_4 + \tau_4 \, \epsilon_5 \, \bar{W}_5 + \mu_3 \, \epsilon_6 \, \bar{W}_6,$$
$$\nabla_\xi \bar{W}_3 = - \tau_2 \, \xi - k_5 \, \epsilon_1 \, \bar{W}_1 - k_6 \, \epsilon_2 \, \bar{W}_2 + k_8 \, \epsilon_4 \, \bar{W}_4$$
$$+ \kappa_9 \, \epsilon_5 \, \bar{W}_5 + \tau_5 \, \epsilon_6 \, \bar{W}_6 + \mu_4 \, \epsilon_7 \, \bar{W}_7,$$
$$\nabla_\xi \bar{W}_4 = - \mu_1 \, \xi - \tau_3 \, \epsilon_1 \, \bar{W}_1 - \kappa_7 \, \epsilon_2 \, \bar{W}_2 - \kappa_8 \, \epsilon_3 \, \bar{W}_3$$
$$+ \kappa_{10} \, \epsilon_5 \, \bar{W}_5 + \kappa_{11} \, \epsilon_6 \, \bar{W}_6 + \tau_6 \, \epsilon_7 \, \bar{W}_7 + \mu_5 \, \epsilon_8 \, \bar{W}_8.$$

Setting $W_i = \epsilon_i \, \bar{W}_i$, we have the following equations

$$\nabla_\xi \xi = h\, \xi + \kappa_1 \, W_1 + \tau_1 \, W_2,$$
$$\nabla_\xi N = -h\, N + \kappa_2 \, W_1 + \kappa_3 \, W_2 + \tau_2 \, W_3 + \mu_1 \, W_4,$$
$$\epsilon_1 \, \nabla_\xi W_1 = - \kappa_2 \, \xi - \kappa_1 \, N + \kappa_4 \, W_2 + \kappa_5 \, W_3 + \tau_3 \, W_4 + \mu_2 \, W_5, \qquad (4.2.2)$$
$$\epsilon_2 \, \nabla_\xi W_2 = - \kappa_3 \, \xi - \tau_1 \, N - \kappa_4 \, W_1 + \kappa_6 \, W_3 + \kappa_7 \, W_4 + \tau_4 \, W_5 + \mu_3 \, W_6,$$
$$\epsilon_3 \, \nabla_\xi W_3 = - \tau_2 \, \xi - \kappa_5 \, W_1 - \kappa_6 \, W_2 + \kappa_8 \, W_4 + \kappa_9 \, W_5 + \tau_5 \, W_6 + \mu_4 \, W_7,$$
$$\epsilon_4 \, \nabla_\xi W_4 = - \mu_1 \, \xi - \tau_3 \, W_1 - \kappa_7 \, W_2 - \kappa_8 \, W_3 + \kappa_{10} \, W_5 + \kappa_{11} W_6$$
$$+ \tau_6 \, W_7 + \mu_5 \, W_8,$$

$$\cdots\cdots\cdots\cdots\cdots\cdots\cdots\cdots\cdots\cdots\cdots\cdots$$

$$\epsilon_{m-1} \, \nabla_\xi W_{m-1} = -\mu_{m-4} \, W_{m-5} - \tau_{m-2} \, W_{m-4} - \kappa_{2m-3} \, W_{m-3}$$
$$- \kappa_{2m-2} \, W_{m-2} + \kappa_{2m} \, W_m,$$
$$\epsilon_m \, \nabla_\xi W_m = -\mu_{m-3} \, W_{m-4} - \tau_{m-1} \, W_{m-3} - \kappa_{2m-1} \, W_{m-2} - \kappa_{2m} \, W_{m-1}.$$

In this case, we call

$$F_3 = \{ \, \xi, \, N, \, W_1, \, \cdots, \, W_m \, \}$$

a *general Frenet frame of Type 3* on M_q^{m+2} along C with respect to a give screen vector bundle $S(TC^\perp)$ and (4.2.2) its *general Frenet equations of Type 3*. The functions $\{\kappa_1, \, \cdots, \, \kappa_{2m}\}$, $\{\tau_1, \, \cdots, \, \tau_{m-1}\}$ and $\{\mu_1, \, \cdots, \, \mu_{m-3}\}$ are the *curvature functions* and the *torsion functions of the first and second kind* of C respectively with respect to the frame F_3.

Next, we also introduce the general Frenet equations of Type 3 with various forms. Let the vector fields $\nabla_\xi \xi - h\xi$ and $\nabla_\xi N + hN - K_2 \, L_1 - K_3 \, L_2$ or equivalently $\nabla_\xi N + hN - \kappa_2 \, W_1 - \kappa_3 \, W_2$ be null. Then there exist four null vector fields $\{L_1, L_2, L_3, L_4\}$ such that $g(L_1, L_2) = 1$, $g(L_3, L_4) = 1$ and all other $g(L_i, L_j) = 0$. By the transformations (4.2.1) for $i = 1$, we have

$$\nabla_\xi L_1 = \frac{1}{\sqrt{2}} \, (\, \nabla_\xi W_2 + \nabla_\xi W_1 \,), \qquad \nabla_\xi L_2 = \frac{1}{\sqrt{2}} \, (\, \nabla_\xi W_2 - \nabla_\xi W_1 \,).$$

Using (4.2.2) in above equation and the following results

$$K_2 = g(\nabla_\xi N, \, L_2) = \frac{1}{\sqrt{2}} \, g(\nabla_\xi N, \, W_2 - W_1) = \frac{1}{\sqrt{2}}(\kappa_3 + \kappa_2),$$

$$K_3 = g(\nabla_\xi N, L_1) = \frac{1}{\sqrt{2}} g(\nabla_\xi N, W_2 + W_1) = \frac{1}{\sqrt{2}}(\kappa_3 - \kappa_2),$$

$$K_4 = g(\nabla_\xi L_1, L_2) = \frac{1}{2} g(\nabla_\xi W_2 + \nabla_\xi W_1, W_2 - W_1) = -\kappa_4,$$

we have

$$\nabla_\xi L_1 = -K_3\xi + K_4 L_1 + K_5 W_3 + K_6 W_4 + \ell_1 W_5 + \ell_2 W_6,$$
$$\nabla_\xi L_2 = -K_2\xi - K_1 N - K_4 L_2 + K_7 W_3 + K_8 W_4 + \ell_3 W_5 + \ell_2 W_6,$$

where $\ell_2 = \frac{\mu_3}{\sqrt{2}}$ and

$$K_5 = \frac{1}{\sqrt{2}}(\kappa_6 - \kappa_5), \quad K_6 = \frac{1}{\sqrt{2}}(\kappa_7 - \tau_3), \quad \ell_1 = \frac{1}{\sqrt{2}}(\tau_4 - \mu_2),$$

$$K_7 = \frac{1}{\sqrt{2}}(\kappa_6 + \kappa_5), \quad K_8 = \frac{1}{\sqrt{2}}(\kappa_7 + \tau_3), \quad \ell_3 = \frac{1}{\sqrt{2}}(\tau_4 + \mu_2).$$

The last two equations are equivalent to the following equations

$$\nabla_\xi L_1 = -K_3\xi + K_4 L_1 + T_5 L_3 + T_6 L_4 + \ell_1 W_5 + \ell_2 W_6,$$
$$\nabla_\xi L_2 = -K_2\xi - K_1 N - K_4 L_2 + T_7 L_3 + T_8 L_4 + \ell_3 W_5 + \ell_2 W_6,$$

where

$$T_5 = \frac{1}{\sqrt{2}}(K_6 + K_5), \qquad T_6 = \frac{1}{\sqrt{2}}(K_6 - K_5),$$

$$T_7 = \frac{1}{\sqrt{2}}(K_8 + K_7), \qquad T_8 = \frac{1}{\sqrt{2}}(K_8 - K_7).$$

Hence we have

$$\nabla_\xi \xi = h\xi + K_1 L_1,$$
$$\nabla_\xi N = -hN + K_2 L_1 + K_3 L_2 + \tau_2 W_3 + \mu_1 W_4,$$
$$\nabla_\xi L_1 = -K_3\xi + K_4 L_1 + K_5 W_3 + K_6 W_4 + \ell_1 W_5 + \ell_2 W_6,$$
$$\nabla_\xi L_2 = -K_2\xi - K_1 N - K_4 L_2 + K_7 W_3 + K_8 W_4 + \ell_3 W_5 + \ell_2 W_6,$$
$$\epsilon_3 \nabla_\xi W_3 = -\tau_2\xi - K_7 L_1 - K_5 L_2 + \kappa_8 W_4 + \kappa_9 W_5 + \tau_5 W_6 + \mu_5 W_7,$$
$$\epsilon_4 \nabla_\xi W_4 = -\mu_1\xi - K_8 L_1 - K_6 L_2 - \kappa_8 W_3 + \kappa_{10} W_5 + \kappa_{11} W_6 + \tau_6 W_7 + \mu_5 W_8,$$
$$\epsilon_5 \nabla_\xi W_5 = -\ell_3 L_1 - \ell_1 L_2 - \kappa_9 W_3 - \kappa_{10} W_4 + \kappa_{12} W_6 + \kappa_{13} W_7 + \tau_7 W_8$$
$$\qquad\qquad + \mu_6 W_9,$$

$$\cdots\cdots\cdots\cdots\cdots\cdots\cdots\cdots\cdots\cdots\cdots\cdots\cdots\cdots \qquad\qquad (4.2.3)$$

$$\epsilon_{m-1}\nabla_\xi W_{m-1} = -\mu_{m-4} W_{m-5} - \tau_{m-2} W_{m-4} - \kappa_{2m-3} W_{m-3}$$
$$\qquad\qquad - \kappa_{2m-2} W_{m-2} + \kappa_{2m} W_m,$$

$$\epsilon_m \nabla_\xi W_m = -\mu_{m-3} W_{m-4} - \tau_{m-1} W_{m-3} - \kappa_{2m-1} W_{m-2} - \kappa_{2m} W_{m-1}.$$

In this case, we also call

$$F_3 = \{\, \xi, N, L_1, L_2, W_3, \cdots, W_m \,\}$$

a general Frenet frame of Type 3 on M_q^{m+2} along C with respect to a given screen vector bundle $S(TC^\perp)$ and (4.2.3) its general Frenet equations of Type 3.

Also, by the transformations (4.2.1) for $i = 3$ we have

$$\nabla_\xi L_3 = \frac{1}{\sqrt{2}}(\nabla_\xi W_4 + \nabla_\xi W_3), \qquad \nabla_\xi L_4 = \frac{1}{\sqrt{2}}(\nabla_\xi W_4 - \nabla_\xi W_3).$$

Using the following relations

$$K_9 = \frac{1}{\sqrt{2}}(T_8 - T_6) = -\frac{1}{\sqrt{2}}(\kappa_5 - \tau_3), \quad K_{11} = -\kappa_8, \quad \ell_6 = \frac{\mu_5}{\sqrt{2}},$$

$$K_{10} = -\frac{1}{\sqrt{2}}(T_8 + T_6) = \frac{1}{\sqrt{2}}(\kappa_7 - \kappa_6), \quad \ell_5 = \frac{1}{\sqrt{2}}(\tau_6 - \mu_4),$$

$$K_{14} = \frac{1}{\sqrt{2}}(T_7 - T_5) = -\frac{1}{\sqrt{2}}(\kappa_5 + \tau_3), \quad \ell_7 = \frac{1}{\sqrt{2}}(\tau_6 - \mu_4),$$

$$K_{15} = -\frac{1}{\sqrt{2}}(T_7 + T_5) = \frac{1}{\sqrt{2}}(\kappa_7 + \kappa_6), \quad K_{12} = \frac{1}{\sqrt{2}}(\kappa_{10} - \kappa_9),$$

$$K_{13} = \frac{1}{\sqrt{2}}(\kappa_{11} - \tau_5), \quad K_{16} = \frac{1}{\sqrt{2}}(\kappa_{10} + \kappa_9), \quad K_{17} = \frac{1}{\sqrt{2}}(\kappa_{11} + \tau_5),$$

we have

$$\nabla_\xi L_3 = -K_9 W_1 - K_{10} W_2 + K_{11} L_3 + K_{12} W_5 + K_{13} W_6 + \ell_5 W_7 + \ell_6 W_8,$$
$$\nabla_\xi L_4 = -T_4 \xi - K_{14} W_1 - K_{15} W_2 - K_{11} L_4 + K_{16} W_5 + K_{17} W_6 + \ell_7 W_7 + \ell_6 W_8.$$

The last two equations are equivalent to the following equations

$$\nabla_\xi L_3 = -T_8 L_1 - T_6 L_2 + K_{11} L_3 + K_{12} W_5 + K_{13} W_6 + \ell_5 W_7 + \ell_6 W_8,$$
$$\nabla_\xi L_4 = -T_4 \xi - T_7 L_1 - T_5 L_2 - K_{11} L_4 + K_{16} W_5 + K_{17} W_6 + \ell_7 W_7 + \ell_6 W_8,$$

where

$$T_5 = \frac{1}{\sqrt{2}}(K_{15} - K_{14}), \qquad T_6 = \frac{1}{\sqrt{2}}(K_{10} - K_9),$$

$$T_7 = \frac{1}{\sqrt{2}}(K_{15} + K_{14}), \qquad T_8 = \frac{1}{\sqrt{2}}(K_{10} + K_9).$$

Hence we have general Frenet equations of the following two forms

$$\nabla_\xi \xi = h\xi + \kappa_1 W_1 + \tau_1 W_2$$
$$\nabla_\xi N = -hN + \kappa_2 W_1 + \kappa_3 W_2 + T_4 L_3$$
$$\epsilon_1 \nabla_\xi W_1 = -\kappa_2 \xi - \kappa_1 N + \kappa_4 W_2 - K_{14} L_3 + K_9 L_4 + \mu_2 W_5, \qquad (4.2.4)$$
$$\epsilon_2 \nabla_\xi W_2 = -\kappa_3 \xi - \tau_1 N - \kappa_4 W_1 + K_{15} L_3 + K_{10} L_4 + \tau_4 W_5 + \mu_3 W_6,$$
$$\nabla_\xi L_3 = -K_9 W_1 - K_{10} W_2 + K_{11} L_3 + K_{12} W_5 + K_{13} W_6 + \ell_5 W_7 + \ell_6 W_8,$$

$$\nabla_\xi L_4 = -T_4 \xi - K_{14} W_1 - K_{15} W_2 - K_{11} L_4 + K_{16} W_5 + K_{17} W_6 + \ell_7 W_7$$
$$+ \ell_6 W_8,$$
$$\epsilon_5 \nabla_\xi W_5 = -\mu_2 W_1 - \tau_4 W_2 - K_{16} L_3 - K_{12} L_4 + \kappa_{12} W_6 + \kappa_{13} W_7 + \tau_7 W_8$$
$$+ \mu_6 W_9,$$

$$\cdots\cdots\cdots\cdots\cdots\cdots\cdots\cdots$$

$$\epsilon_{m-1} \nabla_\xi W_{m-1} = -\mu_{m-4} W_{m-5} - \tau_{m-2} W_{m-4} - \kappa_{2m-3} W_{m-3}$$
$$- \kappa_{2m-2} W_{m-2} + \kappa_{2m} W_m,$$
$$\epsilon_m \nabla_\xi W_m = -\mu_{m-3} W_{m-4} - \tau_{m-1} W_{m-3} - \kappa_{2m-1} W_{m-2} - \kappa_{2m} W_{m-1},$$

and

$$\nabla_\xi \xi = h\xi + K_1 L_1$$
$$\nabla_\xi N = -hN + K_2 L_1 + K_3 L_2 + T_4 L_3$$
$$\nabla_\xi L_1 = -K_3 \xi + K_4 L_1 + T_5 L_3 + T_6 L_4 + \ell_1 W_5 + \ell_2 W_6 \qquad (4.2.5)$$
$$\nabla_\xi L_2 = -K_2 \xi - K_1 N - K_4 L_2 + T_7 L_3 + T_8 L_4 + \ell_3 W_5 + \ell_2 W_6$$
$$\nabla_\xi L_3 = -T_8 L_1 - T_6 L_2 + K_{11} L_3 + K_{12} W_5 + K_{13} W_6 + \ell_5 W_7 + \ell_6 W_8,$$
$$\nabla_\xi L_4 = -T_4 \xi - T_7 L_1 - T_5 L_2 - K_{11} L_4 + K_{16} W_5 + K_{17} W_6 + \ell_7 W_7 + \ell_6 W_8,$$
$$\epsilon_5 \nabla_\xi W_5 = -\mu_2 W_1 - \tau_4 W_2 - K_{16} L_3 - K_{12} L_4 + \kappa_{12} W_6 + \kappa_{13} W_7 + \tau_7 W_8$$
$$+ \mu_6 W_9,$$

$$\cdots\cdots\cdots\cdots\cdots\cdots\cdots\cdots$$

$$\epsilon_{m-1} \nabla_\xi W_{m-1} = -\mu_{m-4} W_{m-5} - \tau_{m-2} W_{m-4} - \kappa_{2m-3} W_{m-3}$$
$$- \kappa_{2m-2} W_{m-2} + \kappa_{2m} W_m,$$
$$\epsilon_m \nabla_\xi W_m = -\mu_{m-3} W_{m-4} - \tau_{m-1} W_{m-3} - \kappa_{2m-1} W_{m-2} - \kappa_{2m} W_{m-1}.$$

In this case, we also call

$$F_3 = \{\, \xi,\, N,\, L_1,\, L_2,\, L_3,\, L_4,\, W_5,\, \cdots,\, W_m \,\}$$

a *general Frenet frame of Type 3* on M_q^{m+2} along C with respect to a given screen vector bundle $S(TC^\perp)$ and (4.2.5) its *general Frenet equations of Type 3*.

Remark. By the same method one can obtain other general Frenet frame and their general Frenet equations of Type 3 with various forms.

Now we consider a general case of a null curve C of M_q^{m+2} ($q \geq 3$). Using the same procedure we have the following equations

$$\nabla_\xi \xi = h\xi + \kappa_1 W_1 + \tau_1 W_2,$$
$$\nabla_\xi N = -hN + \kappa_2 W_1 + \kappa_3 W_2 + \tau_2 W_3 + \mu_1 W_4,$$
$$\epsilon_1 \nabla_\xi W_1 = -\kappa_2 \xi - \kappa_1 N + \kappa_4 W_2 + \kappa_5 W_3 + \tau_3 W_4 + \mu_2 W_5 + \nu_1 W_6,$$
$$\epsilon_2 \nabla_\xi W_2 = -\kappa_3 \xi - \tau_1 N - \kappa_4 W_1 + \kappa_6 W_3 + \kappa_7 W_4 + \tau_4 W_5 + \mu_3 W_6,$$
$$+ \nu_2 W_7 + \eta_1 W_8, \qquad (4.2.6)$$

$$\epsilon_3 \nabla_\xi W_3 = -\tau_2 \xi - \kappa_5 W_1 - \kappa_6 W_2 + \kappa_8 W_4 + \kappa_9 W_5 + \tau_5 W_6 + \mu_4 W_7,$$
$$+ \nu_3 W_6 + \eta_2 W_9 + \sigma_1 W_{10},$$
$$\epsilon_4 \nabla_\xi W_4 = -\mu_1 \xi - \tau_3 W_1 - \kappa_7 W_2 - \kappa_8 W_3 + \kappa_{10} W_5 + \kappa_{11} W_6$$
$$+ \tau_6 W_7 + \mu_5 W_8 + \nu_4 W_9 + \eta_3 W_{10} + \sigma_2 W_{11} + \rho_1 W_{12},$$

$$\cdots\cdots\cdots\cdots\cdots\cdots$$

$$\epsilon_{m-1} \nabla_\xi W_{m-1} = \cdots\cdots - \sigma_{m-10} W_{m-8} - \eta_{m-8} W_{m-7} - \nu_{m-6} W_{m-6}$$
$$- \mu_{m-4} W_{m-5} - \tau_{m-2} W_{m-4} - \kappa_{2m-3} W_{m-3} - \kappa_{2m-2} W_{m-2} + \kappa_{2m} W_m,$$
$$\epsilon_m \nabla_\xi W_m = \cdots\cdots - \sigma_{m-9} W_{m-7} - \eta_{m-7} W_{m-6} - \nu_{m-5} W_{m-5}$$
$$- \mu_{m-3} W_{m-4} - \tau_{m-1} W_{m-3} - \kappa_{2m-1} W_{m-2} - \kappa_{2m} W_{m-1}.$$

In this case, we call

$$F_q = \{\xi, N, W_1, \cdots, W_m\} \qquad (4.2.7)$$

a *general Frenet frame of Type q* on M_q^{m+2} along C with respect to a given screen vector bundle $S(TC^\perp)$ and the equations (4.2.6) its *general Frenet equations of Type q* of C. Also, the functions $\{\kappa_1, \cdots, \kappa_{2m}\}$, $\{\tau_1, \cdots, \tau_{m-1}\}$, $\{\mu_1, \cdots, \mu_{m-3}\}, \cdots$ are the *curvature* and the *torsion functions of the first, second, third,* $\cdots$, $(q-1)$-th kind of C respectively.

Remarks. (i) Let M_q^{m+2} be a real $(m+2)$-dimensional semi-Riemannian manifold of index q (≥ 1) and C be a smooth null curve in M_q^{m+2}. From the discussion so far we know that C has q-type general Frenet equations, named by Type 1, Type 2, ..., Type q, where $q \leq n = \frac{m+2}{2}$.

(ii) One can verify that the general Frenet equations (4.2.6) include all different general Frenet equations of Type 1 (if all torsion functions of all kinds are 0), Type 2 (if all torsion functions of the 2-, 3-, $\cdots$, $(q-1)$-kinds are 0), Type 3 (if all torsion functions of the 3-, 4-, $\cdots$, $(q-1)$-kinds are 0), $\cdots$, Type $(q-1)$ (if all torsion functions of $(q-1)$-kind are 0) and Type q.

(iii) We conclude that if $q = 1$, then M_q^{m+2} has only Type 1 general Frenet equations up to the signs of W_i; if $q = 2$, then M_q^{m+2} has general Frenet equations of two types, named by Type 1 and Type 2, up to the signs of W_i and if $q = n$, then M_q^{m+2} have general Frenet equations of n-types, named by Type 1, Type 2, $\cdots$, Type n, up to the signs of W_i. Hence, we call the equations (4.2.6) the *compound general Frenet equations* of the null curve C and the frame (4.2.7) the *compound general Frenet frame* on M_q^{m+2} along C.

Example 1. Let $\mathbf{R}_3^6$ be the 6-dimensional semi-Riemannian space of index 3 with the semi-Riemannian metric

$$g(x, y) = -x^0 y^0 - x^1 y^1 - x^2 y^2 + x^3 y^3 + x^4 y^4 + x^5 y^5.$$

Suppose C is a null curve in $\mathbf{R}_3^6$ given by the equations

$$C : (A\cos t, \ A\sin t, \ B\sinh t, \ B\cosh t, \ At, \ Bt)$$

where A, B, $t \in R$ such that $(A, B) \neq (0, 0)$. Then,

$$\xi = (-A\sin t, \quad A\cos t, \quad B\cosh t, \quad B\sinh t, \quad A, \quad B),$$
$$\nabla_\xi \xi = (-A\cos t, \quad -A\sin t, \quad B\sinh t, \quad B\cosh t, \quad 0, \quad 0).$$

If we take a spacelike vector field V along C such that

$$V = \begin{cases} (0, \quad 0, \quad 0, \quad 0, \quad 1, \quad 0) & \text{if} \quad A \neq 0 \\ (0, \quad 0, \quad 0, \quad 0, \quad 0, \quad 1) & \text{if} \quad B \neq 0, \end{cases}$$

then $g(\xi, V) = A$ or B and $g(V, V) = 1$. By the relation

$$N = \frac{1}{g(\xi, V)} \left\{ V - \frac{g(V, V)}{2g(\xi, V)} \xi \right\},$$

we obtain the following null transversal vector field

$$N = \begin{cases} \frac{1}{2A^2}(A\sin t, \ -A\cos t, \ -B\cosh t, \ -B\sinh t, \ A, \ -B) & \text{if} \quad A \neq 0 \\ \frac{1}{2B^2}(A\sin t, \ -A\cos t, \ -B\cosh t, \ -B\sinh t, \ -A, \ B) & \text{if} \quad B \neq 0. \end{cases}$$

We need to know the causal character of the vector field $H(t) = \nabla_\xi \xi - h\xi$ along C. By direct calculations we obtain $g(H(t), H(t)) = B^2 - A^2$. Hence $H(t)$ is spacelike, timelike or null according as $B^2 > A^2$, $B^2 < A^2$ or $B^2 = A^2$ respectively.

Choose $A = 1$ and $B = \sqrt{2}$, then $H(t)$ is spacelike and the curve

$$C : (\cos t, \quad \sin t, \quad \sqrt{2}\sinh t, \quad \sqrt{2}\cosh t, \quad t, \quad \sqrt{2}t)$$

falls in the Type 1 with the Frenet frame $F = \{\xi, N, W_1, W_2, W_3, W_4\}$ as follows

$$\xi = (-\sin t, \quad \cos t, \quad \sqrt{2}\cosh t, \quad \sqrt{2}\sinh t, \quad 1, \quad \sqrt{2}),$$
$$N = \frac{1}{2}(\sin t, \ -\cos t, \ -\sqrt{2}\cosh t, \ -\sqrt{2}\sinh t, \quad 1, \ -\sqrt{2}),$$
$$W_1 = (-\cos t, \ -\sin t, \quad \sqrt{2}\sinh t, \quad \sqrt{2}\cosh t, \quad 0, \quad 0),$$
$$W_2 = (-\sqrt{2}\cos t, \ -\sqrt{2}\sin t, \quad \sinh t, \quad \cosh t, \quad 0, \quad 0),$$
$$W_3 = (-\sqrt{2}\sin t, \quad \sqrt{2}\cos t, \quad 0, \quad 0, \quad 0, \quad 1),$$
$$W_4 = (\sqrt{2}\sin t, \ -\sqrt{2}\cos t, \ -\cosh t, \ -\sinh t, \quad 0, \ -2).$$

The general Frenet equations (4.1.1) are given by

$$\nabla_\xi \xi = W_1, \qquad \nabla_\xi N = -\frac{1}{2}W_1, \qquad \nabla_\xi W_1 = \frac{1}{2}\xi - N - \sqrt{2}W_3,$$
$$\nabla_\xi W_2 = -2W_3 - W_4, \quad \nabla_\xi W_3 = -\sqrt{2}W_1 + 2W_2, \quad \nabla_\xi W_4 = -W_2.$$

Next we set $A = \sqrt{2}$ and $B = 1$. The curve

$$C : (\sqrt{2}\cos t, \quad \sqrt{2}\sin t, \quad \sinh t, \quad \cosh t, \quad \sqrt{2}t, \quad t)$$

also falls in Type 1 with the Frenet frame $F = \{\xi,\ N,\ W_1,\ W_2,\ W_3,\ W_4\}$ as follows

$$\xi = (-\sqrt{2}\sin t,\ \sqrt{2}\cos t,\ \cosh t,\ \sinh t,\ \sqrt{2},\ 1),$$

$$N = \frac{1}{2}(\sqrt{2}\sin t,\ -\sqrt{2}\cos t,\ -\cosh t,\ -\sinh t,\ -\sqrt{2},\ 1),$$

$$W_1 = (-\sqrt{2}\cos t,\ -\sqrt{2}\sin t,\ \sinh t,\ \cosh t,\ 0,\ 0),$$

$$W_2 = (-\cos t,\ -\sin t,\ \sqrt{2}\sinh t,\ \sqrt{2}\cosh t,\ 0,\ 0),$$

$$W_3 = (0,\ 0,\ \sqrt{2}\cosh t,\ \sqrt{2}\sinh t,\ 1,\ 0),$$

$$W_4 = (\sin t,\ -\cos t,\ -\sqrt{2}\cosh t,\ -\sqrt{2}\sinh t,\ -2,\ 0).$$

The general Frenet equations (4.1.1) are given by

$$\nabla_\xi \xi = W_1, \qquad \nabla_\xi N = -\frac{1}{2}W_1, \qquad \nabla_\xi W_1 = -\frac{1}{2}\xi + N + \sqrt{2}W_3,$$

$$\nabla_\xi W_2 = 2W_3 + W_4, \qquad \nabla_\xi W_3 = -\sqrt{2}W_1 + 2W_2, \qquad \nabla_\xi W_4 = -W_2.$$

Finally we set $A = B = 1$, then $H(t)$ is null. Therefore, the curve

$$C : (\cos t,\ \sin t,\ \sinh t,\ \cosh t,\ t,\ t)$$

falls in the Type 3 with the Frenet frame $F = \{\xi,\ N,\ L_1,\ L_2,\ L_3,\ L_4\}$ as follows

$$\xi = (-\sin t,\ \cos t,\ \cosh t,\ \sinh t,\ 1,\ 1),$$

$$N = \frac{1}{2}(\sin t,\ -\cos t,\ -\cosh t,\ -\sinh t,\ 1,\ -1),$$

$$L_1 = (-\cos t,\ -\sin t,\ \sinh t,\ \cosh t,\ 0,\ 0),$$

$$L_2 = \frac{1}{2}(\cos t,\ \sin t,\ \sinh t,\ \cosh t,\ 0,\ 0),$$

$$L_3 = (-\sin t,\ \cos t,\ 0,\ 0,\ 0,\ 1),$$

$$L_4 = (0,\ 0,\ \cosh t,\ \sinh t,\ 0,\ 1).$$

The general Frenet equations (4.2.5) are given by

$$\nabla_\xi \xi = L_1, \qquad \nabla_\xi N = -\frac{1}{2}L_1, \qquad \nabla_\xi L_1 = -L_3 + L_4,$$

$$\nabla_\xi L_2 = \frac{1}{2}\xi - N - \frac{1}{2}(L_3 + L_4), \qquad \nabla_\xi L_3 = \frac{1}{2}L_1 - L_2, \qquad \nabla_\xi L_4 = \frac{1}{2}L_1.$$

The general Frenet equations (4.2.3) gives

$$\nabla_\xi \xi = L_1, \qquad \nabla_\xi N = -\frac{1}{2}L_1, \qquad \nabla_\xi L_1 = \sqrt{2}W_3,$$

$$\nabla_\xi L_2 = \frac{1}{2}\xi - N - \frac{1}{\sqrt{2}}W_4, \qquad \nabla_\xi W_3 = \frac{1}{\sqrt{2}}L_2, \qquad \nabla_\xi W_4 = \frac{1}{\sqrt{2}}(L_1 - L_2).$$

Also the general Frenet equations (4.2.2) gives

$$\nabla_\xi \xi = \frac{1}{\sqrt{2}}(W_2 - W_1), \qquad \nabla_\xi N = \frac{1}{2\sqrt{2}}(W_1 - W_2)$$

$$\nabla_\xi W_1 = \frac{1}{2\sqrt{2}}\xi - \frac{1}{\sqrt{2}}N - W_3 - \frac{1}{2}W_4,$$

$$\nabla_\xi W_2 = \frac{1}{2\sqrt{2}}\xi - \frac{1}{\sqrt{2}}N + W_3 - \frac{1}{2}W_4,$$

$$\nabla_\xi W_3 = \frac{1}{2}(W_1 + W_2), \quad \nabla_\xi W_4 = -W_1.$$

4.3   Natural Frenet equations

We consider two Frenet frames F and F^* along neighborhoods $\mathcal{U}$ and $\mathcal{U}^*$ with respect to a given screen vector bundle $S(TC^\perp)$ respectively such that $F \cap F^* \neq \emptyset$. Then the parameter transformations are given by

$$\xi^* = \frac{dt}{dt^*}\xi, \quad N^* = \frac{dt^*}{dt}N, \quad W_\alpha^* = \sum_{\beta=1}^{m} A_\alpha^\beta W_\beta, \quad 1 \le \alpha \le m \qquad (4.3.1)$$

where A_α^β are smooth functions on $\mathcal{U} \cap \mathcal{U}^*$ and the matrix $[A_\alpha^\beta(x)]$ is an element of the semi-orthogonal group $O(q-1, m-q+1)$ for any $x \in \mathcal{U} \cap \mathcal{U}^*$.

Let F and $\bar{F}$ be two Frenet frames with respect to $(t, S(TC^\perp), \mathcal{U})$ and $(\bar{t}, \bar{S}(TC^\perp), \bar{\mathcal{U}})$ respectively. Then the transformations of the screen vector bundle that relate elements of F and $\bar{F}$ on $\mathcal{U} \cap \bar{\mathcal{U}} \neq \emptyset$ are given by

$$\bar{\xi} = \frac{dt}{d\bar{t}}\xi,$$

$$\bar{N} = -\frac{1}{2}\frac{dt}{d\bar{t}}\sum_{\alpha=1}^{m}\epsilon_\alpha(c_\alpha)^2\xi + \frac{d\bar{t}}{dt}N + \sum_{\alpha=1}^{m}c_\alpha W_\alpha, \qquad (4.3.2)$$

$$\bar{W}_\alpha = \sum_{\beta=1}^{m}B_\alpha^\beta\left(W_\beta - \epsilon_\beta\frac{dt}{d\bar{t}}c_\beta\xi\right), \qquad 1 \le \alpha \le m,$$

where c_α and B_α^β are smooth functions on $\mathcal{U} \cap \bar{\mathcal{U}}$ and the $m \times m$ matrix $[B_\alpha^\beta(x)]$ is also an element of the semi-orthogonal group $O(q-1, m-q+1)$ for each $x \in \mathcal{U} \cap \bar{\mathcal{U}}$.

Using the general transformations (4.3.1) that relate elements of F and F^* on $\mathcal{U} \cap \mathcal{U}^*$ and the first and second equations of the compound general Frenet equations (4.2.6) for both F and F^*, we obtain

$$(\kappa_1^*, \tau_1^*, 0, \cdots, 0)\,[A_\alpha^\beta(x)] = (\kappa_1, \tau_1, 0, \cdots, 0)\left(\frac{dt}{dt^*}\right)^2, \qquad (4.3.3)$$

$$(\kappa_2^*, \kappa_3^*, \tau_2^*, \mu_1^*, 0, \cdots, 0)\,[A_\alpha^\beta(x)] = (\kappa_2, \kappa_3, \tau_2, \mu_1, 0, \cdots, 0). \qquad (4.3.4)$$

First, we show that each type of the general Frenet frames always transform to the same type by the transformations of the coordinate neighborhood of C.

Proposition 3.1. *Let C be a null curve of M_q^{m+2} and F and F^* be two general Frenet frames on $\mathcal{U}$ and $\mathcal{U}^*$ induced by the same screen vector bundle*

S(TC⊥) respectively. Then the type of general Frenet equations is invariant to the coordinate transformations.

Proof. In the first case suppose $F^* = F_t^*$, $t \neq 1$ and $F = F_1$. If $\tau_1^* \neq 0$, then we have $\tau_1^* = \kappa_1^*$ and $\tau_1 = 0$. This means from equations (4.3.3) that

$$A_1^2 = -A_2^2, \quad \ldots, \quad A_1^m = -A_2^m.$$

Since W_1^* and W_2^* are the timlike and spacelike vector fields respectively, the first row $(A_1^1, \cdots, A_1^m)$ of $[A_\alpha^\beta(x)]$ is timelike vector field and the second row $(A_2^1, \cdots, A_2^m)$ is a spacelike vector field of $\mathbf{R}_{q-1}^m$ and these vectors are perpendicular to each other. Thus, we have

$$(A_1^1)^2 - 1 = (A_2^1)^2 + 1 = -A_1^1 A_2^1.$$

This implies $A_1^1 = -A_2^1$ which is a contradiction, so it is not possible.

Conversely, suppose $F^* = F_1^*$ and $F = F_t$, $t \neq 1$. If $\tau_1 \neq 0$, then $\tau_1^* = 0$ and $\tau_1 = \kappa_1$. From (4.3.3) we have $A_1^1 = A_1^2$ and $A_1^3 = \cdots = A_1^m = 0$, This means that the first row $(A_1^1, A_1^2, 0, \cdots, 0)$ of the matrix $[A_\alpha^\beta(x)]$ is an null vector. Hence W_1^* is null, so this case is also not possible.

By the similar method, using the Frenet equation (4.2.6), we conclude that the cases $\{F^* = F_t^*, t \neq 1; F = F_1 \text{ and } \tau_1^* = 0, \cdots, \tau_{i-1}^* = 0, \tau_i^* \neq 0\}$ and $\{F^* = F_1^*; F = F_t, t \neq 1 \text{ and } \tau_1 = 0, \cdots, \tau_{i-1} = 0, \tau_i \neq 0\}$ are also not possible.

In the next cases suppose $F^* = F_3^*$, $F = F_2$ and $F^* = F_2^*$, $F = F_3$ respectively. Then we have $A_3^1 = \cdots = A_m^1 = A_3^2 = \cdots = A_m^2 = 0$ and $A_1^3 = \cdots = A_1^m = A_2^3 = \cdots = A_2^m = 0$ if $\tau_1^* = 0$, by the course of the proof of proposition 4.3 in chapter 3 and if $\tau_1^* \neq 0$, by the course of the proof of proposition 4.2 in chapter 3. Due to this fact and the equation (4.3.4), we have

$$\kappa_2^* A_1^1 + \kappa_3^* A_2^1 = \kappa_2, \qquad \kappa_2^* A_1^2 + \kappa_3^* A_2^2 = \kappa_3,$$
$$\tau_2^* A_3^3 + \mu_1^* A_3^4 = \tau_2, \qquad \tau_2^* A_4^3 + \mu_1^* A_4^4 = \mu_1.$$

In case $F^* = F_3^*$, $F = F_2$. If $\mu_1^* \neq 0$, then $\mu_2^* = \tau_2^*$ and $\mu_1 = 0$. Thus we have $A_3^4 = -A_4^4, \cdots, A_3^m = -A_4^m$. Since W_3^* and W_4^* are the timlike and spacelike vector fields respectively, the third row $(0, 0, A_3^3, \cdots, A_3^m)$ of $[A_\alpha^\beta(x)]$ is timelike vector field and the forth row $(0, 0, A_4^3, \cdots, A_4^m)$ is a spacelike vector field of $\mathbf{R}_{q-1}^m$ and these vectors are perpendicular to each other. Thus, we have

$$(A_3^3)^2 - 1 = (A_4^3)^2 + 1 = -A_3^3 A_4^3.$$

From this relation we have the contradictory relation $A_3^3 = -A_4^3$. Hence this case is not possible. In another case $F^* = F_2^*$, $F = F_3$. If $\mu_1 \neq 0$, then $\mu_1^* = 0$ and $\mu_1 = \tau_2$. Then we have $A_3^3 = A_3^4$ and $A_3^5 = \cdots = A_3^m = 0$. This case is also not possible because W_3^* is timelike. Similarly we have the same results for all other cases, which complete the proof.

Now, we introduce another form of Frenet equations of Type 1 different from the general Frenet equations (4.1.1) of Type 1 for null curves as chapter 2 and 3.

Using (4.3.2) and the methods of theorem 1.1 and theorem 1.2 of chapter 2 and theorem 3.1 and theorem 3.2 of chapter 3, we have

Theorem 3.1. *Let C be a null curve of M_q^{m+1} with a general Frenet frame F_1 of Type 1 with respect to the screen vector bundle $S(TC^\perp)$ such that $\kappa_1 \neq 0$. Then there exists a screen vector bundle $\bar{S}(TC)$ which induces another Frenet frame $\bar{F}$ of Type 1 on $\bar{\mathcal{U}}$ such that $\bar{\kappa}_4 = \bar{\kappa}_5 = 0$ on $\mathcal{U} \cap \bar{\mathcal{U}}$.*

Theorem 3.2. *Let C be a Type 1 null curve of M_q^{m+1}. Then there exists a Frenet frame F_1 satisfying the equations*

$$
\begin{aligned}
\nabla_\xi \xi &= h\xi + \kappa_1 W_1, \\
\nabla_\xi N &= -hN + \kappa_2 W_1 + \kappa_3 W_2, \\
\epsilon_1 \nabla_\xi W_1 &= -\kappa_2 \xi - \kappa_1 N, \\
\epsilon_2 \nabla_\xi W_2 &= -\kappa_3 \xi + \kappa_4 W_3, \\
\epsilon_3 \nabla_\xi W_3 &= -\kappa_4 W_2 + \kappa_5 W_4,
\end{aligned}
\qquad (4.3.5)
$$

$$
\cdots\cdots\cdots\cdots
$$

$$
\begin{aligned}
\epsilon_i \nabla_\xi W_i &= -\kappa_{i+1} W_{i-1} + \kappa_{i+2} W_{i+1}, \quad 3 \leq i \leq m-1, \\
\epsilon_m \nabla_\xi W_m &= -\kappa_{m+1} W_{m-1},
\end{aligned}
$$

where $\{\kappa_1, \ldots, \kappa_{m+1}\}$ are smooth functions on $\mathcal{U}$, $\{W_1, \ldots, W_m\}$ is a orthonormal basis of $\Gamma(S(TC^\perp)|_\mathcal{U})$ and $(\epsilon_1, \ldots, \epsilon_m)$ is the signature of (M_q^{m+1}, g).

We call the frame $F_1 = \{\xi, N, W_1, \ldots, W_m\}$ in Theorem 3.2 a *Natural Frenet frame of Type 1* on M along C with respect to the given screen vector bundle $S(TC^\perp)$ and the equations (4.3.5) are called its *Natural Frenet equations of Type 1*. Finally, the functions $\{\kappa_1, \ldots, \kappa_{m+1}\}$ are called *curvature functions* of C with respect to the frame F_1.

Also, we introduce another form of Frenet equations of Type 2 different from the general Frenet equations (4.1.2) of Type 2 for null curves as chapter 2 and 3. Using (4.3.2) and the method of Theorem 3.3 and 3.4 of chapter 3 and the equations (4.1.2) and (4.1.3), we have

Theorem 3.3. *Let C be a null curve of M_q^{m+1} with a general Frenet frame F_2 of Type 2 and a screen vector bundle $S(TC^\perp)$ such that $\tau_1 \neq 0$. Then there exists a screen vector bundle $\bar{S}(TC)$ which induces another Frenet frame $\bar{F}_2$ of Type 2 on $\bar{\mathcal{U}}$ such that $\bar{K}_4 = \bar{K}_7 = \bar{K}_8 = \bar{\ell} = \bar{\kappa}_4 = \bar{\tau}_4 = 0$ on $\mathcal{U} \cap \bar{\mathcal{U}}$.*

Theorem 3.4. *Let C be a null curve of M_q^{m+1} with a general Frenet frame F_2 of Type 2 with a screen vector bundle $S(TC^\perp)$ such that $\kappa_1 \neq 0$ and $\tau_1 = 0$. Then there exists a screen vector bundle $\bar{S}(TC)$ which induces another Frenet frame $\bar{F}_2$ of Type 2 on $\bar{\mathcal{U}}$ such that $\bar{\kappa}_4 = \bar{\kappa}_5 = \bar{\tau}_3 = 0$ on $\mathcal{U} \cap \bar{\mathcal{U}}$.*

Theorem 3.5. *Let C be a Type 2 null curve of M_q^{m+2}. If the vector field C'' is*

non-null, then there exists a Frenet frame $\{\xi, N, W_1, \ldots, W_m\}$ satisfying

$$
\begin{aligned}
\nabla_\xi \xi &= h\xi + \kappa_1 W_1, \\
\nabla_\xi N &= -h\xi + \kappa_2 W_1 + \kappa_3 W_2 + \tau_2 W_3, \\
\epsilon_1 \nabla_\xi W_1 &= -\kappa_2 \xi - \kappa_1 N, \\
\epsilon_2 \nabla_\xi W_2 &= -\kappa_3 \xi + \kappa_4 W_3 + \tau_3 W_4, \\
\epsilon_3 \nabla_\xi W_3 &= -\tau_2 \xi - \kappa_4 W_2 + \kappa_5 W_4 + \tau_4 W_5, \\
\epsilon_4 \nabla_\xi W_4 &= -\tau_3 W_2 - \kappa_5 W_3 + \kappa_6 W_5 + \tau_5 W_6, \\
\epsilon_5 \nabla_\xi W_5 &= -\tau_4 W_3 - \kappa_6 W_4 + \kappa_7 W_6 + \tau_6 W_7,
\end{aligned} \tag{4.3.6}
$$

....

$$
\begin{aligned}
\epsilon_i \nabla_\xi W_i &= -\tau_{i-1} W_{i-2} - \kappa_{i+1} W_{i-1} + \kappa_{i+2} W_{i+1} + \tau_{i+1} W_{i+2}, \; 3 \le i \le m-1, \\
\epsilon_m \nabla_\xi W_m &= -\tau_{m-1} W_{m-2} - \kappa_{m+1} W_{m-1}.
\end{aligned}
$$

If C'' is null, then there exists a Frenet frame $\{\xi, N, L_1, L_2, W_3, \ldots, W_m\}$ satisfying the following equations

$$
\begin{aligned}
\nabla_\xi \xi &= h\xi + K_1 L_1, \\
\nabla_\xi N &= -h N + K_2 L_1 + K_3 L_2 + \tau_2 W_3, \\
\nabla_\xi L_1 &= -K_3 \xi + K_4 W_3 + K_5 W_4, \\
\nabla_\xi L_2 &= -K_2 \xi - K_1 N, \\
\epsilon_3 \nabla_\xi W_3 &= -\tau_2 \xi - K_4 L_2 + \kappa_5 W_4 + \tau_4 W_5, \\
\epsilon_4 \nabla_\xi W_4 &= -K_5 L_2 - \kappa_5 W_3 + \kappa_6 W_5 + \tau_5 W_6, \\
\epsilon_5 \nabla_\xi W_5 &= -\tau_4 W_3 - \kappa_6 W_4 + \kappa_7 W_6 + \tau_6 W_7,
\end{aligned} \tag{4.3.7}
$$

....

$$
\begin{aligned}
\epsilon_i \nabla_\xi W_i &= -\tau_{i-1} W_{i-2} - \kappa_{i+1} W_{i-1} + \kappa_{i+2} W_{i+1} + \tau_{i+1} W_{i+2}, \; 5 \le i \le m-1, \\
\epsilon_m \nabla_\xi W_m &= -\tau_{m-1} W_{m-2} - \kappa_{m+1} W_{m-1}.
\end{aligned}
$$

We call the Frenet frame

$$
F_2 = \{\xi, N, W_1, W_2, \ldots, W_m\} = \{\xi, N, L_1, L_2, W_3, \ldots, W_m\}
$$

in Theorem 3.5 a *Natural Frenet frames of Type 2* on M along C with respect to the given screen vector bundle $S(TC^\perp)$ and the equations (4.3.6) and (4.3.7) are called its *Natural Frenet equations of Type 2* respectively. Finally, the functions $\{\kappa_1, \ldots, \kappa_{m+1}\}$ and $\{K_1, \cdots, K_5, \kappa_5, \ldots, \kappa_{m+1}\}$ are called the *curvature functions* and the functions $\{\tau_1, \ldots, \tau_{m-1}\}$ are called the *torsion functions* of C.

Now we introduce another form of Frenet equations of Type 3 different from the general Frenet equations of Type 3. First of all, we have

Theorem 3.6. *Let C be a null curve of M_q^{m+1} with a general Frenet frame F_3 of Type 3 with respect to the screen vector bundle $S(TC^\perp)$ such that $\tau_1 \mu_1 \neq 0$. Then there exists a screen vector bundle $\bar{S}(TC)$ which induces another Frenet*

frame $\bar{F}_3$ of Type 3 on $\bar{\mathcal{U}}$ such that $\bar{K}_4 = \bar{K}_7 = \bar{K}_8 = \bar{T}_7 = \bar{T}_8 = \bar{\ell}_3 = \bar{\ell}_2 = 0$ on $\mathcal{U} \cap \bar{\mathcal{U}}$.

Proof. Let $\tau_1 \neq 0$. Then, from the equation (4.2.2) and (4.2.4), we have

$$P_1^1 + \dot{P}_2^1 = P_1^2 + P_2^2, \quad P_1^\alpha = -P_2^\alpha, \quad P_\alpha^1 = P_\alpha^2, \quad 3 \le \alpha \le m,$$

because $\bar{\tau}_1 = \bar{\kappa}_1$ and $\tau_1 = \kappa_1$. Using the method of Section 3.3, the matrix $\left[P_\beta^\alpha(x) \right]$ is transformed as

$$T_2(B) = \begin{pmatrix} 0 & P_1^1 + P_2^1 & 0 & \cdots & 0 \\ -P_1^1 + P_1^2 & 0 & 0 & \cdots & 0 \\ 0 & 0 & P_3^3 & \cdots & P_3^m \\ 0 & 0 & P_4^3 & \cdots & P_4^m \\ \vdots & \vdots & \vdots & \cdots & \vdots \\ 0 & 0 & P_m^3 & \cdots & P_m^m \end{pmatrix}, \qquad (4.3.8)$$

where $(P_1^1 + P_2^1)(P_1^1 - P_1^2) = 1$. Also, using the coordinate transformations (4.3.2) that relate elements of F_3 and $\bar{F}_3$ on $\mathcal{U} \cap \bar{\mathcal{U}}$ such that $\tau_1 \neq 0$, we have

$$\bar{N} = -\frac{1}{2}\frac{dt}{d\bar{t}}\sum_{\alpha \ge 1}^m \epsilon_\alpha(c_\alpha)^2 \xi + \frac{d\bar{t}}{dt}N + C_1 L_1 + C_2 L_2 + \sum_{\alpha \ge 3}^m c_\alpha W_\alpha,$$

$$\bar{L}_1 = (B_1^1 + B_2^1)\left\{ L_1 - C_2\frac{dt}{d\bar{t}}\xi \right\}, \quad C_2 = \frac{1}{\sqrt{2}}(c_2 - c_1),$$

$$\bar{L}_2 = (B_1^1 - B_1^2)\left\{ L_2 - C_1\frac{dt}{d\bar{t}}\xi \right\}, \quad C_1 = \frac{1}{\sqrt{2}}(c_1 + c_2), \qquad (4.3.9)$$

$$\bar{W}_\alpha = \sum_{\beta \ge 3}^m B_\alpha^\beta\left\{ W_\beta - \epsilon_\beta c_\beta\frac{dt}{d\bar{t}}\xi \right\}, \quad \alpha, \beta \in \{3, \cdots, m\}.$$

From the fourth equation of the Frenet equations (4.2.3) and (4.2.5), we have

$$\bar{K}_1 = (B_1^1 - B_1^2)K_1\left(\frac{dt}{d\bar{t}}\right)^2,$$

$$\bar{K}_4 = \left\{ K_4 + K_1 C_2\frac{dt}{d\bar{t}} - (B_1^1 - B_1^2)\frac{d(B_1^1 + B_2^1)}{dt} \right\}\frac{dt}{d\bar{t}},$$

$$\bar{K}_7 B_3^3 + \bar{K}_8 B_4^3 + \bar{\ell}_3 B_5^3 + \bar{\ell}_2 B_6^3 = (B_1^1 - B_1^2)\left(K_7 + K_1 c_3\frac{dt}{d\bar{t}}\right)\frac{dt}{d\bar{t}},$$

$$\bar{K}_7 B_3^4 + \bar{K}_8 B_4^4 + \bar{\ell}_3 B_5^4 + \bar{\ell}_2 B_6^4 = (B_1^1 - B_1^2)\left(K_8 + K_1 c_4\frac{dt}{d\bar{t}}\right)\frac{dt}{d\bar{t}},$$

$$\bar{K}_7 B_3^5 + \bar{K}_8 B_4^5 + \bar{\ell}_3 B_5^5 + \bar{\ell}_2 B_6^5 = (B_1^1 - B_1^2)\left(\ell_3 + K_1 c_5\frac{dt}{d\bar{t}}\right)\frac{dt}{d\bar{t}},$$

$$\bar{K}_7 B_3^6 + \bar{K}_8 B_4^6 + \bar{\ell}_3 B_5^6 + \bar{\ell}_2 B_6^6 = (B_1^1 - B_1^2)\left(\ell_2 + K_1 c_6\frac{dt}{d\bar{t}}\right)\frac{dt}{d\bar{t}},$$

$$\bar{K}_7 B_3^\alpha + \bar{K}_8 B_4^\alpha + \bar{\ell}_3 B_5^\alpha + \bar{\ell}_2 B_6^\alpha = K_1 c_\alpha(B_1^1 - B_1^2)\left(\frac{dt}{d\bar{t}}\right)^2.$$

Taking into account that

$$C_2 = -\frac{K_4}{K_1}\frac{d\bar{t}}{dt}; \quad c_3 = -\frac{K_7}{K_1}\frac{d\bar{t}}{dt}; \quad c_4 = -\frac{K_8}{K_1}\frac{d\bar{t}}{dt};$$

$$c_5 = -\frac{\ell_3}{K_1}\frac{d\bar{t}}{dt}; \quad c_6 = -\frac{\ell_2}{K_1}\frac{d\bar{t}}{dt}; \quad c_\alpha = 0, \quad 7 \leq \alpha \leq m,$$

and $B_1^1 - B_1^2 = c\,(\text{constant})$, we have $\bar{K}_4 = \bar{K}_7 = \bar{K}_8 = \bar{\ell}_3 = \bar{\ell}_2 = 0$. On the other hand, since $T_7 = \frac{1}{\sqrt{2}}(K_8 + K_7)$ and $T_8 = \frac{1}{\sqrt{2}}(K_8 - K_7)$, we also have $\bar{T}_7 = \bar{T}_8 = 0$.

Remark. If we take $t = \bar{t}$ in Theorem 3.6, then

$$C_2 = -\frac{K_4}{K_1}; \quad c_3 = -\frac{K_7}{K_1}; \quad c_4 = -\frac{K_8}{K_1};$$

$$c_5 = -\frac{\ell}{K_1}; \quad c_6 = -\frac{\ell_2}{K_1}; \quad c_\alpha = 0, \quad 7 \leq \alpha \leq m,$$

and

$$\bar{N} = -\frac{1}{2}\left(\frac{\epsilon_3\,K_7^2 + \epsilon_4\,K_8^2 + \epsilon_5\,\ell_3^2 + \epsilon_6\,\ell_2^2}{K_1^2}\right)\xi + N - \frac{K_4}{K_1}L_2$$

$$- \frac{K_7}{K_1}W_3 - \frac{K_8}{K_1}W_4 - \frac{\ell_3}{K_1}W_5 - \frac{\ell_2}{K_1}W_6,$$

$$\bar{L}_1 = L_1 + \frac{K_4}{K_1}\xi, \quad \bar{L}_2 = L_2, \quad \bar{W}_3 = W_3 + \frac{K_7}{K_1}\xi,$$

$$\bar{W}_4 = W_4 + \frac{K_8}{K_1}\xi, \quad \bar{W}_5 = W_5 + \frac{\ell_3}{K_1}\xi, \quad \bar{W}_6 = W_6 + \frac{\ell_2}{K_1}\xi,$$

$$\bar{W}_i = W_i, \quad 7 \leq i \leq m.$$

Relabeling $N = \bar{N}$; $L_\alpha = \bar{L}_\alpha$, $\alpha \in \{1, 2\}$; $W_i = \bar{W}_i$, $i \in \{3, 4, 5\}$; $K_4 = \bar{K}_5$, $K_5 = \bar{K}_6$; $\ell = \bar{\ell}_1$; $\kappa_5 = \bar{\kappa}_8$; $\tau_4 = \bar{\kappa}_9$ and $S(TC^\perp) = \bar{S}(TC^\perp)$ and taking only the first five equations in (4.2.3) as follows:

$$\nabla_\xi \xi = h\,\xi + K_1\,L_1,$$
$$\nabla_\xi N = -h\,N + K_2\,L_1 + K_3\,L_2 + \tau_2\,W_3 + \mu_1\,W_4,$$
$$\nabla_\xi L_1 = -K_3\,\xi + K_4\,W_3 + K_5\,W_4 + \ell\,W_5,$$
$$\nabla_\xi L_2 = -K_2\,\xi - K_1\,N,$$
$$\epsilon_3\nabla_\xi W_3 = -\tau_2\,\xi - K_4\,L_2 + \kappa_5\,W_4 + \tau_4\,W_5 + R_8,$$

where $R_8 \in \Gamma(S(TC^\perp))$ is perpendicular to L_1, L_2, W_3, W_4 and W_5. Assume that R_8 is a non-null. Denote $\rho_8 = \|\,R_8\,\|$ and define the new third kind curvature function μ_3 by $\mu_3 = \epsilon_6\,\rho_8$ where ϵ_6 is sign of R_8. If ρ_8 is also non-zero for any t, we set $\bar{W}_6 = \frac{R_8}{\tau_3}$ and $W_6 = \epsilon_6\,\bar{W}_6$, then W_6 is also a non-null unit vector field with the same causality as R_8 along C. Thus we have

$$\epsilon_3\nabla_\xi W_3 = -\tau_2\,\xi - K_4\,L_2 + \kappa_5\,W_4 + \tau_3\,W_5 + \mu_3\,W_6.$$

Using the same method for $\nabla_\xi W_4$, we have

$$\epsilon_4 \nabla_\xi W_4 = -\mu_1\,\xi - K_5\,L_2 - \kappa_5\,W_3 + \kappa_6\,W_5 + \tau_5\,W_6 + \mu_4\,W_7.$$

Repeating above process for all m unit vectors of a semi-orthonormal basis $F_3 = \{L_1,\,L_2,\,W_3,\,\cdots,\,W_m\}$ of $\Gamma(S(TC^\perp))$ and simplifying we obtain the following;

$$
\begin{aligned}
\nabla_\xi \xi &= h\,\xi + K_1\,L_1, \\
\nabla_\xi N &= -h\,N + K_2\,L_1 + K_3\,L_2 + \tau_2\,W_3 + \mu_1\,W_4, \\
\nabla_\xi L_1 &= -K_3\,\xi + K_4\,W_3 + K_5\,W_4 + \ell\,W_5, \\
\nabla_\xi L_2 &= -K_2\,\xi - K_1\,N, \\
\epsilon_3 \nabla_\xi W_3 &= -\tau_2\,\xi - K_4\,L_2 + \kappa_5\,W_4 + \tau_4\,W_5 + \mu_3 W_6, \quad (4.3.10)\\
\epsilon_4 \nabla_\xi W_4 &= -\mu_1\,\xi - K_5\,L_2 - \kappa_5\,W_3 + \kappa_6\,W_5 + \tau_5\,W_6 + \mu_4\,W_7, \\
\epsilon_5 \nabla_\xi W_5 &= -\ell\,L_2 - \tau_4 W_3 - \kappa_6\,W_4 + \kappa_7\,W_6 + \tau_6\,W_7 + \mu_5\,W_8,
\end{aligned}
$$

$$\cdots\cdots\cdots\cdots\cdots\cdots\cdots\cdots\cdots$$

$$
\begin{aligned}
\epsilon_i \nabla_\xi W_i &= -\mu_{i-3}W_{i-3} - \tau_{i-1}W_{i-2} - \kappa_{i+1}W_{i-1} + \kappa_{i+2}W_{i+1} \\
&\quad + \tau_{i+1}W_{i+2} + \mu_i W_{i+3}, \qquad 6 \le i \le m-1, \\
\epsilon_m \nabla_\xi W_m &= -\mu_{m-3}W_{m-3} - \tau_{m-1}W_{m-2} - \kappa_{m+1}W_{m-1},
\end{aligned}
$$

or equivalently, there exists a frame $F_3 = \{W_1,\,\ldots,\,W_m\}$ of $\Gamma(S(TC^\perp))$ satisfying

$$
\begin{aligned}
\nabla_\xi \xi &= h\,\xi + \kappa_1\,W_1 + \tau_1\,W_2, \\
\nabla_\xi N &= -h\,N + \kappa_2\,W_1 + \kappa_3\,W_2 + \tau_2\,W_3 + \mu_1\,W_4, \\
\epsilon_1 \nabla_\xi W_1 &= -\kappa_2\,\xi - \kappa_1\,N + \kappa_4\,W_3 + \tau_3\,W_4 + \mu_2\,W_5, \\
\epsilon_2 \nabla_\xi W_2 &= -\kappa_3\,\xi - \tau_1\,N + \kappa_4\,W_3 + \tau_3\,W_4 + \mu_2\,W_5, \\
\epsilon_3 \nabla_\xi W_3 &= -\tau_2\,\xi - \kappa_4\,W_1 + \kappa_4\,W_2 + \kappa_5\,W_4 + \tau_4\,W_5 + \mu_3\,W_6, \quad (4.3.11)\\
\epsilon_4 \nabla_\xi W_4 &= -\mu_1\,\xi - \tau_3\,W_1 + \tau_3\,W_2 - \kappa_5\,W_3 + \kappa_6\,W_5 + \tau_5\,W_6 + \mu_4\,W_7, \\
\epsilon_5 \nabla_\xi W_5 &= -\mu_2\,W_1 + \mu_2\,W_2 - \tau_4 W_3 - \kappa_6\,W_4 + \kappa_7\,W_6 + \tau_6\,W_7 + \mu_5\,W_8,
\end{aligned}
$$

$$\cdots\cdots\cdots\cdots\cdots\cdots\cdots\cdots\cdots$$

$$
\begin{aligned}
\epsilon_i \nabla_\xi W_i &= -\mu_{i-3}W_{i-3} - \tau_{i-1}W_{i-2} - \kappa_{i+1}W_{i-2} + \kappa_{i+2}W_{i+1} \\
&\quad + \tau_{i+1}W_{i+2} + \mu_i W_{i+3}, \qquad 5 \le i \le m-1, \\
\epsilon_m \nabla_\xi W_m &= -\mu_{m-3}W_{m-3} - \tau_{m-1}W_{m-2} - \kappa_{m+1}W_{m-1}.
\end{aligned}
$$

Also, there is a frame $F_3 = \{L_1,\,\ldots,\,L_4,\,W_5,\,\ldots,\,W_m\}$ of $\Gamma(S(TC^\perp))$ satisfying

$$
\begin{aligned}
\nabla_\xi \xi &= h\xi + K_1 L_1 \\
\nabla_\xi N &= -hN + K_2\,L_1 + K_3\,L_2 + K_4\,L_3 \\
\nabla_\xi L_1 &= -K_3\,\xi + K_5\,L_3 + K_6\,L_4 + \ell\,W_5, \quad (4.3.12)\\
\nabla_\xi L_2 &= -K_2\,\xi - K_1\,N, \\
\nabla_\xi L_3 &= -K_6\,L_2 + K_7\,L_3 + K_8\,W_5 + K_9\,W_6, \\
\nabla_\xi L_4 &= -K_4\,\xi - K_5\,L_2 - K_7\,L_4 + K_{10}\,W_5 + K_{11}\,W_6 + K_{12}\,W_7,
\end{aligned}
$$

$$\epsilon_5 \nabla_\xi W_5 = -\ell L_2 - K_{10} L_3 - K_8 L_4 + \kappa_7 W_6 + \tau_6 W_7 + \mu_5 W_8,$$

$$\cdots\cdots\cdots\cdots\cdots\cdots\cdots\cdots\cdots\cdots\cdots\cdots$$

$$\epsilon_i \nabla_\xi W_i = -\mu_{i-3} W_{i-3} - \tau_{i-1} W_{i-2} - \kappa_{i+1} W_{i-1} + \kappa_{i+2} W_{i+1}$$
$$+ \tau_{i+1} W_{i+2} + \mu_i W_{i+3}, \qquad 3 \le i \le m-1,$$
$$\epsilon_m \nabla_\xi W_m = -\mu_{m-3} W_{m-3} - \tau_{m-1} W_{m-2} - \kappa_{m+1} W_{m-1}.$$

We call the frames F_3 a *Natural Frenet frame of Type 3* on M along C with respect to the given screen vector bundle $S(TC^\perp)$ and the equations (4.3.10) – (4.3.12) are called its *Natural Frenet equations of Type 3*.

Next, using the transformations (4.3.2) such that $\tau_1 = 0$, we have

Theorem 3.7. Let C be a null curve of M_q^{m+1} with a general Frenet frame F_3 of Type 3 and a screen vector bundle $S(TC^\perp)$ such that $\kappa_1 \ne 0$ and $\tau_1 = 0$. Then there exists a screen vector bundle $\bar{S}(TC)$ which induces another Frenet frame $\bar{F}_3$ of Type 3 on $\bar{\mathcal{U}}$ such that $\bar{\kappa}_4 = \bar{\kappa}_5 = \bar{\tau}_3 = \bar{\mu}_2 = \bar{K}_{14} = \bar{K}_9 = 0$.

Proof. By the method of theorem 3.3, from the first equation of the general Frenet equations (4.2.2) such that $\tau_1 = 0$, we have

$$\bar{\kappa}_1 B_1^1 = \kappa_1 \left(\frac{dt}{d\bar{t}}\right)^2; \qquad \bar{\kappa}_1 B_1^\alpha = 0, \quad \alpha \in \{2, \cdots, m\},$$

$B_1^\alpha = B_\alpha^1 = 0 \, (\alpha \ne 1)$ and $B_1^1 = B_1 = \pm 1$. Also, using (4.3.2) and the third equation of the general Frenet equations (4.2.2), we have

$$\bar{\kappa}_4 B_2^2 + \bar{\kappa}_5 B_3^2 + \bar{\tau}_3 B_4^2 + \bar{\mu}_2 B_5^2 = B_1 \left(\kappa_4 + \kappa_1 c_2 \frac{dt}{d\bar{t}}\right) \frac{dt}{d\bar{t}},$$

$$\bar{\kappa}_4 B_2^3 + \bar{\kappa}_5 B_3^3 + \bar{\tau}_3 B_4^3 + \bar{\mu}_2 B_5^3 = B_1 \left(\kappa_5 + \kappa_1 c_3 \frac{dt}{d\bar{t}}\right) \frac{dt}{d\bar{t}},$$

$$\bar{\kappa}_4 B_2^4 + \bar{\kappa}_5 B_3^4 + \bar{\tau}_3 B_4^4 + \bar{\mu}_2 B_5^4 = B_1 \left(\tau_3 + \kappa_1 c_4 \frac{dt}{d\bar{t}}\right) \frac{dt}{d\bar{t}},$$

$$\bar{\kappa}_4 B_2^5 + \bar{\kappa}_5 B_3^5 + \bar{\tau}_3 B_4^5 + \bar{\mu}_2 B_5^5 = B_1 \left(\mu_2 + \kappa_1 c_5 \frac{dt}{d\bar{t}}\right) \frac{dt}{d\bar{t}},$$

$$\bar{\kappa}_4 B_2^\alpha + \bar{\kappa}_5 B_3^\alpha + \bar{\tau}_3 B_4^\alpha + \bar{\mu}_2 B_5^\alpha = B_1 \kappa_1 c_\alpha \left(\frac{dt}{d\bar{t}}\right)^2, \quad \alpha \in \{6, \cdots, m\}.$$

Taking into account that

$$c_2 = -\frac{\kappa_4}{\kappa_1} \frac{d\bar{t}}{dt}; \quad c_3 = -\frac{\kappa_5}{\kappa_1} \frac{d\bar{t}}{dt}; \quad c_4 = -\frac{\tau_3}{\kappa_1} \frac{d\bar{t}}{dt}; \quad c_5 = -\frac{\mu_2}{\kappa_1} \frac{d\bar{t}}{dt};$$

$$c_\alpha = 0, \quad \alpha \in \{6, \cdots, m\}$$

in the last equations, we have $\bar{\kappa}_4 = \bar{\kappa}_5 = \bar{\tau}_3 = \bar{\mu}_2 = 0$. On the other hand, since $K_9 = -\frac{1}{\sqrt{2}}(\kappa_5 - \tau_3)$ and $K_{14} = -\frac{1}{\sqrt{2}}(\kappa_5 + \tau_5)$, we also have $\bar{K}_9 = \bar{K}_{14} = 0$.

Remark. If we take $t = \bar{t}$ in theorem 3.7, then $c_2 = -\frac{\kappa_4}{\kappa_1}$; $c_3 = -\frac{\kappa_5}{\kappa_1}$; $c_4 = -\frac{\tau_3}{\kappa_1}$; $c_5 = -\frac{\mu_2}{\kappa_1}$; $c_\alpha = 0$, $\alpha > 5$ and

$$\bar{N} = -\frac{1}{2}\left(\frac{\epsilon_2\,\kappa_4^2 + \epsilon_3\,\kappa_5^2 + \epsilon_4\,\tau_3^2 + \epsilon_5\,\mu_2^2}{\kappa_1^2}\right)\xi + N$$
$$-\frac{\kappa_4}{\kappa_1}\,W_2 - \frac{\kappa_5}{\kappa_1}\,W_3 - \frac{\tau_3}{\kappa_1}\,W_4 - \frac{\mu_2}{\kappa_1}\,W_5,$$
$$\bar{W}_2 = W_2 + \frac{\kappa_4}{\kappa_1}\,\xi, \quad \bar{W}_3 = W_3 + \frac{\kappa_5}{\kappa_1}\,\xi,$$
$$\bar{W}_4 = W_4 + \frac{\tau_3}{\kappa_1}\,\xi, \quad \bar{W}_5 = W_5 + \frac{\mu_2}{\kappa_1}\,\xi,$$
$$\bar{W}_i = W_i, \quad i \in \{1,\,5,\,\cdots,\,m\}.$$

Relabeling $N = \bar{N}$, $W_1 = \bar{W}_1$, $W_2 = \bar{W}_2$, $\kappa_i = \bar{\kappa}_i$, $i \in \{1, 2, 3\}$, $\tau_2 = \bar{\tau}_2$, $\kappa_4 = \bar{\kappa}_6$ and $S(TC^\perp) = \bar{S}(TC^\perp)$ in the process of the above theorem and taking only the first four equations in (4.2.2) with $\tau_1 = 0$ as follows:

$$\nabla_\xi \xi = h\,\xi + \kappa_1\,W_1,$$
$$\nabla_\xi N = -h\,N + \kappa_2\,W_1 + \kappa_3\,W_2 + \tau_2\,W_3,$$
$$\epsilon_1\,\nabla_\xi W_1 = -\kappa_2\,\xi - \kappa_1\,N,$$
$$\epsilon_2\,\nabla_\xi W_2 = -\kappa_3\,\xi + \kappa_4\,W_3 + R_9,$$

where $R_9 \in \Gamma(S(TC^\perp))$. Thus R_9 is a vector field perpendicular to ξ, N, W_1 and W_2. Assume that R_9 is a null. Using (4.2.1) and define the new second and third kind function τ_3 and μ_2 by $R_3 = \tau_3\,\epsilon_4\,\bar{W}_4 + \mu_2\,\epsilon_5\,\bar{W}_5$, where $\tau_3 = \mu_2$. Thus we have

$$\nabla_\xi \dot{W}_2 = -\kappa_3\,\xi + \kappa_4\,W_3 + \tau_3\,\epsilon_3\,W_4 + \mu_2\,\epsilon_5\,\bar{W}_5.$$

Repeating above process for all the m unit vectors of an orthonormal basis $\{W_1, W_2, W_3, \bar{W}_4, \ldots, \bar{W}_m\}$ of $\Gamma(S(TC^\perp))$, and then setting

$$W_i = \epsilon_i\,\bar{W}_i, \quad i \in \{4, \ldots, m\}$$

and after some simplification we obtain the following;

$$\nabla_\xi \xi = h\xi + \kappa_1 W_1,$$
$$\nabla_\xi N = -hN + \kappa_2 W_1 + \kappa_3 W_2 + \tau_2 W_3,$$
$$\epsilon_1 \nabla_\xi W_1 = -\kappa_2 \xi - \kappa_1 N,$$
$$\epsilon_2 \nabla_\xi W_2 = -\kappa_3 \xi + \kappa_4 W_3 + \tau_3 W_4 + \mu_2 W_5,$$
$$\epsilon_3 \nabla_\xi W_3 = -\tau_2 \xi - \kappa_4 W_2 + \kappa_5 W_4 + \tau_4 W_5 + \mu_3 W_6,$$
$$\epsilon_4 \nabla_\xi W_4 = -\tau_3 W_2 - \kappa_5 W_3 + \kappa_6 W_5 + \tau_5 W_6 + \mu_4 W_7, \qquad (4.3.13)$$
$$\epsilon_5 \nabla_\xi W_5 = -\mu_2 W_2 - \tau_4 W_3 - \kappa_6 W_4 + \kappa_7 W_6 + \tau_6 W_7 + \mu_5 W_8,$$

$$\cdots\cdots\cdots\cdots\cdots\cdots\cdots\cdots$$

$$\epsilon_i \nabla_\xi W_i = -\mu_{i-3} W_{i-3} - \tau_{i-1} W_{i-2} - \kappa_{i+1} W_{i-1} + \kappa_{i+2} W_{i+1} + \tau_{i+1} W_{i+2}$$
$$+ \mu_i W_{i+3}, \quad 4 \le i \le m-1,$$
$$\epsilon_m \nabla_\xi W_m = -\mu_{m-3} W_{m-3} - \tau_{m-1} W_{m-2} - \kappa_{m+1} W_{m-1}.$$

We call the frame $F_3 = \{\xi, N, W_1, \ldots, W_m\}$ a *Natural Frenet frame of Type 3* on M_q^{m+2} along C with respect to the given screen vector bundle $S(TC^\perp)$ and the equations (4.3.13) are called its *Natural Frenet equations of Type 3*. Finally, the functions $\{\kappa_1, \ldots, \kappa_{m+1}\}$, $\{\tau_2, \cdots, \tau_{m-1}\}$ and $\{\mu_2, \cdots, \mu_{m-3}\}$ are called the *curvature functions* and the *torsion functions* of the first and second kind of C respectively with respect to the frame F_3.

Now we consider the general Frenet equations (4.2.6) of a null curve C of M_q^{m+2} ($q \geq 3$). Using the same procedure we have the following equations

Theorem 3.8. *Let C be a Type q null curve of M_q^{m+2}. If C'' is non-null, then there exists a Frenet frame $\{\xi, N, W_1, \ldots, W_m\}$ satisfying*

$$
\begin{aligned}
\nabla_\xi \xi &= h\xi + \kappa_1 W_1, \\
\nabla_\xi N &= -hN + \kappa_2 W_1 + \kappa_3 W_2 + \tau_2 W_3, \\
\epsilon_1 \nabla_\xi W_1 &= -\kappa_2 \xi - \kappa_1 N, \\
\epsilon_2 \nabla_\xi W_2 &= -\kappa_3 \xi + \kappa_4 W_3 + \tau_3 W_4 + \mu_2 W_5, \\
\epsilon_3 \nabla_\xi W_3 &= -\tau_2 \xi - \kappa_4 W_2 + \kappa_5 W_4 + \tau_4 W_5 + \mu_3 W_6 + \nu_1 W_7, \quad (4.3.14)\\
\epsilon_4 \nabla_\xi W_4 &= -\tau_3 W_2 - \kappa_5 W_3 + \kappa_6 W_5 + \tau_5 W_6 + \mu_4 W_7 + \nu_2 W_8 + \eta_1 W_9, \\
\epsilon_5 \nabla_\xi W_5 &= -\mu_2 W_2 - \tau_4 W_3 - \kappa_6 W_4 + \kappa_7 W_6 + \tau_6 W_7 + \mu_5 W_8 + \nu_3 W_9 \\
&\quad + \eta_2 W_{10} + \sigma_1 W_{11},
\end{aligned}
$$

$$
\cdots\cdots\cdots\cdots\cdots\cdots\cdots
$$

$$
\begin{aligned}
\epsilon_{m-1} \nabla_\xi W_{m-1} &= \ldots - \nu_{m-7} W_{m-5} - \mu_{m-4} W_{m-4} - \tau_{m-2} W_{m-3} \\
&\quad - \kappa_m W_{m-2} + \kappa_{m+1} W_m, \\
\epsilon_m \nabla_\xi W_m &= \ldots - \mu_{m-3} W_{m-3} - \tau_{m-1} W_{m-2} - \kappa_{m+1} W_{m-1}.
\end{aligned}
$$

If C'' is null, then there exists a Frenet frame $\{\xi, N, L_1, L_2, W_3, \ldots, W_m\}$ satisfying the following equations

$$
\begin{aligned}
\nabla_\xi \xi &= h\xi + K_1 L_1, \\
\nabla_\xi N &= -hN + K_2 L_1 + K_3 L_2 + \tau_2 W_3 + \mu_1 W_4, \\
\nabla_\xi L_1 &= -K_3 \xi + K_4 W_3 + K_5 W_4 + \ell W_5, \\
\nabla_\xi L_2 &= -K_2 \xi - K_1 N, \\
\epsilon_3 \nabla_\xi W_3 &= -\tau_2 \xi - K_4 L_2 + \kappa_5 W_4 + \tau_4 W_5 + \mu_3 W_6 + \nu_1 W_7, \\
\epsilon_4 \nabla_\xi W_4 &= -\mu_1 \xi - K_5 L_2 - \kappa_5 W_3 + \kappa_6 W_5 + \tau_5 W_6 + \mu_4 W_7 \\
&\quad + \nu_2 W_8 + \eta_1 W_9, \quad (4.3.15)\\
\epsilon_5 \nabla_\xi W_5 &= -\ell L_2 - \tau_4 W_3 - \kappa_6 W_4 + \kappa_7 W_6 + \tau_6 W_7 + \mu_5 W_8 \\
&\quad + \nu_3 W_9 + \eta_2 W_{10} + \sigma_1 W_{11},
\end{aligned}
$$

$$
\cdots\cdots\cdots\cdots\cdots\cdots\cdots
$$

$$
\begin{aligned}
\epsilon_{m-1} \nabla_\xi W_{m-1} &= \ldots - \nu_{m-7} W_{m-5} - \mu_{m-4} W_{m-4} - \tau_{m-2} W_{m-3} \\
&\quad - \kappa_m W_{m-2} + \kappa_{m+1} W_m,
\end{aligned}
$$

$$\epsilon_m \nabla_\xi W_m = \ldots - \mu_{m-3} W_{m-3} - \tau_{m-1} W_{m-2} - \kappa_{m+1} W_{m-1}.$$

In this case, we call

$$F_q = \{\,\xi,\, N,\, W_1,\, \cdots,\, W_m\,\} = \{\xi,\, N,\, L_1,\, L_2,\, W_3,\, \ldots,\, W_m\} \qquad (4.3.16)$$

a *Natural Frenet frame of Type q* on the semi-Riemannian manifold M_q^{m+2} of index q along C with respect to a give screen vector bundle $S(TC^\perp)$ and the equations (4.3.14) and (4.3.15) its *general Frenet equations of Type q*. The functions $\{\kappa_1, \cdots, \kappa_{m+1}\}$, $\{\tau_1, \cdots, \tau_{m-1}\}$, $\{\mu_1, \cdots, \mu_{m-3}\}$, $\cdots$ are called the *curvature functions* and the *torsion functions of the first, second, third, $\cdots$, $(q-1)$-th kind* of C respectively with respect to the frame F_q.

Example 2. Consider a null curve C in $\mathbf{R}_3^6$ given by

$$C : (\,\cos t,\quad \sin t,\quad \sqrt{2}\sinh t,\quad \sqrt{2}\cosh t,\quad t,\quad \sqrt{2}t\,).$$

Then C falls in the Type 1 in Example 1. If we take the Natural Frenet frame $F = \{\xi,\, N,\, W_1,\, W_2,\, W_3,\, W_4\}$ as follows

$$\xi = (-\sin t,\quad \cos t,\quad \sqrt{2}\cosh t,\quad \sqrt{2}\sinh t,\quad 1,\quad \sqrt{2}),$$
$$N = \frac{1}{2}(-5\sin t,\quad 5\cos t,\quad \sqrt{2}\cosh t,\quad \sqrt{2}\sinh t,\quad 3,\quad 3\sqrt{2}),$$
$$W_1 = (-\cos t,\quad -\sin t,\quad \sqrt{2}\sinh t,\quad \sqrt{2}\cosh t,\quad 0,\quad 0),$$
$$W_2 = (-\sqrt{2}\cos t,\quad -\sqrt{2}\sin t,\quad \sinh t,\quad \cosh t,\quad 0,\quad 0),$$
$$W_3 = (0,\quad 0,\quad -2\cosh t,\quad -2\sinh t,\quad -\sqrt{2},\quad -1),$$
$$W_4 = (\cos t,\quad \sin t,\quad -\sqrt{2}\sinh t,\quad -\sqrt{2}\cosh t,\quad 0,\quad 0).$$

Then the Natural Frenet equations (4.3.5) are given by

$$\nabla_\xi \xi = W_1, \qquad \nabla_\xi N = -\frac{3}{2}W_1 + 2\sqrt{2}W_2, \qquad \nabla_\xi W_1 = \frac{3}{2}\xi - N,$$
$$\nabla_\xi W_2 = -2\sqrt{2}\xi - 2W_3, \quad \nabla_\xi W_3 = 2W_2 + 2\sqrt{2}W_4, \quad \nabla_\xi W_4 = -2\sqrt{2}W_3.$$

Example 3. Let C be a null curve in $\mathbf{R}_3^6$ given by

$$C : (\,\cos t,\quad \sin t,\quad \sinh t,\quad \cosh t,\quad t,\quad t\,).$$

We show that C falls in the Type 3 in Example 1. If we take the Natural Frenet frame $F_3 = \{\,\xi,\, N,\, L_1,\, L_2,\, W_3,\, W_4\,\}$ of Type 3 as follows

$$\xi \;=\; (-\sin t,\; \cos t,\; \cosh t,\; \sinh t,\; 1,\; 1),$$
$$N \;=\; \frac{1}{4}(\sin t,\; -\cos t,\; -\cosh t,\; -\sinh t,\; 1,\; 1),$$
$$L_1 \;=\; (-\cos t,\; -\sin t,\; \sinh t,\; \cosh t,\; 0,\; 0),$$

$$L_2 = \frac{1}{2}(\cos t, \ \sin t, \ \sinh t, \ \cosh t, \ \ 0, \ \ 0),$$

$$W_3 = \frac{1}{\sqrt{2}}(-\sin t, \ \cos t, \ -\cosh t, \ -\sinh t, \ \ 0, \ \ 0),$$

$$W_4 = \frac{1}{\sqrt{2}}(0, \ \ 0, \ \ 0, \ \ 0, \ \ -1, \ \ 1).$$

Then the Natural Frenet equations (4.3.10) are given by

$$\nabla_\xi \xi = L_1, \qquad \nabla_\xi N = -\frac{1}{4}L_1, \qquad \nabla_\xi L_1 = -\sqrt{2}\,W_3,$$

$$\nabla_\xi L_2 = \frac{1}{4}\xi + \sqrt{2}\,N, \quad \nabla_\xi W_3 = -\sqrt{2}\,L_2, \quad \nabla_\xi W_4 = 0.$$

4.4 Invariance of Frenet frames

In this section, we examine the dependence of the Frenet frames and the Frenet equations of Type 1, Type 2 and Type 3 on both the parameter transformations and the screen vector bundle transformations of C.

From proposition 3.1 and using the method of this proposition for the Natural Frenet frames, we have

Proposition 4.1. *Let C be a null curve of a semi-Riemannian manifold M_q^{m+2} and F and F^* be two Frenet frames on $\mathcal{U}$ and $\mathcal{U}^*$ induced by the same screen vector bundle $S(TC^\perp)$ respectively. Then the type of Natural Frenet equations is invariant to the transformations of the coordinate neighborhood of C.*

As in theorem 4.1 of chapter 3, we have the following results:

(a) *The vanishing of the first curvature κ_1 on a neighborhood is independent of both the parameter transformations on C and the screen vector bundle transformations.*

(b) *It is possible to find a parameter on C such that $h = 0$ in Frenet equations of all possible types, using the same screen bundle.*

Theorem 4.1. *Let C be a null curve of M_q^{m+2}. Then C is a null geodesic of M_q^{m+2} if and only if the first curvature κ_1 vanishes identically on C.*

Theorem 4.2. *Let $C(p)$ be a Type q non-geodesic null curve of M_q^{m+2}, where p is a distinguished parameter on C. If C'' is non-null, then there exists a Frenet frame $\{\xi, N, W_1, \ldots, W_m\}$ satisfying*

$$\nabla_\xi \xi = \kappa_1 W_1,$$

$$\nabla_\xi N = \kappa_2 W_1 + \kappa_3 W_2 + \tau_2 W_3,$$

$$\epsilon_1 \nabla_\xi W_1 = -\kappa_2 \xi - \kappa_1 N,$$

$$\epsilon_2 \nabla_\xi W_2 = -\kappa_3 \xi + \kappa_4 W_3 + \tau_3 W_4 + \mu_2 W_5,$$
$$\epsilon_3 \nabla_\xi W_3 = -\tau_2 \xi - \kappa_4 W_2 + \kappa_5 W_4 + \tau_4 W_5 + \mu_3 W_6 + \nu_1 W_7,$$
$$\epsilon_4 \nabla_\xi W_4 = -\tau_3 W_2 - \kappa_5 W_3 + \kappa_6 W_5 + \tau_5 W_6 + \mu_4 W_7 + \nu_2 W_8 + \eta_1 W_9,$$
$$\epsilon_5 \nabla_\xi W_5 = -\mu_2 W_2 - \tau_4 W_3 - \kappa_6 W_4 + \kappa_7 W_6 + \tau_6 W_7 + \mu_5 W_8 + \nu_3 W_9$$
$$+ \eta_2 W_{10} + \sigma_1 W_{11},$$

$$\cdots\cdots\cdots\cdots\cdots\cdots\cdots\cdots$$

$$\epsilon_{m-1} \nabla_\xi W_{m-1} = \cdots - \nu_{m-7} W_{m-5} - \mu_{m-4} W_{m-4} - \tau_{m-2} W_{m-3}$$
$$- \kappa_m W_{m-2} + \kappa_{m+1} W_m,$$
$$\epsilon_m \nabla_\xi W_m = \cdots - \mu_{m-3} W_{m-3} - \tau_{m-1} W_{m-2} - \kappa_{m+1} W_{m-1}.$$

If C'' is null, then there exists a Frenet frame $\{\xi, N, L_1, L_2, W_3, \ldots, W_m\}$ satisfying

$$\nabla_\xi \xi = K_1 L_1,$$
$$\nabla_\xi N = K_2 L_1 + K_3 L_2 + \tau_2 W_3 + \mu_1 W_4,$$
$$\nabla_\xi L_1 = -K_3 \xi + K_4 W_3 + K_5 W_4 + \ell W_5,$$
$$\nabla_\xi L_2 = -K_2 \xi - K_1 N,$$
$$\epsilon_3 \nabla_\xi W_3 = -\tau_2 \xi - K_4 L_2 + \kappa_5 W_4 + \tau_4 W_5 + \mu_3 W_6 + \nu_1 W_7,$$
$$\epsilon_4 \nabla_\xi W_4 = -\mu_1 \xi - K_5 L_2 - \kappa_5 W_3 + \kappa_6 W_5 + \tau_5 W_6 + \mu_4 W_7$$
$$+ \nu_2 W_8 + \eta_1 W_9,$$
$$\epsilon_5 \nabla_\xi W_5 = -\ell L_2 - \tau_4 W_3 - \kappa_6 W_4 + \kappa_7 W_6 + \tau_6 W_7 + \mu_5 W_8$$
$$+ \nu_3 W_9 + \eta_2 W_{10} + \sigma_1 W_{11},$$

$$\cdots\cdots\cdots\cdots\cdots\cdots\cdots\cdots$$

$$\epsilon_{m-1} \nabla_\xi W_{m-1} = \cdots - \nu_{m-7} W_{m-5} - \mu_{m-4} W_{m-4} - \tau_{m-2} W_{m-3}$$
$$- \kappa_m W_{m-2} + \kappa_{m+1} W_m,$$
$$\epsilon_m \nabla_\xi W_m = \cdots - \mu_{m-3} W_{m-3} - \tau_{m-1} W_{m-2} - \kappa_{m+1} W_{m-1}.$$

Using (4.3.8) and the method of propositions 4.2 and 4.3 of chapter 3, we have

Proposition 4.2. Let C be a null curve of M_q^{m+2} and F and $\widetilde{F}$ be two general Frenet frames of Type 2 induced by the screen vector bundle $S(TC^\perp)$ such that $\tau_1 \neq 0$. Then, $S(TC^\perp)$ is an orthogonal direct sum of two invariant subspaces $Span\{L_1, L_2\} = Span\{W_1, W_2\}$ and $Span\{W_3, \ldots, W_m\}$ by the transformation of coordinate neighborhoods and the screen vector bundle of C.

By exchanging the form of the general Frenet equations (of the same type) (in case $\tau_1 = \cdots = \tau_{i-1} = 0$ and $\tau_i \neq 0$) and let $\prod_{\alpha=1}^i \kappa_{2\alpha-1} \neq 0$, we have

Proposition 4.3. Let C be a null curve of a semi-Riemannian manifold M_q^{m+2} and F and F^* be two general Frenet frames of Type 2 induced by the screen vector bundle $S(TC^\perp)$ such that $\prod_{\alpha=1}^i \kappa_{2\alpha-1} \neq 0$ and $\tau_1 = \cdots = \tau_{i-1} = 0$; $\tau_i \neq 0$, $1 \leq i \leq m-1$. Then $S(TC^\perp)$ is an orthogonal *direct sum of two invariant subspaces* $Span\{L_i, L_{i+1}\} = Span\{W_i, W_{i+1}\}$ *and* $Span\{W_1, \ldots, \widehat{W}_i, \widehat{W}_{i+1}, \ldots, W_m\}$ *by the transformation of coordinate neighborhoods of* C.

Using the Natural Frenet equations (4.3.5) of Type 1 and the method of proposition 2.1 in chapter 2 and proposition 4.4 in chapter 3, we have

Proposition 4.4. *Let* C *be a null curve of* M_q^{m+2} *and* F *and* F^* *be two Natural Frenet frames of Type 1 on* $\mathcal{U}$ *and* $\mathcal{U}^*$ *respectively induced by the same screen vector bundle* $S(TC^\perp)$. *Suppose* $\prod_{i=1}^{m+1} \kappa_i \neq 0$ *on* $\mathcal{U} \cap \mathcal{U}^* \neq \emptyset$. *Then at any point of* $\mathcal{U} \cap \mathcal{U}^*$ *we have*

$$\kappa_1^* = \kappa_1 A_1 \left(\frac{dt}{dt^*} \right)^2,$$

$$\kappa_2^* = \kappa_2 A_1, \quad \kappa_3^* = \kappa_3 A_2, \tag{4.4.1}$$

$$\kappa_\alpha^* = \kappa_\alpha A_{\alpha-1} \frac{dt}{dt^*}, \quad 4 \leq \alpha \leq m+1, \text{ where } A_\alpha = \pm 1.$$

Hence, κ_2 *and* κ_3 *are invariant functions up to a sign, with respect to the parameter transformations on* C.

Using the general Frenet equations (4.1.1) of Type 1 and the method of proposition 2.2 in chapter 2 and proposition 4.5 in chapter 3, we have

Proposition 4.5. *Let* C *be a null curve of* M_q^{m+2} *and* F *and* F^* *be two general Frenet frames of Type 1 on* $\mathcal{U}$ *and* $\mathcal{U}^*$ *respectively induced by the same screen vector bundle* $S(TC^\perp)$. *Suppose* $\prod_{n=1}^{m} \kappa_{2n-1} \neq 0$ *on* $\mathcal{U} \cap \mathcal{U}^* \neq \emptyset$. *Then at any point of* $\mathcal{U} \cap \mathcal{U}^*$ *we have*

$$\kappa_1^* = \kappa_1 A_1 \left(\frac{dt}{dt^*} \right)^2,$$

$$\kappa_2^* = \kappa_2 A_1, \quad \kappa_3^* = \kappa_3 A_2, \tag{4.4.2}$$

$$\kappa_\alpha^* = \kappa_\alpha A_{\alpha-1} \frac{dt}{dt^*}, \quad 4 \leq \alpha \leq 2m, \text{ where } A_\alpha = \pm 1.$$

Hence, κ_2 *and* κ_3 *are invariant functions up to a sign, with respect to the parameter transformations on* C.

Using the method of proposition 2.3 in chapter 2 and proposition 4.6 in chapter 3 for the different screen vector bundle with $\bar{F}$, we have

Proposition 4.6. *Let* C *be a null curve of* M_q^{m+2} *and* F *and* $\bar{F}$ *be two Natural Frenet frames of Type 1 on* $\mathcal{U}$ *and* $\bar{\mathcal{U}}$ *respectively. Suppose* $\prod_{i=1}^{m+1} \kappa_i \neq 0$ *on* $\mathcal{U} \cap \bar{\mathcal{U}} \neq \emptyset$. *Then, their curvature functions are related by*

$$\bar{\kappa}_1 = \kappa_1 B_1 \left(\frac{dt}{d\bar{t}} \right)^2,$$

$$\bar{\kappa}_2 = \left\{ \kappa_2 + \bar{h} \, c_1 + \frac{dc_1}{d\bar{t}} - \frac{1}{2} \kappa_1 \, c_1^2 \left(\frac{dt}{d\bar{t}} \right)^2 \right\} B_1, \tag{4.4.3}$$

$$\bar{\kappa}_3 = \kappa_3 \, B_2,$$

$$\bar{\kappa}_\alpha = \kappa_\alpha \, B_{\alpha-1} \frac{dt}{d\bar{t}}, \quad \alpha \in \{4, \cdots, m\}, \text{ where } B_i = \pm 1,$$

and $c_2 = \cdots = c_m = 0$. Hence, κ_3 is invariant functions up to a sign, with respect to the transformations of the screen vector bundle of C.

Using proposition 4.6 and the method of proposition 4.3, we have

Theorem 4.3. Let C be a null curve of M_q^{m+2} and F and $\widetilde{F}$ be two Natural Frenet frames of Type 2 induced by $(t, S(TC^\perp), \mathcal{U})$ and $(\widetilde{t}, \widetilde{S}(TC^\perp), \widetilde{\mathcal{U}})$ respectively such that $\prod_{\alpha=1}^{i} \kappa_\alpha \neq 0$ and $\tau_1 = \cdots = \tau_{i-1} = 0$; $\tau_i \neq 0$, $1 \leq i \leq m-1$. Then $S(TC^\perp)$ is an orthogonal direct sum of two invariant subspaces $Span\{L_i, L_{i+1}\} = Span\{W_i, W_{i+1}\}$ and $Span\{W_1, \ldots, \widehat{W}_i, \widehat{W}_{i+1}, \ldots, W_m\}$ by the transformation of the coordinate neighborhoods and the screen vector bundle of C.

Using the method of proposition 4.7 in section 3.3 and (4.3.7), we have

Proposition 4.7. Let C be a null curve of M_q^{m+2} and F and F^* be two Natural Frenet frames of Type 2 on $\mathcal{U}$ and $\mathcal{U}^*$ respectively induced by the same screen vector bundle $S(TC^\perp)$. Suppose $\prod_{i=1}^{m-1} \tau_i \neq 0$ on $\mathcal{U} \cap \mathcal{U}^* \neq \emptyset$. Then at any point of $\mathcal{U} \cap \mathcal{U}^*$ we have

$$K_1^* = K_1 \left(A_1^1 - A_1^2 \right) \left(\frac{dt}{dt^*} \right)^2,$$

$$K_2^* = K_2 \left(A_1^1 - A_1^2 \right), \quad K_3^* = K_3 \left(A_1^1 + A_2^1 \right), \quad \tau_2^* = \tau_2 \, A_3,$$

$$K_4^* = K_4 \, A_3 \left(A_1^1 + A_2^1 \right) \frac{dt}{dt^*}, \quad K_5^* = K_5 \, A_4 \left(A_1^1 + A_2^1 \right) \frac{dt}{dt^*}, \tag{4.4.4}$$

$$\kappa_\alpha^* = \kappa_\alpha \, A_{\alpha-1} \frac{dt}{dt^*}, \quad 5 \leq \alpha \leq m+1, \text{ where } A_i = \pm 1, \ \forall i,$$

$$\tau_\alpha^* = \tau_\alpha \, A_{\alpha+1} \frac{dt}{dt^*}, \quad 4 \leq \alpha \leq m+1, \text{ where } A_i = \pm 1, \ \forall i.$$

Hence, K_2, K_3 and τ_2 are invariant functions up to a sign, with respect to the parameter transformations on C.

Using the general Frenet equations (4.1.3) of Type 2 and the method of proposition 4.8 in chapter 3, we have

Proposition 4.8. Let C be a null curve of M_q^{m+2} and F and F^* be two general Frenet frames of Type 2 on $\mathcal{U}$ and $\mathcal{U}^*$ respectively induced by the same screen vector bundle $S(TC^\perp)$. Suppose $\prod_{i=1}^{m-1} \tau_i \neq 0$ on $\mathcal{U} \cap \mathcal{U}^* \neq \emptyset$. Then at any point of $\mathcal{U} \cap \mathcal{U}^*$

we have

$$K_1^* = K_1 \left(A_1^1 - A_1^2 \right) \left(\frac{dt}{dt^*} \right)^2,$$

$$K_2^* = K_2 \left(A_1^1 - A_1^2 \right), \quad K_3^* = K_3 \left(A_1^1 + A_2^1 \right), \quad \tau_2^* = \tau_2 A_3,$$

$$K_4^* = \left\{ K_4 + \left(A_1^1 + A_2^1 \right) \frac{d(A_1^1 - A_1^2)}{dt} \right\} \frac{dt}{dt^*},$$

$$K_5^* = K_5 A_3 \left(A_1^1 + A_2^1 \right) \frac{dt}{dt^*}, \quad K_6^* = K_6 A_4 \left(A_1^1 + A_2^1 \right) \frac{dt}{dt^*}, \qquad (4.4.5)$$

$$K_7^* = K_7 A_3 \left(A_1^1 - A_1^2 \right) \frac{dt}{dt^*}, \quad K_8^* = K_8 A_4 \left(A_1^1 - A_1^2 \right) \frac{dt}{dt^*},$$

$$\kappa_\alpha^* = \kappa_\alpha A_{2\alpha-3} \frac{dt}{dt^*}, \quad 8 \le \alpha \le m+1, \text{ where } A_\alpha = \pm 1.$$

$$\tau_\alpha^* = \tau_\alpha A_{2\alpha} \frac{dt}{dt^*}, \quad 5 \le \alpha \le m+1, \text{ where } A_\alpha = \pm 1.$$

Hence, K_2, K_3 and τ_2 are invariant functions up to a sign, with respect to the parameter transformations on C.

Also, using the method of proposition 4.9 in chapter 3 for the Natural Frenet equations (4.3.7) of Type 2, we have

Proposition 4.9. Let C be a null curve of M_q^{m+2} and F and $\bar{F}$ be two Natural Frenet frames of Type 2 on $\mathcal{U}$ and $\bar{\mathcal{U}}$ respectively. Suppose $\tau_1 \ne 0$ on $\mathcal{U} \cap \bar{\mathcal{U}} \ne \emptyset$. Then, their curvature functions are related by

$$\bar{K}_1 = K_1 \left(B_1^1 - B_1^2 \right) \left(\frac{dt}{d\bar{t}} \right)^2,$$

$$\bar{K}_2 = \left\{ K_2 + \bar{h} \, C_1 + \frac{dC_1}{d\bar{t}} \right\} \left(B_1^1 - B_1^2 \right), \qquad (4.4.6)$$

$$\bar{K}_3 = K_3 \left(B_1^1 + B_2^1 \right)$$

and $C_2 = c_3 = \cdots = c_m = 0$. Hence, K_3 is invariant functions up to a sign, with respect to the transformations of the screen vector bundle of C.

Now, we let F and F^* be general Frenet frames of Type 3 with four null vector fields $\{L_1, L_2, L_3, L_4\}$ and $\{L_1^*, L_2^*, L_3^*, L_4^*\}$ on $\mathcal{U} \cap \mathcal{U}^*$ respectively, induced by the same screen vector bundle $S(TC^\perp)$. Suppose $\tau_1 \mu_1 \ne 0$ on $\mathcal{U} \cap \mathcal{U}^*$. By the method of proposition 4.2 in chapter 3 and the equation (4.3.3) satisfying $\kappa_1 = \tau_1 \ne 0$, the matrix $\left[A_\beta^\alpha(x)\right]$ is transformed as

$$
\begin{pmatrix}
0 & A_1^1 + A_2^1 & 0 & \cdots & 0 \\
-A_1^1 + A_1^2 & 0 & 0 & \cdots & 0 \\
 & & & & \\
0 & 0 & A_3^3 & \cdots & A_3^m \\
0 & 0 & A_4^3 & \cdots & A_4^m \\
\vdots & \vdots & \vdots & \cdots & \vdots \\
0 & 0 & A_m^3 & \cdots & A_m^m
\end{pmatrix}
$$

and

$$L_1^* = (A_1^1 + A_2^1)L_1, \quad L_2^* = (A_1^1 - A_1^2)L_2, \quad W_\alpha^* = \sum_{\beta \geq 3}^{m} A_\alpha^\beta W_\beta,$$

where $3 \leq \alpha \leq m$. Using this coordinate transformation and (4.3.4), we have

$$
\begin{aligned}
\kappa_2^* A_1^1 + \kappa_3^* A_2^1 &= \kappa_2, & \kappa_2^* A_1^2 + \kappa_3^* A_2^2 &= \kappa_3, \\
\tau_2^* A_3^3 + \mu_1^* A_4^3 &= \tau_2, & \tau_2^* A_3^4 + \mu_1^* A_4^4 &= \mu_1, \\
\tau_2^* A_3^\alpha + \mu_1^* A_4^\alpha &= 0, & 5 \leq \alpha &\leq m.
\end{aligned}
\tag{4.4.7}
$$

Since $\tau_2 = \mu_1 \neq 0$ on $\mathcal{U} \cap \mathcal{U}^*$, it follows from (4.4.7) that

$$A_3^3 + A_4^3 = A_3^4 + A_4^4, \quad A_3^5 = -A_4^5, \quad \cdots, \quad A_3^m = -A_4^m. \tag{4.4.8}$$

Using (4.3.1), (4.4.8) and $L_3 = \frac{1}{\sqrt{2}}(W_3 + W_4)$, we obtain

$$L_3^* = (A_3^3 + A_4^3)L_3 \tag{4.4.9}$$

where $A_3^3 + A_4^3 \neq 0$, otherwise the matrix $[A_\alpha^\beta(x)]$ is singular. By the same method of proposition 4.2 and 4.3 in chapter 3, the matrix $\left[A_\beta^\alpha(x)\right]$ is transformed as

$$
\begin{pmatrix}
0 & A_1^1 + A_2^1 & 0 & 0 & 0 & \cdots & 0 \\
-A_1^1 + A_1^2 & 0 & 0 & 0 & 0 & \cdots & 0 \\
0 & 0 & 0 & A_3^3 + A_4^3 & 0 & \cdots & 0 \\
0 & 0 & -A_3^3 + A_3^4 & 0 & 0 & \cdots & 0 \\
0 & 0 & 0 & 0 & A_5^5 & \cdots & A_5^m \\
\vdots & \vdots & \vdots & \vdots & \vdots & \cdots & \vdots \\
0 & 0 & 0 & 0 & A_m^5 & \cdots & A_m^m
\end{pmatrix}
$$

and

$$L_4^* = (A_3^3 - A_3^4)L_2, \quad W_\alpha^* = \sum_{\beta \geq 5}^{m} A_\alpha^\beta W_\beta, \quad (5 \leq \alpha \leq m), \tag{4.4.10}$$

where $(A_3^3 + A_4^3)(A_3^3 - A_4^3) = 1$. Thus $S(TC^\perp)$ is a orthogonal direct sum of three invariant subspaces $Span\{L_1, L_2\} = Span\{W_1, W_2\}$, $Span\{L_3, L_4\} = Span\{W_3, W_4\}$ and $Span\{W_5, \ldots, W_m\}$ by the transformation of coordinate neighborhoods of C.

Proposition 4.10. *Let C be a null curve of M_q^{m+2} and F be a general Frenet frames of Type 3 on $\mathcal{U}$ induced by the screen vector bundle $S(TC^\perp)$. Suppose $\tau_1\,\mu_1 \neq 0$ on $\mathcal{U}$. Then $S(TC^\perp)$ is a orthogonal direct sum of three invariant subspaces $\mathrm{Span}\,\{L_1, L_2\} = \mathrm{Span}\,\{W_1, W_2\}$, $\mathrm{Span}\,\{L_3, L_4\} = \mathrm{Span}\,\{W_3, W_4\}$ and $\mathrm{Span}\,\{W_5, \cdots, W_m\}$ by the transformation of coordinate neighborhoods of C.*

Similarly, by using the general Frenet equations of Type 3 with another form, we find that the screen distribution $S(TC^\perp)$ is a orthogonal direct sum of three invariant subspaces

$$\mathrm{Span}\,\{L_i, L_{i+1}\} = \mathrm{Span}\,\{W_i, W_{i+1}\}\,;$$
$$\mathrm{Span}\,\{L_j, L_{j+1}\} = \mathrm{Span}\,\{W_j, W_{j+1}\}\,;$$
$$\mathrm{Span}\,\{\,W_1, \dots, \hat{W}_i, \hat{W}_{i+1}, \dots, \hat{W}_j, \hat{W}_{j+1}, \dots, W_m\,\}$$

by the transformation of coordinate neighborhoods of C, where $1 \leq i, j \leq m - 1$, $\{i, i+1\} \neq \{j, j+1\}$ and the matrix $\left[A_\beta^\alpha(x)\right]$ is transformed as

$$\begin{pmatrix} A_1 & & & & & & & & & & \\ & \ddots & & & & & & & & & \mathbf{O} \\ & & A_{i-1} & & & & & & & \\ & & & 0 & D_i & & & & & \\ & & & -E_i & 0 & & & & & \\ & & & & & A_{i+1} & & & & \\ & & & & & & \ddots & & & \\ & & & & & & & A_{j-1} & & \\ & & & & & & & & 0 & D_j \\ \mathbf{O} & & & & & & & & -E_j & 0 \\ & & & & & & & & & & \Delta \end{pmatrix},$$

where

$$\Delta = \begin{pmatrix} A_{j+2}^{j+2} & \cdots & A_{j+2}^m \\ \vdots & \cdots & \vdots \\ A_m^{j+2} & \cdots & A_m^m \end{pmatrix}, \quad D_\alpha = A_\alpha^\alpha + A_{\alpha+1}^\alpha, \quad E_\alpha = A_\alpha^\alpha - A_\alpha^{\alpha+1}.$$

Using this coordinate transformation, we obtain

$$L_i^* = D_i\,L_i, \quad L_{i+1}^* = E_i\,L_{i+1}, \quad L_j^* = D_j\,L_j, \quad L_{j+1}^* = E_j\,L_{j+1},$$
$$W_\alpha^* = \sum_{\beta \geq 1}^m A_\alpha^\beta\,W_\beta, \quad \alpha,\,\beta \in \{1, \dots, \hat{i}, i\,\hat{+}\,1, \dots, \hat{j}, j\,\hat{+}\,1, \dots, m\}.$$

Proposition 4.11. *Let C be a null curve of M_q^{m+2} and F and F^* be two general Frenet frames of Type 3 on $\mathcal{U} \cap \mathcal{U}^*$ induced by the screen vector bundle $S(TC^\perp)$*

such that $\prod_{\alpha=1}^{i} \kappa_{2\alpha-1} \neq 0$ and $\tau_1 = \cdots = \tau_{i-1} = 0$, $\tau_i \neq 0$, $1 \leq i \leq m-1$; $\mu_1 = \cdots = \mu_{j-1} = 0$, $\mu_j \neq 0$, $1 \leq j \leq m-3$. Then $S(TC^\perp)$ is orthogonal direct sum of 3 invariant subspaces $Span\{L_i, L_{i+1}\} = Span\{W_i, W_{i+1}\}$, $Span\{L_j, L_{j+1}\} = Span\{W_j, W_{j+1}\}$ and $Span\{W_1, \cdots, \widehat{W}_i, \widehat{W}_{i+1}, \cdots, \widehat{W}_j, \widehat{W}_{j+1}, \cdots, W_m\}$ by the transformation of the coordinate neighborhoods of C.

Next, by using the general Frenet equations of Type r (≥ 2), we find that the screen distribution $S(TC^\perp)$ is a orthogonal direct sum of r-th invariant subspaces $U_{\sigma(1)}, U_{\sigma(2)}, \cdots, U_{\sigma(r)}$ by the transformation of coordinate neighborhoods of C and the matrix $\left[A_\beta^\alpha(x) \right]$ is transformed as

$$
\begin{pmatrix}
\ddots & & & & & & & \\
& \Delta_{\sigma(1)} & & & & & & \\
& & \ddots & & & & \mathbf{O} & \\
& & & \Delta_{\sigma(2)} & & & & \\
& & & & \ddots & & & \\
& & & & & \Delta_{\sigma(3)} & & \\
& & & & & & \ddots & \\
& & & & & & & \Delta_{\sigma(r)} \\
& \mathbf{O} & & & & & & & \ddots \\
& & & & & & & & & \Delta
\end{pmatrix},
$$

where

$$
\Delta = \begin{pmatrix}
A_{\sigma(r)+2}^{\sigma(r)+2} & \cdots & A_{\sigma(r)+2}^m \\
\vdots & \cdots & \vdots \\
A_m^{\sigma(r)+2} & \cdots & A_m^m
\end{pmatrix}, \quad
\Delta_{\sigma(k)} = \begin{pmatrix}
0 & D_{\sigma(k)} \\
-E_{\sigma(k)} & 0
\end{pmatrix}.
$$

Using this coordinate transformation, we obtain

$$
L_{\sigma(k)}^* = D_{\sigma(k)} L_{\sigma(k)}, \quad L_{\sigma(k)+1}^* = E_{\sigma(k)} L_{\sigma(k)+1}, \quad k = 1, \cdots, r,
$$

$$
W_\alpha^* = \sum_{\beta \geq 1}^m A_\alpha^\beta W_\beta, \quad \alpha, \beta \in \{1, \dots, \widehat{\sigma(k)}, \widehat{\sigma(k)} + 1, \dots, m\}.
$$

Proposition 4.12. Let C be a null curve of M_q^{m+2} and F be a general Frenet frames of Type r (≥ 2) on $\mathcal{U}$ induced by the screen vector bundle $S(TC^\perp)$. Suppose $\tau_{\sigma(1)} \mu_{\sigma(2)} \nu_{\sigma(3)} \cdots \neq 0$ on $\mathcal{U}$. Then $S(TC^\perp)$ is a orthogonal direct sum of r-th invariant subspaces $U_{\sigma(1)}, \cdots, U_{\sigma(r)}$ by the transformation of coordinate neighborhoods of C.

Using propositions 4.9 and 4.12 and the method of theorem 4.3, we have

Theorem 4.4. *Let C be a null curve of M_q^{m+2} and F be a Natural Frenet frames of Type r (≥ 2) on $\mathcal{U}$ induced by the screen vector bundle $S(RC^{\perp})$. Suppose $\tau_{\sigma(1)} \mu_{\sigma(2)} \nu_{\sigma(3)} \cdots \neq 0$ on $\mathcal{U}$. Then $S(TC^{\perp})$ is a orthogonal direct sum of r-th invariant subspaces $U_{\sigma(1)}, \cdots, U_{\sigma(r)}$ by the transformation of coordinate neighborhoods and the screen vector bundle of C.*

For the Type 3 Frenet frames, using the parameter transformations (4.4.9) and (4.4.10) that relate elements of F_3 and F_3^* on $\mathcal{U} \cap \mathcal{U}^*$ such that $\tau_1 \mu_1 \neq 0$ and using the Natural Frenet equations (4.3.12) of Type 3, we have

Proposition 4.13. *Let C be a null curve of M_q^{m+2} and F and F^* be two Natural Frenet frames of Type 3 on $\mathcal{U}$ and $\mathcal{U}^*$ respectively induced by the same screen vector bundle $S(TC^{\perp})$. Let $\tau_1 \neq 0$ and $\prod_{i=1}^{m-3} \mu_i \neq 0$ on $\mathcal{U} \cap \mathcal{U}^* \neq \emptyset$. Then,*

$$K_1^* = K_1 \left(A_1^1 - A_1^2\right) \left(\frac{dt}{dt^*}\right)^2,$$

$$K_2^* = K_2 \left(A_1^1 - A_1^2\right), \quad K_3^* = K_3 \left(A_1^1 + A_2^1\right), \quad K_4^* = K_4 \left(A_3^3 - A_3^4\right),$$

$$K_5^* = K_5 \left(A_1^1 + A_2^1\right)(A_3^3 - A_3^4)\frac{dt}{dt^*}, \quad K_6^* = K_6 \left(A_1^1 + A_2^1\right)(A_3^3 + A_4^3)\frac{dt}{dt^*},$$

$$\ell^* = \ell A_5 (A_1^1 + A_2^1)\frac{dt}{dt^*}, \tag{4.4.11}$$

$$K_7^* = K_7 \frac{dt}{dt^*}, \quad K_8^* = K_8 A_5 \frac{dt}{dt^*}, \quad K_9^* = K_9 A_6 \frac{dt}{dt^*},$$

$$K_{10}^* = K_{10} A_5 \frac{dt}{dt^*}, \quad K_{11}^* = K_{11} A_6 \frac{dt}{dt^*}, \quad K_{12}^* = K_{12} A_7 \frac{dt}{dt^*},$$

$$\kappa_\alpha^* = \kappa_\alpha A_{\alpha-1} \frac{dt}{dt^*}, \ 7 \leq \alpha \leq m+1, \quad \tau_\alpha^* = \tau_\alpha A_{\alpha+1} \frac{dt}{dt^*}, \ 6 \leq \alpha \leq m+1,$$

$$\mu_\alpha^* = \mu_\alpha A_{\alpha-3} \frac{dt}{dt^*}, \ 5 \leq \alpha \leq m+1, \quad \text{where } A_i = \pm 1, \ \forall i.$$

Hence, K_2, K_3 and K_4 are invariant functions up to a sign, with respect to the parameter transformations on C.

Proof. From (4.4.6) and the first equations of the Natural Frenet equations (4.3.12) of Type 3, we have $K_1^* = K_1 \left(A_1^1 - A_1^2\right) \left(\frac{dt}{dt^*}\right)^2$. Also, from (4.4.6) and the second equations of (4.3.12), we have $K_2^* = K_2 \left(A_1^1 - A_1^2\right)$, $K_3^* = K_3 \left(A_1^1 + A_2^1\right)$ and $K_4^* = K_4 \left(A_3^3 - A_3^4\right)$. From the third equations of (4.3.12), we have $K_5^* = K_5 \left(A_1^1 + A_2^1\right)(A_3^3 - A_3^4)\frac{dt}{dt^*}$, $K_6^* = K_6 \left(A_1^1 + A_2^1\right)(A_3^3 + A_4^3)\frac{dt}{dt^*}$ and $\ell^* A_5^5 = \ell(A_1^1 + A_2^1)\frac{dt}{dt^*}$; $\ell^* A_5^\alpha = 0$ for all $\alpha \in \{6, \cdots, m\}$. Since $\mu_2 \neq 0$, this implies $\ell \neq 0$, therefore, $\ell_2^* \neq 0$ on $\mathcal{U} \cap \mathcal{U}^*$. Thus $A_5^6 = \ldots = A_5^m = 0$. Since $[A_\alpha^\beta]$ is a Lorentzian matrix, we infer that $A_5^5 = A_5 = \pm 1$ and $A_6^5 = \ldots = A_m^5 = 0$. Then from the fifth equation of (4.3.12), we have $K_7^* = K_7 \frac{dt}{dt^*}$, $K_8^* = K_8 A_5 \frac{dt}{dt^*}$ and $K_9^* A_6^6 = K_9 \frac{dt}{dt^*}$; $K_9^* A_6^\alpha = 0$ for all $\alpha \in \{7, \cdots, m\}$. Since $\mu_3 \neq 0$, this implies

$K_9 \neq 0$, therefore, $K_9^* \neq 0$ on $\mathcal{U} \cap \mathcal{U}^*$. Thus $A_6^7 = \ldots = A_6^m = 0$. Since $[A_\alpha^\beta]$ is a Lorentzian matrix, we infer that $A_6^6 = A_6 = \pm 1$ and $A_7^6 = \ldots = A_m^6 = 0$. Repeating this process for all other equations of (4.3.12), we obtain all the relations in (4.4.11), which completes the proof.

Using the general Frenet equations (4.2.5) of Type 3 and the method of proposition 4.11, we have

Proposition 4.14. Let C be a null curve of M_q^{m+2} and F and F^* be two general Frenet frames of Type 3 on $\mathcal{U}$ and $\mathcal{U}^*$ respectively induced by the same screen vector bundle $S(TC^\perp)$. Suppose $\tau_1 \mu_1 \neq 0$ on $\mathcal{U} \cap \mathcal{U}^* \neq \emptyset$. Then at any point of $\mathcal{U} \cap \mathcal{U}^*$ we have

$$K_1^* = K_1 (A_1^1 - A_1^2) \left(\frac{dt}{dt^*} \right)^2,$$

$$K_2^* = K_2 (A_1^1 - A_1^2), \quad K_3^* = K_3 (A_1^1 + A_2^1), \quad T_4^* = T_4 (A_3^3 - A_3^4),$$

$$T_5^* = T_5 (A_1^1 + A_2^1)(A_3^3 - A_3^4) \frac{dt}{dt^*}, \quad T_6^* = T_6 (A_1^1 + A_2^1)(A_3^3 + A_4^3) \frac{dt}{dt^*},$$

$$T_7^* = T_7 (A_1^1 - A_1^2)(A_3^3 - A_3^4) \frac{dt}{dt^*}, \quad T_8^* - T_8 (A_1^1 - A_1^2)(A_3^3 + A_4^3) \frac{dt}{dt^*},$$

$$K_4^* = \left\{ K_4 + (A_1^1 + A_2^1) \frac{d(A_1^1 - A_1^2)}{dt} \right\} \frac{dt}{dt^*},$$

$$K_{11}^* = \left\{ K_{11} + (A_3^3 + A_4^3) \frac{d(A_3^3 - A_3^4)}{dt} \right\} \frac{dt}{dt^*}.$$

Hence, K_2, K_3 and T_4 are invariant functions up to a sign, with respect to the parameter transformations on C.

Also, using the transformations (4.3.2) that relate elements of F_2 and $\bar{F}_3$ on $\mathcal{U} \cap \bar{\mathcal{U}}$ such that $\tau_1 \mu_1 \neq 0$ and by the method of section 3.3, we have

$$\bar{L}_1 = (B_1^1 + B_2^1) \left(L_1 - C_2 \frac{dt}{d\bar{t}} \xi \right), \quad \bar{L}_2 = (B_1^1 - B_1^2) \left(L_2 - C_1 \frac{dt}{d\bar{t}} \xi \right),$$

$$\bar{L}_3 = (B_3^3 + B_4^3) \left(L_3 - C_4 \frac{dt}{d\bar{t}} \xi \right), \quad \bar{L}_4 = (B_3^3 - B_3^4) \left(L_4 - C_3 \frac{dt}{d\bar{t}} \xi \right),$$

$$\bar{W}_\alpha = \sum_{\beta \geq 5}^{m} B_\alpha^\beta \left(W_\beta - \epsilon_\beta c_\beta \frac{dt}{d\bar{t}} \xi \right), \tag{4.4.12}$$

By the method of proposition 4.6 for the Frenet equations (4.3.12) of Type 3, we have

Proposition 4.15. Let C be a null curve of M_q^{m+2} and F and $\bar{F}$ be two Natural Frenet frames of Type 3 on $\mathcal{U}$ and $\bar{\mathcal{U}}$ respectively. Suppose $\tau_1 \mu_1 \neq 0$ on $\mathcal{U} \cap \bar{\mathcal{U}} \neq \emptyset$.

Then, their curvature functions are related by

$$\bar{K}_1 = K_1 \left(B_1^1 - B_1^2 \right) \left(\frac{dt}{d\bar{t}} \right)^2,$$

$$\bar{K}_2 = \left\{ K_2 + \bar{h}\, C_1 + \frac{dC_1}{d\bar{t}} \right\} \left(B_1^1 - B_1^2 \right), \qquad (4.4.13)$$

$$\bar{K}_3 = K_3 \left(B_1^1 + B_2^1 \right)$$

$$\bar{K}_4 = K_4 \left(B_3^3 - B_3^4 \right)$$

and $C_2 = C_3 = C_4 = c_5 = \cdots = c_m = 0$. *Hence, K_3 and K_4 are invariant functions up to a sign, with respect to the transformations of the screen vector bundle of C.*

4.5 Geometry of null curves in M_q^6

Let C be a null curve in a 6-dimensional semi-Riemannian manifold M_q^6. Suppose $F = \{\xi, N, W_1, \cdots, W_4\}$ and $\bar{F} = \{\bar{\xi}, \bar{N}, \bar{W}_1, \cdots, \bar{W}_4\}$ are two general Frenet frames of C with their respective screen vector bundle. We know from propositions 4.1 that they both belong to one of Type 1, Type 2 and Type 3.

Lemma 5.1. *Let C be a null curve of M_q^6 with two general Frenet frames F and $\bar{F}$ of Type 1 such that $\kappa_1 \neq 0$. Then, their curvature functions are related by*

$$\bar{\kappa}_1 = \kappa_1 B_1 \left(\frac{dt}{d\bar{t}} \right)^2;$$

$$\bar{\kappa}_2 = B_1 \left\{ \kappa_2 + \bar{h} c_1 + \frac{d\kappa_1}{d\bar{t}} - \frac{1}{2}\kappa_1 \left(\frac{dt}{d\bar{t}} \right)^2 \sum_{\alpha=1}^{4} (c_\alpha)^2 - (c_2\kappa_4 + c_3\kappa_5)\frac{dt}{d\bar{t}} \right\};$$

$$\bar{\kappa}_3 = \kappa_3 B_2^2 + \bar{h} \sum_{\alpha=2}^{4} B_2^\alpha c_\alpha + \sum_{\alpha=2}^{4} B_2^\alpha \frac{dc_\alpha}{dt} + (c_1\kappa_4 - c_3\kappa_6 - c_4\kappa_7)B_2^2 \frac{dt}{d\bar{t}}$$

$$+ (c_1\kappa_5 + c_2\kappa_6 - c_4\kappa_8)B_2^4 \frac{dt}{d\bar{t}} + (c_2\kappa_7 + c_3\kappa_8)B_2^4 \frac{dt}{d\bar{t}};$$

$$\bar{\kappa}_4 = B_1 \left\{ B_2^2 \left(\kappa_4 + \kappa_1 \frac{dt}{d\bar{t}} c_2 \right) + B_2^3 \left(\kappa_5 + \kappa_1 \frac{dt}{d\bar{t}} c_3 \right) + B_2^4 \kappa_1 \frac{dt}{d\bar{t}} c_4 \right\} \frac{dt}{d\bar{t}};$$

$$\bar{\kappa}_5 = B_1 \left\{ B_3^2 \left(\kappa_4 + \kappa_1 \frac{dt}{d\bar{t}} c_2 \right) + B_3^3 \left(\kappa_5 + \kappa_1 \frac{dt}{d\bar{t}} c_3 \right) + B_3^4 \kappa_1 \frac{dt}{d\bar{t}} c_4 \right\} \frac{dt}{d\bar{t}};$$

$$\bar{\kappa}_6 = \left| \begin{pmatrix} B_2^2 & B_2^3 & \kappa_8 \\ B_3^2 & B_3^3 & -\kappa_7 \\ B_2^4 & B_3^4 & \kappa_6 \end{pmatrix} \right| \frac{dt}{d\bar{t}} + \sum_{\alpha=2}^{4} B_3^\alpha \frac{dB_2^\alpha}{d\bar{t}};$$

$$\bar{\kappa}_7 = \left| \begin{pmatrix} B_2^2 & -\kappa_8 & B_4^2 \\ B_2^3 & \kappa_7 & B_4^3 \\ B_2^4 & -\kappa_6 & B_4^4 \end{pmatrix} \right| \frac{dt}{d\bar{t}} + \sum_{\alpha=2}^{4} B_4^\alpha \frac{dB_2^\alpha}{d\bar{t}};$$

$$\bar{\kappa}_8 = \left| \begin{pmatrix} \kappa_8 & B_3^2 & B_4^2 \\ -\kappa_7 & B_3^3 & B_4^3 \\ \kappa_6 & B_3^4 & B_4^4 \end{pmatrix} \right| \frac{dt}{d\bar{t}} + \sum_{\alpha=2}^{4} B_4^\alpha \frac{dB_3^\alpha}{d\bar{t}};$$

$$B_4^2 \left(\kappa_4 + \kappa_1 \frac{dt}{d\bar{t}} c_2 \right) + B_4^3 \left(\kappa_5 + \kappa_1 \frac{dt}{d\bar{t}} c_3 \right) + B_4^4 \kappa_1 \frac{dt}{d\bar{t}} c_4 = 0.$$

Proof. For the Type 1 it follows from (4.3.2) that $B_1^i = B_i^1 = 0 \, (i \neq 1)$; $B_1^1 = B_1 = \pm 1$. Therefore the general transformations relating the elements of F and $\bar{F}$ on $\mathcal{U} \cap \bar{\mathcal{U}}$ are given by

$$\bar{\xi} = \frac{dt}{d\bar{t}} \xi,$$

$$\bar{N} = -\frac{1}{2} \frac{dt}{d\bar{t}} \sum_{i=1}^{4} \epsilon_i \, (c_i)^2 \, \xi + \frac{d\bar{t}}{dt} N + \sum_{i=1}^{4} c_i \, W_i, \qquad (4.5.1)$$

$$\bar{W}_1 = B_1 \left(W_1 + \frac{dt}{d\bar{t}} c_1 \, \xi \right),$$

$$\bar{W}_\alpha = \sum_{\beta=2}^{4} B_\alpha^\beta \left(W_\beta - \frac{dt}{d\bar{t}} c_\beta \, \xi \right), \quad \alpha \in \{\, 2, 3, 4 \,\}.$$

The relations in lemma follow by straightforward calculations from the general Frenet equations of Type 1 and the use of (4.5.1).

Theorem 5.1. *Let C be a null curve of M_q^6 with a general Frenet frame F of Type 1 and a screen vector bundle $S(TC^\perp)$ on $\mathcal{U} \subset C$ such that $\kappa_1 \neq 0$ on $\mathcal{U}$. Then there exists a screen vector bundle $\bar{S}(TC)$ which induces another Frenet frame $\bar{F}$ of Type 1 on $\mathcal{U}$ such that $\bar{\kappa}_4 = \bar{\kappa}_5 = 0$.*

Proof. Define the following vector fields in terms of the elements of F on $\mathcal{U}$

$$\bar{N} = -\frac{1}{2} \left(\frac{\kappa_4^2 + \kappa_5^2}{\kappa_1^2} \right) \xi + N - \frac{\kappa_4}{\kappa_1} W_2 - \frac{\kappa_5}{\kappa_1} W_3,$$

$$\bar{W}_2 = W_2 + \frac{\kappa_4}{\kappa_1} \xi, \quad \bar{W}_3 = W_3 + \frac{\kappa_5}{\kappa_1} \xi, \qquad (4.5.2)$$

$$\bar{W}_i = W_i, \quad i \in \{\, 1, 4 \,\}.$$

Let $\mathcal{U}^*$ be another coordinate neighborhood with parameter t^* on C such that $\mathcal{U} \cap \mathcal{U}^* \neq \emptyset$. By proposition 4.5 we have the following on $\mathcal{U} \cap \mathcal{U}^*$

$$\kappa_1^* = \kappa_1 A_1 \left(\frac{dt}{dt^*} \right)^2; \quad \kappa_4^* = \kappa_4 A_3 \frac{dt}{dt^*},$$

$$\kappa_5^* = \kappa_5 A_4 \frac{dt}{dt^*}, \quad W_i^* = A_i W_i, \quad i \in \{\, 1, 2, 3, 4 \,\}. \qquad (4.5.3)$$

Define $\{\,\bar{N}^*, \bar{W}_1^*, \dots, \bar{W}_4^*\,\}$ by (4.5.2) but on $\mathcal{U}^*$ with respect to F^*, induced by the same $S(TC^\perp)$ on $\mathcal{U}^*$. Then by using (4.3.1), (4.3.2), (4.5.2) and (4.5.3) we obtain

$$\bar{N}^* = \frac{dt^*}{dt}\,\bar{N}, \quad \bar{W}_i^* = A_i\,\bar{W}_i\,, \quad i \in \{\,1,2,3,4\,\}.$$

Hence there exists a vector bundle $\bar{S}(TC^\perp)$ spanned by $\{\,\bar{W}_1, \dots, \bar{W}_4\,\}$ on $\mathcal{U}$ given by (4.5.2). Moreover, it is easy to check that this vector bundle is complementary to TC in $TC^\perp$. The null transversal vector, with respect to $S(TC^\perp)$, is locally represented by $\bar{N}$ from (4.5.2). Finally taking into account that $t = \bar{t}$ and $c_2 = -\frac{\kappa_4}{\kappa_1}$; $c_3 = -\frac{\kappa_5}{\kappa_1}$ in the fourth and the fifth equations of Lemma 4, we obtain $\bar{\kappa}_4 = \bar{\kappa}_5 = 0$ which completes the proof.

Example 4. Let C be a null curve in $\mathbf{R}_3^6$ given by

$$C : (\,\cos t, \quad \sin t, \quad \sqrt{2}\sinh t, \quad \sqrt{2}\cosh t, \quad t, \quad \sqrt{2}t\,).$$

Then C falls in Type 1 with the general Frenet frame

$$F = \{\,\xi,\, N,\, W_1,\, W_2,\, W_3,\, W_4\,\}, \quad \text{where}$$

$$\xi = (\,-\sin t, \quad \cos t, \quad \sqrt{2}\cosh t, \quad \sqrt{2}\sinh t, \quad 1, \quad \sqrt{2}\,),$$

$$N = \frac{1}{2}\,(\,\sin t, \quad -\cos t, \quad -\sqrt{2}\cosh t, \quad -\sqrt{2}\sinh t, \quad 1, \quad -\sqrt{2}\,),$$

$$W_1 = (\,-\cos t, \quad -\sin t, \quad \sqrt{2}\sinh t, \quad \sqrt{2}\cosh t, \quad 0, \quad 0\,),$$

$$W_2 = (\,-\sqrt{2}\cos t, \quad -\sqrt{2}\sin t, \quad \sinh t, \quad \cosh t, \quad 0, \quad 0\,),$$

$$W_3 = (\,-\sqrt{2}\sin t, \quad \sqrt{2}\cos t, \quad 0, \quad 0, \quad 0, \quad 1\,),$$

$$W_4 = (\,\sqrt{2}\sin t, \quad -\sqrt{2}\cos t, \quad -\cosh t, \quad -\sinh t, \quad 0, \quad -2\,).$$

The general Frenet equations (4.1.1) and the curvature functions are given by

$$\nabla_\xi \xi = W_1, \quad \nabla_\xi N = -\frac{1}{2}W_1, \quad \nabla_\xi W_1 = \frac{1}{2}\xi - N - \sqrt{2}W_3,$$

$$\nabla_\xi W_2 = -2W_3 - W_4, \quad \nabla_\xi W_3 = -\sqrt{2}W_1 + 2W_2, \quad \nabla_\xi W_4 = -W_2,$$

$$h = 0, \quad \kappa_1 = 1, \quad \kappa_2 = -\frac{1}{2}, \quad \kappa_3 = 0, \quad \kappa_4 = 0,$$

$$\kappa_5 = -\sqrt{2}, \quad \kappa_6 = 2, \quad \kappa_7 = 1, \quad \kappa_8 = 0.$$

Since $\kappa_1 = 1$, $\kappa_4 = 0$ and $\kappa_5 = -\sqrt{2}$, let

$$\bar{N} = -\xi + N + \sqrt{2}W_3,$$
$$\bar{W}_3 = W_3 - \sqrt{2}\xi,$$
$$\bar{W}_i = W_i, \quad i \in \{\,1,\,2,\,4\,\},$$

then $\bar{\kappa}_4 = -g(\nabla_\xi \bar{W}_1, \bar{W}_2) = 0$ and $\bar{\kappa}_5 = -g(\nabla_\xi \bar{W}_1, \bar{W}_3) = 0$.

We assume that the transformations (4.3.2) are *diagonal transformations*, that is, they satisfy $B_i^j = B_j^i = 0$ $(i \neq j)$. For this case, it follows from the last equation of lemma 5.1 that $c_4 = 0$. Using this we obtain

Theorem 5.2. *Let C be a null curve of M_q^6 with Frenet frame of Type 1 such that $\kappa_1 \neq 0$. Then, there exist a null 2-surface which is invariant with respect to both the parameter transformations on C and the diagonal screen vector bundle transformations.*

Proof. Let C^* be an integral curve of the vector field W_4. Since, by Lemma 5.1, $c_4 = 0$ for a diagonal screen vector bundle transformation, the 2-surface $S = C \times C^*$ is always invariant with respect to this particular class of screen transformations. S can neither be Lorentz nor definite because its two base vectors $\{\ell, W_4\}$ contain a single null vector ℓ. Therefore, S must be null. This completes the proof.

Now we consider the case when F and $\bar{F}$ are both of Type 2. Using the equations (4.2.6) and the method of the proposition 4.6, we have

Lemma 5.2. *Let C be a null curve of M_q^6 with two general Frenet frames F and $\bar{F}$ of Type 2 such that $\kappa_1 \neq 0$. Then their curvature functions are related by*

$$\bar{K}_1 = K_1 H \left(\frac{dt}{d\bar{t}}\right)^2,$$

$$\bar{K}_2 = H \left\{ K_2 + \bar{h}\, C_1 + \frac{dC_1}{d\bar{t}} - K_1 C_1 C_2 \left(\frac{dt}{d\bar{t}}\right)^2 \frac{K_1}{2} \left(\frac{dt}{d\bar{t}}\right)^2 \sum_{i=3}^{4} \epsilon_i\, (c_i)^2 \right.$$
$$\left. -(C_1\, k_4 + c_3\, K_7 + c_4\, K_8)\, \frac{dt}{d\bar{t}} \right\},$$

$$\bar{K}_3 = G \left\{ K_3 + \bar{h}\, C_2 + \frac{dC_2}{d\bar{t}} + (C_2\, k_4 - c_3\, K_5 - c_4\, K_6)\, \frac{dt}{d\bar{t}} \right\},$$

$$\bar{K}_4 = \left\{ k_4 + K_1\, C_2\, \frac{dt}{d\bar{t}} - H\, \frac{dG}{dt} \right\} \frac{dt}{d\bar{t}},$$

$$\bar{K}_5 = G \left\{ K_5\, B_3^3 + K_6\, B_3^4 \right\} \frac{dt}{d\bar{t}}, \qquad\qquad (4.5.4)$$

$$\bar{K}_6 = G \left\{ K_5\, B_4^3 + K_6\, B_4^4 \right\} \frac{dt}{d\bar{t}},$$

$$\bar{K}_7 = H \left\{ \left(K_7 + K_1\, \frac{dt}{d\bar{t}}\, c_3 \right) B_3^3 + \left(K_8 + K_1\, \frac{dt}{d\bar{t}}\, c_4 \right) B_3^4 \right\} \frac{dt}{d\bar{t}},$$

$$\bar{K}_8 = H \left\{ \left(K_7 + K_1\, \frac{dt}{d\bar{t}}\, c_3 \right) B_4^3 + \left(K_8 + K_1\, \frac{dt}{d\bar{t}}\, c_4 \right) B_4^4 \right\} \frac{dt}{d\bar{t}}.$$

Proof. The matrix $[\, B_j^i\, (x)\,]$, in the relations (4.3.2), is made up of two 2×2 matrices (a Lorentz and an orthogonal). Therefore, using (4.3.2), the general transformations are given by

$$\bar{\xi} = \frac{dt}{d\bar{t}}\xi,$$

$$\bar{N} = -\frac{1}{2}\frac{dt}{d\bar{t}}\sum_{i=1}^{4}\epsilon_i(c_i)^2\xi + \frac{d\bar{t}}{dt}N + C_1L_1 + C_2L_2 + c_3W_3 + c_4W_4,$$

$$\bar{L}_1 = G\left\{L_1 - \frac{dt}{d\bar{t}}C_2\,\xi\right\}, \quad \bar{L}_2 = H\left\{L_2 - \frac{dt}{d\bar{t}}C_1\,\xi\right\}, \tag{4.5.5}$$

$$\bar{W}_\alpha = \sum_{\beta=3}^{4} B_\alpha^\beta\left(W_\beta - \frac{dt}{d\bar{t}}c_\beta\,\xi\right).$$

Straightforward calculations from above relations and the use of (4.1.3) implies (4.5.4), which proves this lemma.

By a procedure same as for the theorem 5.1, one can prove the following:

Theorem 5.3. *Let C be a null curve of M_q^6 with screen bundle space $S(TC^\perp)$ and a general Frenet frame F of Type 2 such that $\kappa_1 \neq 0$ on $\mathcal{U}$. Then there exists a screen vector bundle $\bar{S}(TC^\perp)$ which induces another Frenet frame F on $\mathcal{U}$ such that $\bar{K}_7 = \bar{K}_8 = 0$ on $\mathcal{U}$.*

Next we consider the case when F and $\bar{F}$ are both of Type 3. Using the equations (4.2.5), we have the following general result for M_q^6.

$$\bar{\xi} = \frac{dt}{d\bar{t}}\xi,$$

$$\bar{N} = -(C_1C_2 + C_3C_4)\frac{dt}{d\bar{t}}\xi + \frac{d\bar{t}}{dt}N + \sum_{\alpha=1}^{4} C_\alpha L_\alpha, \tag{4.5.6}$$

$$\bar{L}_1 = H_1\left\{L_1 - C_2\frac{dt}{d\bar{t}}\xi\right\}, \qquad \bar{L}_2 = G_1\left\{L_2 - C_1\frac{dt}{d\bar{t}}\xi\right\},$$

$$\bar{L}_3 = H_3\left\{L_3 - C_4\frac{dt}{d\bar{t}}\xi\right\}, \qquad \bar{L}_4 = G_3\left\{L_4 - C_3\frac{dt}{d\bar{t}}\xi\right\},$$

where $G_1 = B_1^1 - B_1^2$, $H_1 = B_1^1 + B_2^1$, $G_3 = B_3^3 - B_3^4$, $H_3 = B_3^3 + B_4^3$, $C_1 = \frac{1}{\sqrt{2}}(c_2 - c_1)$, $C_2 = \frac{1}{\sqrt{2}}(c_2 + c_1)$, $C_3 = \frac{1}{\sqrt{2}}(c_4 - c_3)$ and $C_4 = \frac{1}{\sqrt{2}}(c_4 + c_3)$.

By the procedure same as before, one can prove the following:

Lemma 5.3. *Let C be a null curve of M_q^6 with two general Frenet frames F and $\bar{F}$ of Type 3 such that $\kappa_1 \neq 0$. Then the functions G_3 and H_3 are constant and their curvature functions are related by*

$$\bar{K}_1 = K_1 H_1 \left(\frac{dt}{d\bar{t}}\right)^2,$$

$$\bar{K}_2 = H_1 \left\{ K_2 + \bar{h} C_1 + \frac{dC_1}{d\bar{t}} - K_1(C_1 C_2 + C_3 C_4) \left(\frac{dt}{d\bar{t}} \right)^2 \right.$$
$$\left. -(C_1 K_4 - C_3 T_8 - C_4 T_7) \frac{dt}{d\bar{t}} \right\},$$

$$\bar{K}_3 = G_1 \left\{ K_3 + \bar{h} C_2 + \frac{dC_2}{d\bar{t}} + (C_2 K_4 - C_3 T_6) \frac{dt}{d\bar{t}} \right\},$$

$$\bar{T}_4 = H_3 \left\{ T_4 + \bar{h} C_3 + \frac{dC_3}{d\bar{t}} + (C_1 T_5 + C_2 T_7 + C_3 K_{11}) \frac{dt}{d\bar{t}} \right\},$$

$$0 = \bar{h} C_4 + \frac{dC_4}{d\bar{t}} + (C_1 T_6 + C_2 T_8 - C_4 K_{11}) \frac{dt}{d\bar{t}},$$

$$\bar{K}_4 = \left\{ K_4 - K_1 C_2 \frac{dt}{d\bar{t}} + H_1 \frac{dG_1}{dt} \right\} \frac{dt}{d\bar{t}}, \qquad (4.5.7)$$

$$\bar{T}_5 = G_1 H_3 T_5 \frac{dt}{d\bar{t}}, \qquad \bar{T}_6 = G_1 G_3 T_6 \frac{dt}{d\bar{t}},$$

$$\bar{T}_7 = H_1 H_3 \left\{ T_7 + K_1 C_3 \frac{dt}{d\bar{t}} \right\} \frac{dt}{d\bar{t}},$$

$$\bar{T}_8 = H_1 G_3 \left\{ T_8 - K_1 C_4 \frac{dt}{d\bar{t}} \right\} \frac{dt}{d\bar{t}}, \qquad \bar{K}_{11} = K_{11} \frac{dt}{d\bar{t}}. \qquad (4.5.8)$$

By the method of theorem 5.3, one can prove the following:

Theorem 5.4. Let C be a null curve of M_q^6 with screen bundle space $S(TC^\perp)$ and a general Frenet frame F of Type 3 such that $\kappa_1 \neq 0$ on $\mathcal{U}$. Then there exists a screen vector bundle $\bar{S}(TC^\perp)$ which induces another Frenet frame F' on $\mathcal{U}$ such that $\bar{T}_7 = \bar{T}_8 = 0$ on $\mathcal{U}$.

Example 5. Let C be a null curve in $\mathbf{R}_3^6$ given by the equations

$$C : (\cos t, \quad \sin t, \quad \sinh t, \quad \cosh t, \quad t, \quad t).$$

Then C falls in Type 3 with the Frenet frame $F = \{\xi, N, L_1, L_2, L_3, L_4\}$ as follows

$$\xi = (-\sin t, \cos t, \cosh t, \sinh t, \quad 1, \quad 1),$$
$$N = \frac{1}{2}(\sin t, \quad -\cos t, \quad -\cosh t, \quad -\sinh t, \quad 1, \quad -1),$$
$$L_1 = (-\cos t, \quad -\sin t, \quad \sinh t, \quad \cosh t, \quad 0, \quad 0),$$
$$L_2 = \frac{1}{2}(\cos t, \quad \sin t, \quad \sinh t, \quad \cosh t, \quad 0, \quad 0),$$
$$L_3 = (-\sin t, \quad \cos t, \quad 0, \quad 0, \quad 0, \quad 1),$$
$$L_4 = (0, \quad 0, \quad \cosh t, \quad \sinh t, \quad 0, \quad 1).$$

The general Frenet equations (4.2.5) and their curvature and torsion functions are:

$$\nabla_\xi \xi = L_1, \qquad \nabla_\xi N = -\frac{1}{2} L_1, \qquad \nabla_\xi L_1 = -L_3 + L_4,$$

$$\nabla_\xi L_2 = \frac{1}{2}\xi - N - \frac{1}{2}(L_3 + L_4), \quad \nabla_\xi L_3 = \frac{1}{2}L_1 - L_2, \quad \nabla_\xi L_4 = \frac{1}{2}L_1,$$

$$h = 0, \ K_1 = 1, \ K_2 = -\frac{1}{2}, \ K_3 = 0, \ K_4 = 0,$$

$$T_4 = 0, \ T_5 = -1, \ T_6 = 1, \ T_7 = T_8 = -\frac{1}{2}, \ K_{11} = 0.$$

Since $K_1 = 1$ and $T_7 = T_8 = -\frac{1}{2}$, let $\bar{L}_i = L_i$, $i \in \{1, 2\}$ and

$$
\begin{aligned}
\bar{N} &= \frac{T_7 T_8}{(K_1)^2}\xi + N - \frac{T_7}{K_1}L_3 + \frac{T_8}{K_1}L_4 \\
&= \frac{1}{4}\xi + N + \frac{1}{2}L_3 - \frac{1}{2}L_4, \\
\bar{L}_3 &= L_3 - \frac{T_8}{K_1}\xi = L_3 + \frac{1}{2}\xi, \\
\bar{L}_4 &= L_4 + \frac{T_7}{K_1}\xi = L_4 - \frac{1}{2}\xi,
\end{aligned}
$$

then we have $\bar{T}_7 = g(\nabla_\xi \bar{L}_1, \bar{L}_3) = 0$ and $\bar{T}_8 = g(\nabla_\xi \bar{L}_2, \bar{L}_4) = 0$.

4.6   Fundamental theorem of null curves in $\mathbf{R}_q^{m+2}$

Let $\mathbf{R}_q^{m+2}$ be the $(m + 2)$-dimensional semi-Euclidean space of index q with the semi-Euclidean metric

$$g(x, y) = -\left(\sum_{i=0}^{q-1} x^i y^i\right) + \left(\sum_{\alpha=q}^{m+1} x^\alpha y^\alpha\right).$$

Suppose C is a null curve in $\mathbf{R}_q^{m+2}$ locally given by the equations

$$x^A = x^A(t), \quad t \in \mathbf{I} \subset \mathbf{R}, \quad A \in \{0, 1, \ldots, (m+1)\}.$$

First, we define in $\mathbf{R}_q^{m+2}$ the natural quasi-orthonormal basis

$$
\begin{aligned}
L_1 &= \left(\frac{1}{\sqrt{2}}, 0, 0, \cdots, 0, \frac{1}{\sqrt{2}}\right), & L_1^* &= \left(-\frac{1}{\sqrt{2}}, 0, 0, \cdots, 0, \frac{1}{\sqrt{2}}\right), \\
E_1 &= (0, 1, 0, \cdots, 0, 0), & E_2 &= (0, 0, 1, 0, \cdots, 0),
\end{aligned}
$$

$$\cdots\cdots\cdots\cdots\cdots\cdots\qquad\qquad\cdots\cdots\cdots\cdots\cdots\cdots\qquad\qquad (4.6.1)$$

$$E_{m-1} = (0, \cdots, 0, 1, 0, 0), \quad E_m = (0, \cdots, 0, 0, 1, 0),$$

where $\{L_1, L_1^*\}$ are null vectors such that $g(L_1, L_1^*) = 1$.

In case $q = 1$, $E_1, \cdots, E_m$ are spacelike vector fields. In case $q = 2$, E_1 is a timelike vector field and $E_2, \cdots, E_m$ are spacelike vector fields. In case $q = 3$, E_1 and E_2 are timelike vector fields and $E_3, \cdots, E_m$ are spacelike vector fields, etc. It is easy to see that

$$L_1^A L_1^{*B} + L_1^B L_1^{*A} + \sum_{\alpha=1}^{m} \epsilon_\alpha E_\alpha^A E_\alpha^B = h^{AB}, \qquad (4.6.2)$$

for any $A, B \in \{0, \ldots, (m+1)\}$, where we put

$$h^{AB} = \begin{cases} -1, & A = B \in \{0, \cdots, q-1\}; \\ 1, & A = B \in \{q, \cdots, (m+1)\}; \\ 0, & A \neq B. \end{cases}$$

Next, we define in $\mathbf{R}_q^{m+2}$ $(2 \leq q \leq n)$ the quasi-orthonormal basis

$$L_1 = \left(\frac{1}{\sqrt{2}}, 0, 0, \cdots, 0, 0, \frac{1}{\sqrt{2}} \right), \quad L_1^* = \left(-\frac{1}{\sqrt{2}}, 0, 0, \cdots, 0, 0, \frac{1}{\sqrt{2}} \right),$$

$$L_2 = \left(0, \frac{1}{\sqrt{2}}, 0, \cdots, 0, \frac{1}{\sqrt{2}}, 0 \right), \quad L_2^* = \left(0, -\frac{1}{\sqrt{2}}, 0, \cdots, 0, \frac{1}{\sqrt{2}}, 0 \right),$$

$$E_2 = (0, 0, 1, 0, \cdots, 0, 0), \qquad E_3 = (0, 0, 0, 1, \cdots, 0, 0) \qquad (4.6.3)$$

$$\cdots\cdots\cdots\cdots\cdots\cdots \qquad\qquad \cdots\cdots\cdots\cdots\cdots\cdots$$

$$E_{m-2} = (0, 0, \cdots, 0, 1, 0, 0, 0), \qquad E_{m-1} = (0, 0, \cdots, 0, 1, 0, 0),$$

where $\{L_1, L_2, L_1^*, L_2^*\}$ are null vector fields such that

$$g(L_i, L_j^*) = \delta_{ij}, \qquad g(L_i, L_j) = 0, \qquad g(L_i^*, L_j^*) = 0$$

and $\{E_2, \cdots, E_{q-2}\}$ and $\{E_{q-1}, \cdots, E_{m-1}\}$ are orthonormal timelike and spacelike vector fields respectively. In this case also we find

$$\sum_{i=1}^{2} (L_i^A L_i^{*B} + L_i^B L_i^{*A}) + \sum_{\alpha=2}^{m-1} \epsilon_\alpha E_\alpha^A E_\alpha^B = h^{AB}, \qquad (4.6.4)$$

for any $A, B \in \{0, \ldots, (m+1)\}$. Let $E_1 = \frac{L_2 - L_2^*}{\sqrt{2}}$ and $E_m = \frac{L_2 + L_2^*}{\sqrt{2}}$, then $\{E_1, E_m\}$ are orthonormal timelike and spacelike vector fields respectively which is orthogonal to $\{E_2, \cdots, E_{m-1}\}$. The equation (4.6.4) becomes

$$L_1^A L_1^{*B} + L_1^B L_1^{*A} + \sum_{\alpha=1}^{m} \epsilon_\alpha E_\alpha^A E_\alpha^B = h^{AB}. \qquad (4.6.5)$$

In the last case, we define in $\mathbf{R}_k^{m+2}$ the quasi-orthonormal basis

$$L_1 = \left(\frac{1}{\sqrt{2}}, 0, 0, 0, \cdots, 0, 0, \cdots, 0, 0, \frac{1}{\sqrt{2}} \right),$$

$$L_1^* = \left(-\frac{1}{\sqrt{2}}, 0, 0, 0, \cdots, 0, 0, \cdots, 0, 0, \frac{1}{\sqrt{2}}\right),$$

$$L_2 = \left(0, \frac{1}{\sqrt{2}}, 0, 0, \cdots, 0, 0, \cdots, 0, \frac{1}{\sqrt{2}}, 0\right),$$

$$L_2^* = \left(0, -\frac{1}{\sqrt{2}}, 0, 0, \cdots, 0, 0, \cdots, 0, \frac{1}{\sqrt{2}}, 0\right), \qquad (4.6.6)$$

$$L_3 = \left(0, 0, \frac{1}{\sqrt{2}}, 0, \cdots, 0, 0, \cdots, 0, \frac{1}{\sqrt{2}}, 0, 0\right),$$

$$L_3^* = \left(0, 0, -\frac{1}{\sqrt{2}}, 0, \cdots, 0, 0, \cdots, 0, \frac{1}{\sqrt{2}}, 0, 0\right),$$

$$\cdots\cdots\cdots\cdots\cdots$$

$$L_k = \left(0, \cdots, 0, \frac{1}{\sqrt{2}}, \frac{1}{\sqrt{2}}, 0, \cdots, 0\right),$$

$$L_k^* = \left(0, \cdots, 0, -\frac{1}{\sqrt{2}}, \frac{1}{\sqrt{2}}, 0, \cdots, 0\right),$$

where $\{L_1, \cdots, L_k, L_1^*, \cdots, L_k^*\}$ are null vector fields such that

$$g(L_i, L_j^*) = \delta_{ij}, \qquad g(L_i, L_j) = 0, \qquad g(L_i^*, L_j^*) = 0.$$

In this case also we find

$$\sum_{i=1}^{k} (L_i^A L_i^{*B} + L_i^B L_i^{*A}) = h^{AB}, \qquad (4.6.7)$$

for any $A, B \in \{0, \ldots, (m+1)\}$. Let

$$E_1 = \frac{L_2 - L_2^*}{\sqrt{2}}, \qquad E_m = \frac{L_2 + L_2^*}{\sqrt{2}},$$

$$E_2 = \frac{L_3 - L_3^*}{\sqrt{2}}, \qquad E_{m-1} = \frac{L_2 + L_2^*}{\sqrt{2}},$$

$$\cdots\cdots\cdots \qquad\qquad \cdots\cdots\cdots$$

$$E_{k-1} = \frac{L_k - L_k^*}{\sqrt{2}}, \qquad E_{m-k+2} = \frac{L_k + L_k^*}{\sqrt{2}},$$

then $\{E_1, \cdots, E_{k-1}\}$ and $\{E_{m-k+2}, \cdots, E_m\}$ are mutually orthogonal orthonormal sets of timelike and spacelike vector fields respectively. The equation (4.6.6) becomes

$$L_1^A L_1^{*B} + L_1^B L_1^{*A} + \sum_{\alpha=1}^{4} \epsilon_\alpha E_\alpha^A E_\alpha^B = h^{AB}. \qquad (4.6.8)$$

We are now in a position to state the fundamental existence and uniqueness theorem for null curves of semi-Euclidean space $\mathbf{R}_q^{m+2}$.

Theorem 6.1. Let $\kappa_i, \tau_i, \mu_i, \nu_i, \eta_i, \cdots : [-\varepsilon, \varepsilon] \to \mathbf{R}$ be everywhere continuous functions, $x_o = (x_o^i)$ a fixed point of $\mathbf{R}_q^{m+2}$ and $\{L_i, L_i^*, E_\alpha\}$ the quasi-orthonormal basis in (4.6.1), (4.6.3) and (4.6.6) respectively. Then there exists a unique null curve $C : [-\varepsilon, \varepsilon] \to \mathbf{R}_q^{m+2}$ such that $x^i = x^i(t)$, $C(0) = x_o$ and $\{\kappa_i, \tau_i, \mu_i, \nu_i, \eta_i, \cdots\}$ are curvature and torsion functions with respect to a general Frenet frame of Type q $(1 \le q \le n)$ $F = \left\{\frac{d}{dp}, N, W_1, \cdots, W_m\right\}$ satisfying

$$\frac{d}{dp} = L_1, \quad N(0) = L_1^*, \quad W_\alpha(0) = E_\alpha, \quad \alpha \in \{1, \cdots, m\}.$$

Proof. Note that $\nabla_\mu X$ is just X' for any vector field X defined on $\mathcal{U}$. Using the equations (4.6.8) we consider the system of differential equation

$$\xi' = h\,\xi + \kappa_1\,W_1 + \tau_1\,W_2,$$
$$N' = -h\,N + \kappa_2\,W_1 + \kappa_3\,W_2 + \tau_2\,W_3 + \mu_1\,W_4,$$
$$\epsilon_1\,W_1' = -\kappa_2\,\xi - \kappa_1\,N + \kappa_4\,W_2 + \kappa_5\,W_3 + \tau_3\,W_4 + \mu_2\,W_5 + \nu_1\,W_6,$$
$$\epsilon_2\,W_2' = -\kappa_3\,\xi - \tau_1\,N - \kappa_4\,W_1 + \kappa_6\,W_3 + \kappa_7\,W_4 + \tau_4\,W_5 + \mu_3\,W_6,$$
$$\qquad\qquad + \nu_2\,W_7 + \eta_1\,W_8, \tag{4.6.9}$$
$$\epsilon_3\,W_3' = -\tau_2\,\xi - \kappa_5\,W_1 - \kappa_6\,W_2 + \kappa_8\,W_4 + \kappa_9\,W_5 + \tau_5\,W_6 + \mu_4\,W_7,$$
$$\qquad\qquad + \nu_3\,W_6 + \eta_2\,W_9 + \sigma_1\,W_{10},$$
$$\epsilon_4\,W_4' = -\mu_1\,\xi - \tau_3\,W_1 - \kappa_7\,W_2 - \kappa_8\,W_3 + \kappa_{10}\,W_5 + \kappa_{11}\,W_6$$
$$\qquad\qquad + \tau_6\,W_7 + \mu_5\,W_8 + \nu_4\,W_9 + \eta_3\,W_{10} + \sigma_2\,W_{11} + \rho_1\,W_{12},$$

$$\cdots\cdots\cdots\cdots\cdots\cdots\cdots\cdots\cdots$$

$$\epsilon_{m-1}\,W_{m-1}' = \cdots - \mu_{m-6}\,W_{m-5} - \tau_{m-4}\,W_{m-4} - \kappa_{2m-3}\,W_{m-3}$$
$$\qquad\qquad - \kappa_{2m-2}\,W_{m-2} + \kappa_{2m}\,W_m,$$
$$\epsilon_m\,W_m' = \cdots - \mu_{m-5}\,W_{m-4} - \tau_{m-3}\,W_{m-3} - \kappa_{2m-1}\,W_{m-2} - \kappa_{2m}\,W_{m-1}.$$

Then there exists a unique solution $\{\xi(t), N(t), W_1(t), \cdots, W_m(t)\}$ satisfying the initial conditions $\xi(0) = L_1$, $N(0) = L_1^*$, $W_\alpha(0) = E_\alpha$, $\alpha \in \{1, \ldots, m\}$. Now we claim that $\{\xi(t), N(t), W_1(t), \cdots, W_m(t)\}$ is a quasi-orthonormal basis such that $\{\xi, N\}$ and $\{W_1, \cdots, W_m\}$ are null and non-null vectors respectively for $t \in [-\varepsilon, \varepsilon]$. To this end, by direct calculations using (4.6.9), we obtain

$$\frac{d}{dp}\left\{\xi^A N^B + \xi^B N^A + \sum_{i=1}^m \epsilon_i\,W_i^A\,W_i^B\right\} = 0. \tag{4.6.10}$$

As for $t = 0$ we have (4.6.2), (4.6.5) and (4.6.8), from (4.6.10) it follows that

$$\xi^A N^B + \xi^B N^A + \sum_{i=1}^m \epsilon_i\,W_i^A\,W_i^B = h^{AB}. \tag{4.6.11}$$

Let

$$N_2 = \frac{W_1 + W_m}{\sqrt{2}}, \qquad\qquad N_2^* = \frac{W_1 - W_m}{\sqrt{2}},$$

$$N_3 = \frac{W_2 + W_{m-1}}{\sqrt{2}}, \qquad\qquad N_3^* = \frac{W_2 - W_{m-1}}{\sqrt{2}},$$

$$\cdots\cdots\cdots\cdots\cdots\cdots \qquad\qquad \cdots\cdots\cdots\cdots\cdots\cdots$$

$$N_q = \frac{W_{q-1} + W_{m-q+1}}{\sqrt{2}}, \qquad N_q^* = \frac{W_{q-1} - W_{m-q+1}}{\sqrt{2}},$$

then $\{N_i, N_i^*\}$ are null vector fields such that

$$g(N_i(t), N_j^*(t)) = \delta_{ij}, \quad g(N_i(t), N_j(t)) = 0, \quad g(N_i^*(t), N_j^*(t)) = 0.$$

Further on, construct the field of frames

$$W_0 = \frac{1}{\sqrt{2}}(\xi - N), \qquad W_{m+1} = \frac{1}{\sqrt{2}}(\xi + N), \tag{4.6.12}$$

where W_0 is a timelike vector field and W_{m+1} is a spacelike one. Then (4.6.11) becomes

$$\sum_{i=0}^{m+1} \epsilon_i W_i^A W_i^B = h^{AB}. \tag{4.6.13}$$

We define for each $p \in [-\varepsilon, \varepsilon]$ the matrix $D(p) = (d^{AB}(p))$ such that

$$d^{ab} = W_b^a; \ d^{ai} = \sqrt{-1} W_i^a; \ d^{hb} = -\sqrt{-1} W_b^h; \ d^{hi} = W_i^h,$$

for any $a, b \in \{0, \cdots, q-1\}$, $h, i \in \{q, \cdots, (m+1)\}$. By using (4.6.13) it is easy to check that $D(p)D(p)^t = I_{m+2}$, which implies that $\{W_1, \cdots, W_m\}$ is an orthonormal basis for any $p \in [-\varepsilon, \varepsilon]$. Then, from (4.6.11), we conclude that $\{\xi, N, W_1, \cdots, W_m\}$ is a quasi-orthonormal basis for any $p \in [-\varepsilon, \varepsilon]$. The null curve is obtained by integrating the system

$$\frac{dx^i}{dp} = \mu(t), \qquad x^i(0) = x_o^i.$$

Taking into account of (4.6.11) we see the frame $F = \{\mu, N, W_1, \ldots, W_4\}$ is a general Frenet frame of Type q for C with the curvature and torsion functions $\{\kappa_i, \tau_i, \mu_i, \nu_i, \eta_i, \cdots\}$. This completes the proof of theorem.

Exercises

Find the Frenet frames and the curvature functions of

(1) A null curve C in $\mathbf{R}_3^6$ given by

$$x^0 = \cos t, \quad x^1 = \sin t, \quad x^2 = \sinh t, \quad x^3 = \cosh t,$$
$$x^4 = \sqrt{2} \sin t, \quad x^5 = \sqrt{2} \cos t, \quad t \in \mathbf{R}.$$

(2) A null curve C in $\mathbf{R}_3^6$ given by

$$x^0 = \frac{1}{3} t^3 - 2t, \quad x^1 = \frac{1}{2} t^2, \quad x^2 = t \sinh t,$$
$$x^3 = t \cosh t, \quad x^4 = \sqrt{2} t, \quad x^5 = \frac{1}{3} t^3 - t, \quad t \in \mathbf{R}.$$

(3) A null curve C in $\mathbf{R}_3^7$ given by

$$x^0 = \frac{1}{3} t^3 - 2t, \quad x^1 = \frac{1}{2} t^2, \quad x^2 = \sqrt{2} t \sinh t, \quad x^3 = \sqrt{2} t \cosh t,$$
$$x^4 = t \cos t, \quad x^5 = t \sin t, \quad x^6 = \frac{1}{3} t^3 - t, \quad t \in \mathbf{R}.$$

(4) A null curve C in $\mathbf{R}_3^7$ given by

$$x^0 = \sin t, \quad x^1 = \cos t, \quad x^2 = \sinh t, \quad x^3 = \cosh t,$$
$$x^4 = \cos t, \quad x^5 = \sin t, \quad x^6 = t, \quad t \in \mathbf{R}.$$

(5) A null curve C in $\mathbf{R}_3^8$ given by

$$x^0 = t, \quad x^1 = \frac{1}{2} t^2, \quad x^2 = \frac{1}{6} t^3 + t, \quad x^3 = \frac{1}{2} t^2,$$
$$x^4 = \frac{1}{6} t^3, \quad x^5 = t, \quad x^6 = t \sin t, \quad x^7 = t \cos t, \quad t \in \mathbf{R}.$$

(6) A null curve C in $\mathbf{R}_4^8$ given by

$$x^0 = t, \quad x^1 = \frac{1}{2} t^2, \quad x^2 = t^2 + t, \quad x^3 = \sqrt{3} t \cosh t,$$
$$x^4 = \sqrt{3} t \sinh t, \quad x^5 = t \cos t, \quad x^6 = t \sin t, \quad x^7 = \frac{1}{2} t^2 + 2, \quad t \in \mathbf{R}.$$

Chapter 5

Geometry of null Cartan curves

As we have seen in previous chapters that, contrary to the Riemannian case, the general Frenet equations of a null curve C are not unique as they depend on the parameter of C and the choice of a screen distribution. We have also seen (in chapter 2, section 3) that, following Bonnor [18], one can find a unique Cartan Frenet frame, with respect to a distinguished parameter, for a non-geodesic null curve of some low dimensional Lorentzian manifolds. In this chapter, we first generalize the results of Bonnor for a null curve in M_1^{m+2} and, then, present a complete classification of null helices in 3, 4 and 5 dimensional Lorentzian space forms. Finally, we briefly present latest results on null Cartan curves in M_2^{m+2}.

5.1 Null Cartan curves in Lorentzian manifolds

The objective is to find a Frenet frame with a minimal number of curvature functions and such that they are invariant under Lorentzian transformations. For this purpose, we use a Natural Frenet frame $\{\xi, N, W_1, \ldots, W_m\}$, with their Natural Frenet equations given by (2.1.4) of chapter 2 in proving the following theorem:

Theorem 1.1. *Let C be a null curve of an orientable Lorentzian manifold (M_1^{m+2}, g) parameterized by the distinguished parameter p such that $\mathcal{A} = \{C'(p), C''(p), \ldots, C^{(m+2)}(p)\}$ is a basis of $T_{C(p)}M$ for all p. Then there exists exactly one Frenet frame $F = \{\xi, N, W_1, \ldots, W_m\}$, satisfying the equations*

$$\nabla_\xi \xi = W_1,$$
$$\nabla_\xi N = \sigma_1 W_1 + \sigma_2 W_2,$$
$$\nabla_\xi W_1 = -\sigma_1 \xi - N,$$
$$\nabla_\xi W_2 = -\sigma_2 \xi + \sigma_3 W_3$$

$$\cdots\cdots\cdots\cdots\cdots$$

$$(5.1.1)$$

$$\nabla_\xi W_i = -\sigma_i W_{i-1} + \sigma_{i+1} W_{i+1}, \quad i \in \{3, \dots, m-1\},$$
$$\nabla_\xi W_m = -\sigma_m W_{m-1},$$

and fulfilling the following two conditions:

(i) $\{C'(p), C''(p), \dots, C^{(i+2)}(p)\}$ and $\{\xi, N, W_1, \dots, W_i\}$ have the same orientation for $2 \leq i \leq m-1$,

(ii) $\{\xi, N, W_1, \dots, W_m\}$ is positively oriented and $\sigma_i > 0, \forall i \geq 2$.

Proof. Using the notations of chapter 2, without any loss of generality we assume that a null curve C, of (M_1^{m+2}, g), is parameterized by a special parameter p such that $g(C'', C'') = 1$ and $h = 0$. Now choose $\xi = C'$ and $W_1 = C''$ so that $\kappa_1 = 1$. It is easy to show that the null transversal bundle N is generated by

$$N = -C^{(3)} - \kappa_2 C', \quad \text{where} \quad \kappa_2 = g(\nabla_\xi N, W_1) = \frac{1}{2} g(C^{(3)}, C^{(3)}).$$

Using above in the second equation of (2.1.4) and after calculating $\nabla_\xi N$ we obtain

$$\kappa_3 = \pm \sqrt{g(C^{(4)}, C^{(4)}) - g(C^{(3)}, C^{(3)})^2}$$
$$W_2 = \mp \frac{1}{\kappa_3} \left(C^{(4)} + g(C^{(3)}, C^{(3)}) C'' + g(C^{(4)}, C^{(3)}) C' \right).$$

A direct computation and renaming the curvature functions ($\sigma_1 = \kappa_2, \sigma_2 = \kappa_3$ and $\sigma_3 = \kappa_4$), we obtain first 4 equations of (5.1.1). If $m > 3$, then, by a similar reasoning we obtain all the other equations of (5.1.1) which completes the proof.

Note. The above theorem 1.1 was first proved by Ferrández-Giménez-Lucas [35] by using Frenet equations which are identical to our Natural Frenet equations (2.1.4). Following their terminology we call the Frenet frame of the equations (5.1.1), its curvature functions and the corresponding curve C the *Cartan frame* (see Cartan [20]), the *Cartan curvatures* and the *null Cartan curve* respectively.

In order not to repeat, we state (see [35, pages 4850-4854] for their proofs) the following results directly coming from the theorem 1.1.

Proposition 1.1. Let C be a null Cartan Curve in M_1^{m+2}. Then the Cartan curvatures $\{\sigma_1, \dots, \sigma_m\}$ are given by

$$\sigma_1 = \frac{1}{2} g(C^{(3)}, C^{(3)}), \quad \sigma_2^2 = -D_4, \quad \sigma_i^2 = \frac{D_i D_{i+2}}{D_{i+1}^2}.$$

Moreover, $\sigma_2 < 0$, $\sigma_i > 0$ for all $i \in \{3, \dots, m-1\}$, and $\sigma_m < 0$ according to $\mathcal{A}$ is positive or negatively oriented respectively, where

$$D_i = \begin{vmatrix} g(C', C') & g(C', C'') & \cdots & g(C', C^{(i)}) \\ g(C'', C') & g(C'', C'') & \cdots & g(C'', C^{(i)}) \\ \vdots & \vdots & & \vdots \\ g(C^{(i)}, C') & g(C^{(i)}, C'') & \cdots & g(C^{(i)}, C^{(i)}) \end{vmatrix}. \qquad (5.1.2)$$

Corollary 1. *The Cartan curvatures of a curve C in M_1^{m+2} are invariant under Lorentzian transformations.*

For physical reasons we consider a Cartan null curve C in a 4-dimensional space-time manifold (M_1^4, g), with a Cartan frame $\{\xi, N, W_1, W_2\}$ with respect to a distinguished parameter p so that its Cartan equations are

$$\dot{\xi} \equiv \nabla_\xi \xi = W_1, \tag{5.1.3}$$

$$\dot{N} \equiv \nabla_\xi N = \sigma_1 W_1 + \sigma_2 W_2, \tag{5.1.4}$$

$$\dot{W}_1 \equiv \nabla_\xi W_1 = -\sigma_1 \xi - N, \tag{5.1.5}$$

$$\dot{W}_2 \equiv \nabla_\xi W_2 = -\sigma_2 \xi. \tag{5.1.6}$$

As it is known, for regular curves in $\mathbf{R}^3$, there exists a Frenet frame (T, N, B) of unit tangent, normal and binormal vector fields such that N and B are expressed in terms of T and its covariant derivatives along T. Following the same idea, taking $\sigma_2 \neq 0$ we now express $\{N, W_1, W_2\}$ in terms of ξ and its covariant derivatives. The equation (5.1.3) provides

$$W_1 = \dot{\xi}. \tag{5.1.7}$$

Next, taking covariant differentiation of (5.1.7) with respect to ξ and then comparing the result with (5.1.5), we obtain

$$N = -\sigma_1 \xi - \ddot{\xi}, \tag{5.1.8}$$

where $\ddot{\xi} = \nabla_\xi \nabla_\xi \xi$. Finally, differentiating covariantly (5.1.8) with respect to ξ and then comparing with (5.1.4) we obtain

$$W_2 = -\frac{1}{\sigma_2}\left\{\sigma_1 \xi + 2\sigma_1 \dot{\xi} + \dddot{\xi}\right\}, \quad \text{where} \quad \dddot{\xi} = \nabla_\xi \nabla_\xi \nabla_\xi \xi. \tag{5.1.9}$$

Definition 1.1. *A space form is a complete connected semi-Riemannian manifold of constant curvature.*

The simplest *Lorentzian space form* is a flat (means zero scalar curvature) Minkowski space $\mathbf{R}_1^{m+2}$ for which, proceeding exactly as in section 5 of chapter 2 but using the Cartan equations (5.1.1) instead of the general Frenet frame equations (2.5.6), one can prove the following *fundamental theorem of null Cartan curves.*

Theorem 1.2. *Let $\sigma_1, \ldots, \sigma_m : [-\varepsilon, \varepsilon] \to \mathbf{R}$ be everywhere differentiable functions with $\sigma_2 < 0$ and $\sigma_i > 0$ for all $i \in \{3, \ldots, m-1\}$. Let x be a point in $\mathbf{R}_1^{m+2}$ and consider $\{\xi^0, N^0, W_0^0, \ldots, W_m^0\}$ a positively oriented quasi-orthonormal basis of $\mathbf{R}_1^{m+2}$. Then there exists a unique null Cartan curve C in $\mathbf{R}_1^{m+2}$, with $C(0) = x$, whose Cartan Frame $\{\xi, N, W_1, \ldots, W_m\}$ satisfies*

$$\xi(0) = \xi^0, \quad N(0) = N^0, \quad W_i(0) = W_i^0, \quad i \in \{1, \ldots, m\}.$$

Theorem 1.3. *If two null Cartan curves C and $\bar{C}$ in $\mathbf{R}_1^{m+2}$ have same Cartan curvatures $\{\sigma_1, \ldots, \sigma_m\}$, where $\sigma_i : [-\varepsilon, \varepsilon] \to \mathbf{R}$ are differentiable functions, then there exists a Lorentzian transformation of $\mathbf{R}_1^{m+2}$ which maps C into $\bar{C}$.*

Next Lorentzian space form is the spacetime of positive scalar curvature, called *De Sitter spacetime* and denoted by $\mathbf{S}_1^{m+2}$, which has the topology $\mathbf{R}^1 \times \mathbf{S}^{m+1}$ and can be visualized as a hypersurface in $\mathbf{R}_1^{m+3}$. Here $\mathbf{S}^{m+1}$ denotes a $(m+1)$-sphere. In particular, a 4-dimensional De Sitter spacetime is physically important space [56, page 124] which can be seen as the hyperboloid

$$\alpha^2 = -t^2 + u^2 + x^2 + y^2 + z^2$$

in a 5-dimensional Minkowski space $\mathbf{R}_1^5$ with metric

$$ds^2 = -dt^2 + du^2 + dx^2 + dy^2 + dz^2.$$

Let $C(p)$ be a null curve in a De Sitter space $\mathbf{S}_1^{m+2} \subset \mathbf{R}_1^{m+3}$, with respect to a special parameter p. Denote by D_p the covariant derivative along C in $\mathbf{S}_1^{m+2}$. Then, for any vector field X along C, we have

$$D_p X = X' + g(X, C')C, \tag{5.1.10}$$

where g is the canonical metric in $\mathbf{R}_1^{m+3}$. Using (5.1.10) for a Cartan null curve with a Cartan frame $\{\xi, N, W_1, \ldots, W_m\}$, it is straightforward to see that the equations (5.1.1) can be replaced by the following equations:

$$\begin{aligned}
\nabla_\xi \xi &= W_1, \\
\nabla_\xi N &= -C + \sigma_1 W_1 + \sigma_2 W_2, \\
\nabla_\xi W_1 &= -\sigma_1 \xi - N, \\
\nabla_\xi W_2 &= -\sigma_2 \xi + \sigma_3 W_3, \\
&\quad\cdots\cdots\cdots\cdots\cdots \\
\nabla_\xi W_i &= -\sigma_i W_{i-1} + \sigma_{i+1} W_{i+1}, \quad i \in \{3, \ldots, m-1\}, \\
\nabla_\xi W_m &= -\sigma_m W_{m-1}.
\end{aligned} \tag{5.1.11}$$

Using similar procedure, for a null curve in $\mathbf{S}_1^{m+2}$, the fundamental theorem 1.2 and its corollary 2 can be replaced as follows:

Theorem 1.4. *Let $\sigma_1, \ldots, \sigma_m : [-\varepsilon, \varepsilon] \to \mathbf{R}$ be everywhere differentiable functions with $\sigma_2 < 0$ and $\sigma_i > 0$ for all $i \in \{3, \ldots, m-1\}$. Let x be a point in $\mathbf{S}_1^{m+2}$ and consider $\{\xi^0, N^0, W_0^0, \ldots, W_m^0\}$ a positively oriented quasi-orthonormal basis of $T_x \mathbf{S}_1^{m+2}$. Then there exists a unique null Cartan curve C in $\mathbf{S}_1^{m+2}$, with $C(0) = x$, whose Cartan Frame $\{\xi, N, W_1, \ldots, W_m\}$ satisfies*

$$\xi(0) = \xi^0, \quad N(0) = N^0, \quad W_i(0) = W_i^0, \quad i \in \{1, \ldots, m\}.$$

Theorem 1.5. *If two null Cartan curves C and $\bar{C}$ in $\mathbf{S}_1^{m+2}$ have same Cartan curvatures $\{\sigma_1, \ldots, \sigma_m\}$, where $\sigma_i : [-\varepsilon, \varepsilon] \to \mathbf{R}$ are differentiable functions, then* there exists a *Lorentzian transformation of* $\mathbf{S}_1^{m+2}$ *which maps* C *into* $\bar{C}$.

Finally, a Lorentzian space form of constant negative curvature is known as *anti-De Sitter spacetime*, denoted by $\mathbf{H}_1^{m+2}$ and is topologically $\mathbf{S}^1 \times \mathbf{R}^{m+1}$. It can be regarded as a hypersurface of revolution in $\mathbf{R}_2^{m+3} = \mathbf{R}_2^2 \times \mathbf{R}^{m+1}$. In particular, a 4-dimensional anti-De Sitter spacetime is also physically important space [56, page 131] which can be seen as the hyperboloid

$$\alpha^2 = -u^2 - v^2 + x^2 + y^2 + z^2$$

in a 5-dimensional semi-Euclidean space $\mathbf{R}_2^5$ with metric

$$-du^2 - dv^2 + dx^2 + dy^2 + dz^2 = ds^2.$$

Its universal covering space of constant curvature $c = -1$ (constructed by unwrapping the circle S^1) has the metric

$$ds^2 = -\cosh^2 r \, dt'^2 + dr^2 + \sinh^2 r \left(d\theta^2 + \sin^2 \theta \, d\phi^2 \right),$$

where the coordinates (t', r, θ, ϕ) cover the entire spacetime [56, page 131].

Let $C(p)$ be a null curve in an anti-De Sitter space $\mathbf{H}_1^{m+2} \subset \mathbf{R}_2^{m+3}$, with respect to a special parameter p. Denote by D_p the covariant derivative along C in $\mathbf{H}_1^{m+2}$. Proceeding as in the case of De Sitter space, for a Cartan null curve in $\mathbf{H}_1^{m+2}$, the equation (5.1.10) will be replaced by $D_p X = X' - g(X, C')C$, where g is now the canonical metric in $\mathbf{R}_2^{m+3}$ and the same Cartan equations as (5.1.11) with $-C$ in the second equation replaced by C, due to negative constant scalar curvature. Furthermore, proceeding as in the case of De Sitter space, for a null curve in $\mathbf{H}_1^{m+2}$, the fundamental theorem 1.3 and its corollary 3 can be replaced as follows:

Theorem 1.6. *Let* $\sigma_1, \ldots, \sigma_m : [-\varepsilon, \varepsilon] \to \mathbf{R}$ *be everywhere differentiable functions with* $\sigma_2 < 0$ *and* $\sigma_i > 0$ *for all* $i \in \{3, \ldots, m-1\}$. *Let* x *be a point in* $\mathbf{H}_1^{m+2}$ *and consider* $\{\xi^0, N^0, W_0^0, \ldots, W_m^0\}$ *a positively oriented quasi-orthonormal basis of* $T_x \mathbf{H}_1^{m+2}$. *Then there exists a unique Cartan curve* C *in* $\mathbf{H}_1^{m+2}$, *with* $C(0) = x$, *whose null Cartan Frame* $\{\xi, N, W_1, \ldots, W_m\}$ *satisfies*

$$\xi(0) = \xi^0, \quad N(0) = N^0, \quad W_i(0) = W_i^0, \quad i \in \{1, \ldots, m\}.$$

Theorem 1.7. *If two null Cartan curves* C *and* $\bar{C}$ *in* $\mathbf{H}_1^{m+2}$ *have same Cartan curvatures* $\{\sigma_1, \ldots, \sigma_m\}$, *where* $\sigma_i : [-\varepsilon, \varepsilon] \to \mathbf{R}$ *are differentiable functions, then there exists a Lorentzian transformation of* $\mathbf{H}_1^{m+2}$ *which maps* C *into* $\bar{C}$.

Definition 1.2. *A curve is called a helix if it has constant Cartan curvatures.*

If C is a helix, then it satisfies the following differential equation:

$$C^{m+3} = a_1 C' + a_2 C^{(3)} + \ldots + a_s C^{(m+1)}, \quad m + 2 = 2s, \qquad (5.1.12)$$
$$C^{m+3} = a_1 C'' + a_2 C^{(4)} + \ldots + a_s C^{(m+1)}, \quad m + 2 = 2s + 1, \qquad (5.1.13)$$

where the coefficients are given by

$$
a_i = \sigma_2^2 \sum_{\substack{4 \le j_1 \cdots < j_{s-i-1} \le m \\ j_r - j_{r-1} \ge 2\, \forall r}} \sigma_{j_1}^2 \cdots \sigma_{j_{s-i-1}}^2 - 2\sigma_1 \sum_{\substack{3 \le j_1 < \cdots < j_{s-i} \le m \\ j_r - j_{r-1} \ge 2\, \forall r}} \sigma_{j_1}^2 \cdots \sigma_{j_{s-i}}^2,
$$

$$
a_{s-1} = \sigma_2^2 - 2\sigma_1 \sum_{j=3}^{m} k_j^2 - \sum_{\substack{3 \le j_1 < j_2 \le m \\ j_2 - j_1 \ge 2}} \sigma_{j_1}^2 \sigma_{j_2}^2,
$$

$$
a_s = -2\sigma_1 - \sum_{j=3}^{m} k_j^2.
$$

5.2 Null Cartan helices

The study on helices in 3-dimensional Euclidean spaces goes back to the year 1802 when Lancret (see details in [106]) stated that a curve is a *generalized helix* if and only if the ratio of curvature to torsion is constant. Much later on, in 1931, Hayden [54, 55] defined a generalized helix in an odd n-dimensional Euclidean space as a curve for which the ratios k_{2i}/k_{2i-1} are constant, where k_1, k_2, ..., k_{n-1} are the curvature functions of the respective Frenet equations (also, see Barros [6]). Hayden proved that a curve is a generalized helix if there exists a parallel vector field lying in the osculating space of the curve making constant angles with the tangent and the principal normals. In this section we present up-to-date works on *null Cartan helices*, briefly called null helices (see definition 2 in section 1), in low dimensional Lorentzian spaces of constant curvature, starting from a 3-dimensional Minkowski space. In section 3, we provide latest information on *null generalized helices*, references for further study on the physical significance of generalized helices and characterization theorems for null generalized helices in $\mathbf{R}_1^3$ and $\mathbf{R}_1^5$.

Null helices in $\mathbf{R}_1^3$. J. Bertrand studied a pair of curves in a 3-dimensional Euclidean space which possess common normal direction. Such a pair of curves is called a spacelike Bertrand pair. *Bertrand curves* are characterized as follows:

Proposition ([34, page 41]). *A curve C in a Euclidean 3-space, parameterized by the arclength, is a Bertrand curve if and only if C is a plane curve or curves whose curvature κ and torsion τ are in linear relation: $a\kappa + b\tau = 1$ for some constants a and b. The product of torsion of Bertrand pair is constant.*

Recall from section 3 of chapter 2 that a curve C, together with a Frenet frame $F = \{\xi, N, W\}$ with respect to a distinguished parameter p, is called a *framed null curve*. Using the usual terminology, the spacelike unit vector field W and the transversal null vector field N, of F, will be called *principal normal* and the *binormal* vector fields respectively.

Extending above result to null curves in $\mathbf{R}_1^3$, recently, Honda-Inoguchi [57] and Inoguchi-Lee [64] have done some work on a pair of null curves $(C, \bar{C})$, called a *null Bertrand pair* and their relation with null helices in R_1^3, which we now describe.

Definition 2.1 ([64]). *Let $(C, \bar{C})$ be a pair of framed null curves in $\mathbf{R}_1^3$, with distinguished parameters p and $\bar{p}$ respectively. This pair is said to be a null Bertrand pair if their principal normal vector fields are linearly dependent.*

The curve $\bar{C}$ is called a *Bertrand mate* of C and vice versa. A framed null curve is said to be a *null Bertrand curve* if it admits a Bertrand mate. By above definition 2.1, there exists a functional relation $\bar{p} = \bar{p}(p)$ for a null Bertrand pair $(C, \bar{C})$ such that $\bar{W}(\bar{p}(p)) = \epsilon W(p)$, $\epsilon = \pm 1$, i.e., the normal lines coincide at their respective points. To show that the null Bertrand curves have been used in characterizing null helices, we assume that each framed null curve is a null Cartan curve (to assure its uniqueness) with the following type of Cartan equations:

$$\begin{aligned}
\nabla_\xi \xi &= W, \\
\nabla_\xi N &= \tau W, \\
\nabla_\xi W &= -\tau \xi - N.
\end{aligned} \tag{5.2.1}$$

having Frenet frame $F = \{\xi, N, W\}$ and a single Cartan curvature τ. We reproduce complete proofs (given in [64] and [57] using a different approach) of the following characterization theorems for null helices in $\mathbf{R}_1^3$:

Theorem 2.1 ([64]). *Let $C(p)$ be a null Cartan curve in $\mathbf{R}_1^3$, where p is a special parameter. Then C admits a Bertrand mate $\bar{C}$ if and only if C and $\bar{C}$ have same nonzero constant curvatures. Moreover, $\bar{C}$ is congruent to C.*

Proof. Suppose C admits a Bertrand mate $\bar{C}(\bar{p})$, with respect to a special parameter $\bar{p}$. Let $\{\xi, N, W\}$ and $\{\bar{\xi}, \bar{N}, \bar{W}\}$ be their respective Cartan frames. Since the normal lines coincide at corresponding points, we may write

$$\bar{C} = C + \alpha W, \tag{5.2.2}$$

for some function $\alpha(p) \neq 0$. Differentiating (5.2.2) with respect to p and using the Cartan equations (5.2.1), we get

$$\frac{d\bar{p}}{dp} \bar{\xi} = (1 - \alpha\tau)\xi - \alpha N + \alpha' W. \tag{5.2.3}$$

For null Bertrand curves, the condition $g(\bar{\xi}, W) = 0$ holds. Using this in (5.2.3) implies that $\alpha' = 0$ and so α is a constant. This, together with null $\bar{\xi}$, means that $\alpha(1 - \alpha\tau) = 0$, so $\tau = \frac{1}{\alpha} = $ constant. Using this value of τ in (5.2.3) we get

$$\bar{\xi} = \mu N, \quad \mu = -\alpha \frac{dp}{d\bar{p}} \neq 0. \tag{5.2.4}$$

Differentiating (5.2.4) with respect to p and using Cartan frame (5.2.1) we get

$$\frac{d\bar{p}}{dp} \bar{W} = \mu' N + \mu\tau W.$$

Since $\bar{W} = \pm W$ we get $\frac{d\bar{p}}{dp} = \pm 1$ and $\mu = \pm \alpha =$ constant, so the Frenet frame of $\bar{C}$ is $\{\mp \alpha N, \mp \alpha^{-1}\xi, -W\}$. Thus $\bar{C}$ has same constant curvature $\tau = 1/\alpha$.

Conversely, let C be a null Cartan curve with constant curvature $\tau = 1/\alpha \neq 0$. Then $\bar{C} = C + \alpha W$ is a null curve with a special parameter p and framed by

$$\bar{\xi} = -\alpha N, \quad \bar{N} = -\alpha^{-1}\xi, \quad \bar{W} = -W.$$

Thus $(C, \bar{C})$ is a Bertrand pair and $\bar{C}$ also has constant Cartan curvature $\tau = 1/\alpha$. The congruence of C and $\bar{C}$ trivially follows, which completes the proof.

Example. Let $C(p) = (p, \cos p, \sin p)$ be a null helix with Frenet frame

$$F = \left\{ C', \ N = \frac{1}{2}(-1, -\sin p, -\cos p), \ W = (0, -\cos p, -\sin p) \right\}.$$

It is easy to see that its torsion $\tau = \frac{1}{2}$. Define $\bar{C} = C + 2W$ with $\bar{p} = p$. Then, $\bar{C} = (p, -\cos p, -\sin p)$. Therefore, $\bar{C}$ is congruent to the original curve C.

Now we discuss results on pairs of Cartan null curves which have common binormal directions. As a comparison, we first state the following classical result:

Proposition ([48, page 161]). *Let $C(s)$ be a curve in $\mathbf{R}^3$, with an arc length parameter s. Take a curve $\bar{C}$ parameterized by arc length parameter such that the binormal directions of $\bar{C}$ and C coincide. Then both curves are plane curves.*

Let $C(p)$ and $\bar{C}(\bar{p})$ be two Cartan null curves with special parameters p and $\bar{p}$ and Frenet frames $\{\xi, N, W\}$ and $\{\bar{\xi}, \bar{N}, \bar{W}\}$ respectively. Suppose

$$\bar{C} = C + u\xi + vN + wW,$$

for some functions u, v and w. Then, $\bar{C}$ is called an *associated Cartan null curve* of C, with reference coordinates (u, v, w). Differentiating $\bar{C}$ we get

$$\frac{d\bar{p}}{dp}\bar{\xi} = (1 - u' - \tau w)\xi + (v' - u)N + (w' + u + \tau v)W. \tag{5.2.5}$$

Since $\bar{\xi}$ is null, we obtain

$$(w' + u + \tau v)^2 + 2(1 + u' - \tau w)(v' - w) = 0. \tag{5.2.6}$$

The null curves behave differently as stated in the following result:

Theorem 2.2 ([57]). *(1) Let $C(p)$ be a null Cartan curve in $\mathbf{R}_1^3$, where p is a special parameter. Assume there exists a null Cartan curve $\bar{C}(\bar{p})$ parameterized by special parameter $\bar{p}$ such that the binormal directions of $\bar{C}$ and C coincide. Then at the corresponding point, their Cartan curvatures coincide.*

(2) Let $C(p)$ be a Cartan null curve in $\mathbf{R}_1^3$, where p is a special parameter. Then there exists a null Cartan curve $\bar{C}(\bar{p})$, with a special parameter $\bar{p}$

such that the binormal directions and the respective curvatures of C and $\bar{C}$ coincide.

Proof. (1) Assume a pair $(C(p), \bar{C}(\bar{p}))$ of Cartan null curves with their special parameters p and $\bar{p}$ respectively such that their binormal directions coincide. Then, $\bar{C}$ is parameterized as

$$\bar{C}(\bar{p}(p)) = C(p) + v(p)N(p) \tag{5.2.7}$$

for some function $v(p) \neq 0$ and some parameterization $\bar{p} = \bar{p}(p)$. Hence, $\bar{C}$ is an associated curve to C with the reference coordinates $(0, v, 0)$. Without any loss of generality, we assume that

$$\bar{N} = \alpha N \tag{5.2.8}$$

for some function $\alpha(p) \neq 0$. Then, using (5.2.5) we obtain

$$\frac{d\bar{p}}{dp}\bar{\xi} = \xi + v'N + \tau vW. \tag{5.2.9}$$

Also (5.2.6) and $g(\bar{\xi}, \bar{N}) = 1$ implies

$$2v' + (v\tau)^2 = 0, \quad \text{and} \quad \alpha = \frac{d\bar{p}}{dp}. \tag{5.2.10}$$

Thus, from (5.2.10) the function v is completely determined by

$$\frac{1}{v} = \frac{1}{2}\int (\tau(p))^2 dp + c, \quad \text{for some constant} \quad c.$$

Using $W \times N = -N$, the principal normal $\bar{W}$ of $\bar{C}$ can be expressed as

$$\bar{W} = \bar{\xi} \times \bar{N} = W - v\tau N. \tag{5.2.11}$$

Differentiating (5.2.8) by p and using (5.2.11) we get

$$\alpha\bar{\tau}(W - v\tau N) = \frac{da}{dp}N + \alpha\tau W.$$

Comparing both sides of this equation, we get

$$\bar{\tau} = \tau, \quad \frac{da}{dp} = -\alpha v\tau^2 \implies a = a_0 v^2, \ a_0 \in \mathbf{R}^+.$$

Differentiating (5.2.11) we get $a^2 = v\tau'$. Thus, the Cartan curvatures at the corresponding points coincide.

(2) For every null Cartan curve $C(p)$, define a new curve $\bar{C}(\bar{p})$ by

$$\bar{C} = C + vN, \quad \frac{1}{v} = \frac{1}{2}\int (\tau(p))^2 dp + c,$$

where c is a nonzero constant. Define a function

$$\bar{p} = a_0 \int v^2 dp, \quad a_o \text{ is a nonzero constant.}$$

Then $\bar{C}$ is a Cartan null curve, with special parameter $\bar{p}$ framed by

$$\left\{ \bar{\xi} = \frac{d}{d\bar{p}}\bar{C}, \quad \bar{N} = \frac{d\bar{p}}{dp}N, \quad \bar{W} = \bar{\xi} \times \bar{N} \right\}.$$

Clearly, C and $\bar{C}$ have common binormal directions, which completes the proof.

The following corollary (proof is immediate from the theorem 2.2) characterizes null helices of zero Cartan curvature (such helices are called *null cubics*):

Corollary 2.1. *Let $(C(p), \bar{C}(\bar{p}))$ be a pair of null Cartan curves, with their special parameters p and $\bar{p}$ respectively. Suppose their binormal directions coincide and $g(\bar{C} - C, \xi)$ is a constant. Then C is a null cubic.*

Null helices in a 4-dimensional Minkowski space $\mathbf{R}_1^4$. Bonnor [18] was the first who followed Cartan's [20] 3-dimensional case and introduced a Cartan frame for null curves in $\mathbf{R}_1^4$. He proved the fundamental existence and congruence theorems. Consider a Frenet frame $\{\xi, N, W_1, W_2\}$, with respect to a distinguished parameter p, of a framed null curve $C(p)$. We assume that each framed null curve is a Cartan null curve (to assure its uniqueness) and the following Cartan equations:

$$\begin{aligned}
\nabla_\xi \xi &= W_1, \\
\nabla_\xi N &= \sigma_1 W_1 + \sigma_2 W_2, \\
\nabla_\xi W_1 &= -\sigma_1 \xi - N, \\
\nabla_\xi W_2 &= -\sigma_2 \xi,
\end{aligned} \qquad (5.2.12)$$

having Cartan curvatures σ_1 and σ_2. Null Cartan curves which have constant σ_1 and σ_2 (not both zero) are called null helices. Their equations are [18]

$$\sigma_2 \neq 0: \quad C(p) = \sqrt{\frac{1}{a^2 + b^2}} \left(\frac{1}{a}\sinh ap, \frac{1}{a}\cosh ap, \frac{1}{b}\sin bp, \frac{1}{b}\cos bp \right),$$

$$a = \left[(\sigma_1^2 + \sigma_2^2)^{1/2} - \sigma_1 \right]^{1/2}, \quad b = \left[(\sigma_1^2 + \sigma_2^2)^{1/2} + \sigma_1 \right]^{1/2},$$

$$\sigma_2 = 0, \ \sigma_1 > 0: C(p) = b^{-2}(bp, \ \sin bp, \ 0, \ \cos bp), \quad b = \sqrt{2\sigma_1}.$$

$$\sigma_2 = 0, \ \sigma_1 < 0: C(p) = \frac{1}{a^2}(\sinh ap, \cosh bp, 0, ap), \quad a = \sqrt{-\sigma_1}.$$

Recently, Cöken and Ciftci [22] have followed the 3-dimensional notion of Bertrand curves and proved a characterization theorem for null helices in $\mathbf{R}_1^4$, which we now describe. We will follow the terminology used for the 3-dimensional case previously discussed. Accordingly, the spacelike unit vector field W_1 is called the principal normal vector field.

Definition 2.2. *Let $(C, \bar{C})$ be a pair of framed null curves in $\mathbf{R}_1^4$, with distinguished parameters p and $\bar{p}$ respectively. This pair is said to be a null*

Bertrand pair if their principal normal vector fields are linearly dependent.

The curve $\bar{C}$ is called a *Bertrand mate* of C and vice versa. A framed null curve is said to be a *null Bertrand curve* if it admits a Bertrand mate. By above definition 2.2, there exists a functional relation $\bar{p} = \bar{p}(p)$ for a null Bertrand pair $(C, \bar{C})$ such that $\bar{W}(\bar{p}(p)) = \epsilon W(p)$, $\epsilon = \pm 1$, i.e., the normal lines coincide at their corresponding points. Now we show that the null Bertrand curves have been used to characterize null helices in $\mathbf{R}_1^4$.

Lemma 2.1. *The distance between corresponding points of a Bertrand null curve and its Bertrand mate is a constant.*

Proof. Let $(C(p), \bar{C}(\bar{p}))$ be a null Bertrand pair with their corresponding Cartan frames $\{\xi, N, W_1, W_2\}$ and $\{\bar{\xi}, \bar{N}, \bar{W}_1, \bar{W}_2\}$ respectively. Then, since the normal lines coincide, we may write $\bar{C} = C + \alpha W_1$, for some function $\alpha(p) \neq 0$. Differentiating this relation with respect to p we get

$$\frac{d\bar{p}}{dp}\bar{\xi} = (1 - \alpha\tau)\xi - \alpha N + \alpha' W_1.$$

For null Bertrand curves, the condition $g(\bar{\xi}, W_1) = 0$ holds. Using this in above implies that $\alpha' = 0$ and so α is a constant. Thus, the norm $\|\bar{C} - C\|$ is also a constant which proves the lemma.

Theorem 2.3 ([22]). *A null Cartan curve in $\mathbf{R}_1^4$ is a Bertrand null curve if and only if σ_1 is nonzero constant and σ_2 is zero.*

Proof. Suppose $(C(p), \bar{C}(\bar{p}))$ is a null Bertrand pair. Then, Lemma 2.1 allows us to write

$$\bar{C} = C + \alpha W_1, \tag{5.2.13}$$

where α is the constant distance between the two curves. Differentiating (5.2.13) with respect to p we get

$$\frac{d\bar{p}}{dp}\bar{\xi} = (1 - \alpha\tau)\xi - \alpha N. \tag{5.2.14}$$

This, together with null $\bar{\xi}$, means that $\alpha(1 - \alpha\tau) = 0$, so $\sigma_1 = \frac{1}{\alpha} = $ constant. Inserting this value in (5.2.14) yields

$$\bar{\xi} = -\left(\frac{d\bar{p}}{dp}\right)^{-1}\frac{1}{\sigma_1}. \tag{5.2.15}$$

Differentiating (5.2.15) with respect to p and using Cartan frame (5.2.12) we get

$$\frac{d\bar{p}}{dp}\bar{W} = -\left(\frac{d\bar{p}}{dp}\right)^{-1}W_1 + \left(\frac{d\bar{p}}{dp}\right)^{-2}\frac{d^2\bar{p}}{dp^2}\frac{1}{\sigma_1}N - \left(\frac{d\bar{p}}{dp}\right)^{-1}\frac{\sigma_2}{\sigma_1}W_2.$$

Since $\bar{W} = \pm W$ we get $\frac{d^2 \bar{\rho}}{dp^2} = 0$ and $\sigma_2 = 0$.

Conversely, let C be a null Cartan curve with σ_1 and σ_2 as in the hypothesis. Then consider a curve $\bar{C}$ with coordinate function

$$\bar{C} = C + \frac{1}{\sigma_1} W_1, \tag{5.2.16}$$

which after differentiating with respect to p gives $(\bar{C})' = (\sigma_1)^{-1} N$. This shows that $\bar{C}$ is null. Moreover, $\bar{C}$ is a Cartan curve, since differentiating the last equation gives $(\bar{C})'' = W_1$. Thus the special parameters of C and $\bar{C}$ are the same and the normal vector of C equals to (minus) the normal vector of $\bar{C}$ at the same parameter values. These two facts with (5.2.13) implies that the normal lines of C and $\bar{C}$ coincide at their corresponding points, which completes the proof.

Note. It follows from above Theorem 2.3 that the only Bertrand null curves in $\mathbf{R}_1^4$ are null helices with $\sigma_2 = 0$.

Proposition 2.1 ([22]). *A null Cartan curve C in $\mathbf{R}_1^4$ is a 3-dimensional null helix if and only if there exists a fixed direction V such that*

$$g(\xi, V) = a, \quad g(N, V) = b, \tag{5.2.17}$$

where a, b are nonzero constants and $\{\xi, N, W_1, W_2\}$ is the Cartan frame of $\acute{C}(p)$.

Proof. Suppose V is a fixed direction satisfying (5.2.17). Taking derivatives of both these equations with respect to the p, we get $\sigma_1 = -\frac{b}{a}$ and $\sigma_2 = 0$.

Conversely, assume that C is a null helix and $\sigma_2 = 0$. Then, if we set $V = \sigma_1 \xi - N$, it is easy to see that V is a fixed direction and the two relations of (5.2.17) holds, which completes the proof.

Using the Cartan equations (5.1.1) and definition 1.2, Ferrández et al. [35] have found a complete classification of null helices in three low dimensional spacetimes of constant curvature, namely, the 5-dimensional Minkowski space $\mathbf{R}_1^5$, the de Sitter space $\mathbf{S}_1^4$ and the anti-de Sitter space $\mathbf{H}_1^4$, by means of thirteen specific problems. They concluded that in $\mathbf{R}_1^5$ there are three different families of helices, in $\mathbf{S}_1^4$ there is only one type of helices and in $\mathbf{H}_1^4$ there are up to nine distinct types of helices. For the benefit of readers, we reproduce all problems taken from their paper [35].

Null helices in $\mathbf{R}_1^5$. Using the Cartan equations (5.1.1) and (5.1.13) for $m = 3$, a null helix $C(p)$ in $\mathbf{R}_1^5$ will satisfy the following differential equation

$$C^{(6)} + (2\sigma_1 + \sigma_3^2) C^{(4)} - (\sigma_2^2 - 2\Sigma_1 \Sigma_3^2) C'' = 0,$$

where σ_1, σ_2 and σ_3 are its surviving Cartan curvatures.

Problem 1 (Type 1). Let a, b and c be three nonzero constants and $C_{(a,\,b,\,c)}(p)$, $\frac{1}{b^2} < c^2 < \frac{1}{a^2}$, be a null curve of $\mathbf{R}_1^5$ defined by

$$C_{(a,\,b,\,c)}(p) = (cp,\ A\sin ap,\ A\cos ap,\ B\sin bp,\ B\cos bp),$$

where $A = \frac{1}{a}\sqrt{\frac{c^2 b^2 - 1}{b^2 - a^2}}$ and $B = \frac{1}{b}\sqrt{\frac{1 - c^2 a^2}{b^2 - a^2}}$.

Then it is easy to see that $C(p)$ is a helix whose Cartan curvatures are:

$$\sigma_1 = \frac{1}{2}(b^2 + a^2)(1 - b^2 c^2),$$
$$(\sigma_2)^2 = -a^2 b^2 (a^2 c^2 - 1)(b^2 c^2 - 1),$$
$$(\sigma_3)^2 = (abc)^2.$$

Problem 2 (Type 2). Let a, b and c be three nonzero constants and $C_{(a,\ b,\ c)}(p)$, $0 < a^2 c^2 < 1$, be a null curve of $\mathbf{R}_1^5$ defined by

$$C_{(a,\ b,\ c)}(p) = (\ A \sinh ap, \ A \cosh ap, \ B \sin bp, \ B \cos bp, \ cp\ ),$$

where $A = \frac{1}{a}\sqrt{\frac{1 + c^2 b^2}{a^2 + b^2}}$ and $B = \frac{1}{b}\sqrt{\frac{1 - c^2 a^2}{a^2 + b^2}}$.

Then $C(p)$ is a helix whose Cartan curvatures are:

$$\sigma_1 = \frac{1}{2}(b^2 - a^2)(1 + b^2 c^2),$$
$$(\sigma_2)^2 = -a^2 b^2 (a^2 c^2 - 1)(b^2 c^2 + 1),$$
$$(\sigma_3)^2 = (abc)^2.$$

Problem 3 (Type 3). Let a and b be two nonzero constants and $C_{(a,\ b)}(p)$, $0 < 2b^2 < 1$, be a null curve of $\mathbf{R}_1^5$ defined by

$$C_{(a,\ b)}(p) = \left(\frac{3}{2}\left(\frac{1 - 2b^2}{(ab)^2}\right)p + \frac{1}{6}b^2 p^3 + p^3, \ \frac{b^2}{\sqrt{2}}p^2, \ \frac{3}{2}\left(\frac{1 - 2b^2}{(ab)^2}\right)p,\right.$$
$$\left.-\frac{1}{6}b^2 p^3 + p^3, \ \left(\frac{\sqrt{1 - 2b^2}}{a^2}\right)\sin ap, \ \left(\frac{\sqrt{1 - 2b^2}}{a^2}\right)\cos ap\right).$$

Then $C(p)$ is a helix whose Cartan curvatures are:

$$\sigma_1 = \frac{1}{2}a^2(1 - 2b^2),$$
$$(\sigma_2)^2 = 2 a^4 b^2 (1 - 2b^2),$$
$$(\sigma_3)^2 = 2 a^2 b^2.$$

Following is the classification theorem of the family of null helices in $\mathbf{R}_1^5$.

Theorem 2.4 ([35]). *Let $C(p)$ be a null curve immersed in $\mathbf{R}_1^5$. Then C is a helix if and only if it is congruent to a helix of type 1 or type 2 or type 3.*

Proof. Suppose σ_1, $\sigma_2 \neq 0$ and $\sigma_3 \neq 0$ are the constant curvatures of C. By the congruence theorem 2.3. it suffices to find a helix (of one of above type) with these curvatures. We consider three cases as follows:

Case 1. Set $\alpha = 2\sigma_1 + \sigma_3^2$, $\beta = \sqrt{(2\sigma_1 - \sigma_3^2)^2 + 4\sigma_2^2}$ and $\omega = \frac{\sigma_3^2}{2\sigma_1\sigma_3^2 - \sigma_2^2}$. Assume that $\alpha > \beta$. Let the helix $C_{(a,\,b,\,c)}$ of type 1 be determined by

$$a^2 = \frac{1}{2}(\alpha - \beta), \quad b^2 = \frac{1}{2}(\alpha + \beta), \quad c^2 = \omega.$$

Then, a straightforward computation shows that $\frac{1}{b^2} < c^2 < \frac{1}{a^2}$ and that the Cartan curvatures of $C_{(a,\,b,\,c)}$ are σ_1, σ_2 and σ_3.

Case 2. Suppose $\alpha < \beta$. Let the helix $C_{(a,\,b,\,c)}$ of type 2 be determined by

$$a^2 = \frac{1}{2}(\beta - \alpha), \quad b^2 = \frac{1}{2}(\alpha + \beta), \quad c^2 = \omega.$$

In this case $0 < a^2 c^2 < 1$ and the Cartan curvatures of C are σ_1, σ_2 and σ_3.

Case 3. Finally, if $\alpha = \beta$, then we take the helix $C_{(a,\,b)}$ determined by

$$a^2 = \frac{\sigma_2^2 + \sigma_3^4}{\sigma_3^2}, \quad b^2 = \frac{\sigma_3^4}{2(\sigma_2^2 + \sigma_3^4)}, \quad 0 < 2b^2 < 1$$

with Cartan curvatures σ_1, σ_2 and σ_3. This completes the proof.

Null helices in De-Sitter space $\mathbf{S}_1^4 \subset \mathbf{R}_1^5$. Using the Cartan equations (5.1.11), for $m = 3$, one can show that a null helix $C(p)$ in $\mathbf{S}_1^4$ satisfies the following differential equation

$$C^{(5)} + 2\,\sigma_1 C^{(3)} - (1 + \sigma_2^2)C' = 0,$$

whose general solution is given by

$$C_{(a,\,b)}\,p = A_1 \sinh ap + A_2 \cosh ap + A_3 \sin bp + A_4 \cos bp + A_5,$$

where A_1, A_2, A_3, A_4 and A_5 are constant vectors in $\mathbf{R}_1^5$, σ_1 and σ_2 are surviving curvatures and a, b are nonzero constants.

Problem 4. Let C be a null curve in $\mathbf{S}_1^4 \subset \mathbf{R}_1^5$ defined by $C_{(a,\,b)}(p) =$

$$\sqrt{\frac{1}{a^2 + b^2}}\left(\frac{1}{a}\sinh ap, \ \frac{1}{a}\cosh ap, \ \frac{1}{b}\sin bp, \ \frac{1}{b}\cos bp, \ \sqrt{\frac{a^4 - 1}{a^2} + \frac{b^4 - 1}{b^2}}\right),$$

where a and b are nonzero constants such that $a^2 b^2 > 1$. Then, we obtain

$$\sigma = \frac{b^2 - a^2}{2}, \qquad \sigma_2^2 = a^2 b^2 - 1, \tag{5.2.18}$$

which shows that the curve $C_{(a,\,b)}(p)$ is a null helix.

Following is the classification theorem of the family of null helices in $\mathbf{S}_1^4$.

Theorem 2.5 ([35]). *Let $C(p)$ be a null curve immersed in $\mathbf{S}_1^4$. Then C is a helix if it is congruent to one of the family described in problem 4.*

Proof. Let $a = \sqrt{\sigma_1 + \sqrt{\sigma_1^2 + \sigma_2^2 + 1}}$ and $b = \sqrt{-\sigma_1 + \sqrt{\sigma_1^2 + \sigma_2^2 + 1}}$ be two constants. Then from (5.1.11) it is easy to see that the curve C, of problem 4 has Cartan curvatures σ_1 and σ_2. Thus, theorem 2.4 follows from the congruence theorem 1.5 of previous section.

Null helices in anti-De Sitter space $\mathbf{H}_1^4 \subset \mathbf{R}_1^5$. In a similar way, using the Cartan equations (5.1.11), for $m = 3$ and $-C$ replaced by C, one can show that a null helix $C(p)$ in $\mathbf{H}_1^4$ satisfies the following differential equation

$$C^{(5)} + 2\,\sigma_1 C^{(3)} + (1 - \sigma_2^2)C' = 0.$$

Problem 5 (Type A_1). Let $C_a(p), 0 < a^2 < 1$, be a null curve in $\mathbf{H}_1^4$ given by

$$C_a(p) \;=\; \left(\frac{p}{2a} \cosh ap, \;\; \frac{1}{a^2}(\cosh ap - \frac{1}{2}\sinh ap), \right.$$

$$\left. \frac{1}{a^2}(\sinh ap - \frac{1}{2}ap\cosh ap), \;\; \frac{p}{2a}\sinh ap, \;\; \frac{\sqrt{(1-a^4)}}{a^2} \right).$$

Then C_a is a helix with Cartan curvatures $\sigma_1 = -a^2$ and $\sigma_2^2 = 1 - a^4$.

Problem 6 (Type A_2). Let $C_b, \; 0 < b^2 < 1$ be a null curve in $\mathbf{H}_1^4$ given by

$$C_b(p) \;=\; \left(\frac{1}{b^2}(\sin bp - \frac{1}{2}bp\cos bp), \;\; \frac{1}{b^2}(\cos bp + \frac{1}{2}\sin bp), \right.$$

$$\left. -\frac{p}{2b}\cos bp, \;\; \frac{p}{2b}\sin bp, \;\; \frac{\sqrt{(1-b^4)}}{b^2} \right).$$

Then C_b is a helix with Cartan curvatures $\sigma_1 = b^2$ and $\sigma_2^2 = 1 - b^4$.

Problem 7 (Type A_3). Let $C_a(p), \; a^2 = 1$, be a null curve in $\mathbf{H}_1^4$ given by

$$C_a(p) = \left(1 - \frac{p^4}{24}, \;\; \frac{a(p^3 + p)}{2\sqrt{3}}, \;\; \frac{p^4}{24}, \;\; \frac{a(p^3 - p)}{2\sqrt{3}}, \;\; \frac{p^2}{2} \right).$$

Then C_a is a helix with Cartan curvatures $\sigma_1 = 0$ and $\sigma_2^2 = 1$.

Problem 8 (Type B_1). Let $C_{(a,\, b)}(p), \; 0 < a^2 < b^2$ and $a^2 b^2 < 1$, be a null curve in $\mathbf{H}_1^4$ defined by

$$C_{(a,\, b)}(p) = \sqrt{\frac{1}{b^2 - a^2}} \left(\frac{1}{a}\sin ap, \;\; \frac{1}{a}\cos ap, \;\; \frac{1}{b}\sin bp, \;\; \frac{1}{b}\cos bp, \;\; c \right),$$

where $c = \sqrt{\frac{1+a^4}{a^2} - \frac{1+b^4}{b^2}}$. Then $C_{(a,\, b)}$ is a helix with Cartan curvatures

$$\sigma_1 = \frac{1}{2}(a^2 + b^2), \quad \sigma_2^2 = 1 - a^2 b^2.$$

Problem 9 (Type B_2). Let $C_{(a,\, b)}(p), \; 0 < b^2 < a^2$ and $a^2 b^2 < 1$, be a null curve in $\mathbf{H}_1^4$ defined by

$$C_{(a,\, b)}(p) = \sqrt{\frac{1}{a^2 - b^2}} \left(\frac{1}{a}\sinh ap, \;\; \frac{1}{b}\cosh bp, \;\; \frac{1}{b}\sinh bp, \;\; \frac{1}{a}\cosh ap, \;\; d \right),$$

where $d = \sqrt{\frac{1+b^4}{b^2} - \frac{1+a^4}{a^2}}$. Then $C_{(a,\,b)}$ is a helix with Cartan curvatures

$$\sigma_1 = -\frac{1}{2}(a^2 + b^2), \quad \sigma_2^2 = 1 - a^2 b^2.$$

Problem 10 (Type B_3). Let $C_a(p)$, $a \neq 0$, be a null curve in $\mathbf{H}_1^4$ given by

$$C_a(p) = \left(\frac{2 + 2a^4 + a^2 p^2}{2a^2 \sqrt{1 + a^4}}, \ \frac{p}{a}, \ \frac{p^2}{2\sqrt{1 + a^4}}, \ \frac{1}{a^2} \sin ap, \ \frac{1}{a^2} \cos ap \right).$$

Then C_a is a helix with Cartan curvatures $\sigma_1 = \frac{a^2}{2}$ and $\sigma_2^2 = 1$.

Problem 11 (Type B_4). Let $C_b(p)$, $b \neq 0$, be a null curve in $\mathbf{H}_1^4$ given by

$$C_b(p) = \left(\frac{2 + 2b^4 + b^2 p^2}{2b^2 \sqrt{1 + b^4}}, \ \frac{1}{b^2} \sinh bp, \ \frac{1}{b^2} \cosh bp, \ \frac{p^2}{2\sqrt{1 + b^4}}, \ \frac{p}{b} \right).$$

Then C_b is a helix with Cartan curvatures $\sigma_1 = -\frac{b^2}{2}$ and $\sigma_2^2 = 1$.

Problem 12 (Type B_5). Let $C_{(a,\,b)}(p)$, $ab \neq 0$, be a null curve in $\mathbf{H}_1^4$ given by

$$C_{(a,\,b)}(p) = \sqrt{\frac{1}{a^2 + b^2}} \left(c, \ \frac{1}{a} \sinh ap, \ \frac{1}{a} \cosh ap, \ \frac{1}{b} \sin bp, \ \frac{1}{b} \cos bp, \ c \right),$$

where $c = \sqrt{\frac{1+a^4}{a^2} + \frac{1+b^4}{b^2}}$. Then $C_{(a,\,b)}$ is a helix with Cartan curvatures

$$\sigma_1 = \frac{1}{2}(b^2 - a^2), \quad \sigma_2^2 = 1 + a^2 b^2.$$

Problem 13 (Type D). Let $C_{(a,\,b)}(p)$ be a null curve in $\mathbf{H}_1^4$ given by

$$C_{(a,\,b)}(p) = \frac{1}{2ab(c)} (2ab \cosh ap \sin bp + d \sinh ap \cos ap,$$

$$-2apb \cosh ap \cos bp, \quad \sinh ap \sin bp,$$

$$c \sinh ap \cos bp, \quad c \sinh ap \sin bp, \quad 2ab\sqrt{1 - c^2}).$$

where $a^2 + b^2 < 1$ and set $c = a^2 + b^2$ and $d = a^2 - b^2$. Then $C_{(a,\,b)}$ is a helix with Cartan curvatures
$$\sigma_1 = b^2 - a^2, \quad \sigma_2^2 = 1 - (a^2 + b^2).$$

Following is the classification theorem of the family of null helices in $\mathbf{H}_1^4$.

Theorem 2.6 ([35]). *Let $C(p)$ be a null curve immersed in $\mathbf{H}_1^4$. Then C is a helix if and only if it is congruent to one helix of the family of problems 5-13.*

Proof. The proof is similar as in the previous cases, with the following table for all the 9 possibilities:

Type A_1 $\sigma_1^2 + \sigma_2^2 = 1$, $\sigma_1 < 0$, $0 < \sigma_2^2 < 1$, $a^2 = -\sigma_1$

Type A_2 $\quad$ $\sigma_1^2 + \sigma_2^2 = 1,\ \sigma_1 > 0,\ 0 < \sigma_2^2 < 1,\ b^2 = \sigma_1$

Type A_3 $\quad$ $\sigma_1 = 0,\ \sigma_2^2 = 1$

Type B_1 $\quad$ $\sigma_1^2 + \sigma_2^2 > 1,\ \sigma_1 > 0,\ 0 < \sigma_2^2 < 1,\ a^2 = \sigma_1 - c,\ b^2 = \sigma_1 + c$

Type B_2 $\quad$ $\sigma_1^2 + \sigma_2^2 > 1,\ \sigma_1 < 0,\ 0 < \sigma_2^2 < 1,\ a^2 = -\sigma_1 - c,\ b^2 = -\sigma_1 - c$

Type B_3 $\quad$ $\sigma_1^2 + \sigma_2^2 > 1,\ \sigma_1 > 0,\ \sigma_2^2 = 1,\ b^2 = 2\sigma_1$

Type B_4 $\quad$ $\sigma_1^2 + \sigma_2^2 > 1,\ \sigma_1 > 0,\ \sigma_2^2 = 1,\ a^2 = -2\sigma_1$

Type B_5 $\quad$ $\sigma_2^2 = 1,\ a^2 = -\sigma_1 + c,\ b^2 = \sigma_1 + c$

Type D $\quad$ $\sigma_1^2 + \sigma_2^2 < 1,\ a^2 = \frac{1}{2}(-\sigma_1 + \sqrt{1 - \sigma_2^2}),\ b^2 = \frac{1}{2}(\sigma_1 + \sqrt{1 - \sigma_2^2})$

where we set $c = \sqrt{\sigma_1^2 + \sigma_2^2 - 1}$.

5.3 Brief notes and research problems

(a) Pseudo-spherical null curves in $\mathbf{R}_1^4$. By a pseudo-spherical null curve we mean a null curve that completely lies on a pseudo-sphere $\mathcal{S}_1^3(r)$, say of radius $r > 0$ and of center A given by $\mathcal{S}_1^3(r)$ [82]

$$\mathcal{S}_1^3(r) = \left\{ X \in \mathbf{R}_1^4 \ :\ g(X - A,\ X - A) = r^2 \right\}.$$

Just as the Euclidean case, we define the *osculating pseudo-sphere* as follows:

Definition 3.1. *Let C be a null Cartan curve in $\mathbf{R}_1^4$. Then the pseudo-sphere having 5 point contact with C is called the osculating pseudo-sphere of C.*

We assume that $C(p)$ be a null Cartan curve in $\mathbf{R}_1^4$ with respect to a special parameter p having Cartan equations (5.2.12) and Cartan curvatures $\sigma_1,\ \sigma_2$.

Lemma ([22]). *Let $C(p)$ be a null Cartan curve in $\mathbf{R}_1^4$. The center point of the osculating pseudo-sphere at a point $C(p_0)$ is*

$$A(p_0) = C(p_0) + \frac{1}{\sigma_2} W_2(p_0).$$

Proof. For any point $C(p_0)$ the position vector $A(p_0) - C(p_0)$ can be written as a linear combination of the Cartan frame $F = \{\xi,\ N,\ W_1,\ W_2\}$ as follows:

$$A(p_0) - C(p_0) = a_1 \xi + a_2 N + a_3 W_1 + a_4 W_2,$$

where $a_i(p_0)$ for $1 \le i \le 4$ are differentiable functions. Consider the function

$$f(p) = g\left(A(p) - C(p),\ A(p) - C(p)\right) - r^2,$$

where r is the radius of the osculating-sphere. Then, all the derivatives of f at the point p_0 vanish due to the definition of the osculating pseudo-sphere at p_0. Now a straightforward computation leads to

$$a_1(p_0) = a_2(p_0) = a_3(p_0) = 0 \quad \text{and} \quad a_4(p_0) = (\sigma_2(p_0))^{-1}.$$

Thus the result follows from above relations, which completes the proof.

Theorem ([22]). Let $C(p)$ be a null Cartan curve in $\mathbf{R}_1^4$. Then C is a pseudo-spherical curve if and only if σ_2 is a nonzero constant.

Proof. Suppose C lies on $\mathcal{S}_1^3(r)$. Then the osculating pseudo-spheres at all points of C are exactly $\mathbf{S}_1^3(r)$ and so r and σ_2 are non-zero constants.

Conversely, if σ_2 is a nonzero constant, then, all the osculating pseudo-spheres have the same radius. Using above lemma, we consider the function

$$A(p) = C(p) + \frac{1}{\sigma_2} W_2(p).$$

giving the central point of the osculating pseudo-sphere. Since the derivative of this function is zero everywhere, this function is constant. Consequently, the curve C lies on $\mathbf{S}_1^3(r)$ as the equation $g\left(A(p) - C(p), \ A(p) - C(p)\right) = r^2$ holds for all p, which completes the proof.

Note. Null curves with nonzero constant curvature σ_2 do lie on a pseudo-sphere but they may not be its geodesics, since null geodesics of pseudo-spheres are null straight lines of $\mathbf{R}_1^4$. Null helices with nonzero σ_2 are pseudo-spherical, but the other null helices and the null cubics are not. In generally, there is no 3-dimensional null curve that lies on a pseudo-sphere in Minkowski space (see [95]). Similarly, one can show that the hyperbolic space

$$\mathbf{H}^3(r) = \{X \in \mathbf{R}_1^4 \ : \ g(X, \ X) = -r^2\}$$

may not contain any null curve.

(b) Null generalized helices in $\mathbf{R}_1^{m+2}$. Recall, Lancret (see details in [106]) stated that a curve in a 3-dimensional Euclidean space is a *generalized helix* if and only if the ratio of curvature to torsion is constant. Also see the works of Hayden [54, 55] on generalized helices in odd n-dimensional Euclidean spaces. The physical interest of non-null generalized helices comes from an interplay between geometry and integrable Hamiltonian systems (see [79, 80]). Related to null curves, in chapter 6 we have presented latest works of Barros-Ferrández-Lucus-Meroño [8] and Ferrández-Giménez-Lucas [40] on physical applications of null generalized helices in $\mathbf{R}_1^{m+2}$. In this note, we use theorem 1.1 of section 1, the Cartan equations (5.1.1), its Cartan frame $\{\xi, N, W_1, \ldots, W_m\}$ and reproduce the latest geometric results (for some details on the proofs see [37]) as follows:

Definition 3.2 ([37]). *A null Cartan curve $C(p)$ in $\mathbf{R}_1^{m+2}(m = 2q + 1)$ is a generalized helix if there exists a constant vector $v \neq 0$ such that all the products $g(\xi(p), v) \neq 0$, $g(N(p), v)$ and $g(W_{2i+1}(p), v) \neq 0, 1 \leq i \leq q - 1$ are constants.*

The vector v can be spacelike, timelike or lightlike and the direction of v will be called the axis of v. Without any loss of generality we assume that a non-null v is unitary. A straightforward computation from the Cartan equations (5.1.1) provides the following result called *the Lancret theorem for null curves*:

Theorem 3.1. *Let $C(p)$ be a null Cartan curve in $\mathbf{R}_1^{m+2}(m = 2q + 1)$. Then, the following statements are equivalent:*

(i) *there exist constants $\{a, a_1, \ldots, a_q\}(a_i \neq 0)$ such that*

$$\sigma_1(p) = a, \quad \sigma_{2i+1}(p) = a_i\sigma_{2i}, \quad 1 \leq i \leq q.$$

(ii) *C is a generalized helix.*

In particular, for a generalized helix $C(p)$ in $\mathbf{R}_1^3$ the condition $g(C'(p), v)$ is constant means that the tangent indicatrix lies in a plane, or in other words, there exists a nonzero vector v in $\mathbf{R}_1^3$ which is orthogonal to the acceleration vector field of C. The following characterization result is immediate from Theorem 4.1.

Proposition 3.1. *A Cartan null curve in $\mathbf{R}_1^3$ is a generalized helix if and only if it is a Cartan helix.*

Following are exactly three types of null helices, up to congruence in $\mathbf{R}_1^3$.

$$C(p) = \left(\frac{1}{a}, \frac{1}{a^2}\sin ap, \frac{1}{a^2}\cos ap\right); \quad \sigma_1 = \frac{a^2}{2} > 0, \quad \text{timelike axis}:$$

$$C(p) = \left(\frac{1}{b^2}\sinh bp, \frac{1}{b^2}\cosh bp, -\frac{1}{b}\right); \quad \sigma_1 = -\frac{b^2}{2} < 0, \quad \text{spacelike axis}:$$

$$C(p) = \left(\frac{p^3}{4} + \frac{p}{3}, \frac{p^2}{2}, \frac{p^3}{4} - \frac{p}{3}\right); \quad \sigma_1 = 0, \quad \text{null axis}.$$

Now we deal with the 5-dimensional case. As per definition 4.1, a null Cartan curve $C(p)$ in $\mathbf{R}_1^5$ is a generalized helix if $g(\xi(p), v)$ and $g(N, v)$ are constants. We first consider the case with non-null axis. Set $g(v, v) = \epsilon = \pm 1$.

Let Σ denote the hyperplane orthogonal to the non-null v, P the projection map onto Σ and $\bar{\alpha} = P(C)$ the non-null projection curve. Then one can write

$$C(p) = \bar{\alpha}(p) + \bar{\mu}v, \quad \bar{\mu} \text{ is a non-constant differentiable function.} \tag{5.3.1}$$

Clearly, the $\bar{\alpha}$ and v have same causal structure. Let $\alpha(s)$ represent the arc-length parameterization of $\bar{\alpha}$ with curvatures k_1, k_2 and k_3. Following is immediate:

Lemma 3.1. *Let $C(p)$ be a null Cartan curve in $\mathbf{R}_1^5$ and $\alpha(s)$ be the orthogonal projection of C onto a non-degenerate hyperplane Σ, where p and s are the special and the arc-length parameters of C and α respectively. Then p and s are linearly related if and only if k_1 is constant. Moreover, in this case, σ_1 is constant if and only if k_2 is constant.*

Using above lemma, we have the following characterization result:

Theorem 3.2. *A null Cartan curve C in $\mathbf{R}_1^5$ is a generalized helix with non-null axis if and only if it is a null geodesic of a Lorentzian cylinder constructed on a non-degenerate curve in $\mathbf{R}^4$ or $\mathbf{R}_1^4$ with constant curvature and torsion.*

Now we assume null axis for which $v \in \Sigma$ is null. Thus, Σ is a lightlike orthogonal hyperplane of $\mathbf{R}_1^5$ and we have the following decomposition:

$$T_p\mathbf{R}_1^5 = Span\{v\} \perp S(T_x\Sigma) \oplus tr(T_x\Sigma), \quad \forall x \in \Sigma$$

where $\perp$ denotes orthogonal direct sum, $S(T_x\Sigma)$ is a chosen screen of Σ and $tr(T_x\Sigma)$ is a screen transversal part of Σ at every x (details may seen in [28, pages 77-79]). Since $g(\xi, v) = a$ is non-zero constant, we set $\bar{\xi} = \frac{1}{a}\xi$ a transversal section along C which satisfies $g(\bar{\xi}, v) = 1$. Then the projection $\bar{\alpha}$ of C with respect to $\bar{\xi}$ is given by $\bar{\alpha}(p) = C(p) - g(C(p), v)\bar{\xi}(p)$. Now $g(C'(p), v) = a$ implies that $g(C(p), v) = a(p + b)$ for some constant b. Thus, for this case, the relation (5.3.1) is replaced to

$$C(p) = \bar{\alpha}(p) + (p + b)\xi(p). \tag{5.3.2}$$

Let $\{T, V_1, V_2\}$ be an orthonormal basis of the spacelike screen $S(T_x\Sigma)$. Using the Gram-Schmidt method for the non-degenerate case (see details in [37, page 8245]) we have the following Frenet equations of a spacelike curve in Σ.

$$\begin{aligned}
T' &= \rho_1 V_1, \\
V_1' &= -\rho_1 T + \rho_2 V_2, \\
V_2' &= -\rho_2 V_1 + \rho_3 v, \\
v' &= 0.
\end{aligned} \tag{5.3.3}$$

The following two theorems are straightforward using above material for the case of null axis (details may be seen in [37, page 8248-8250]).

Theorem 3.3. *If a null Cartan curve C in $\mathbf{R}_1^5$ is a generalized helix with null axis v and Σ its lightlike orthogonal hyperplane, then, the curvatures $\{\rho_1, \rho_2, \rho_3\}$ of the projected spacelike projected curve $\bar{\alpha}$ satisfying*

$$\rho_1 = \sqrt{\frac{\Sigma_1(p)}{p}}, \quad \rho_2 = \frac{\Sigma_2(p)}{2\sqrt{\Sigma_1(p)p}}, \quad \rho_3 = \frac{r\Sigma_2(p)}{\sqrt{2p}}$$

for some constant r. Conversely, if $\bar{\alpha}$ is a spacelike curve in a lightlike hyperplane Σ of $\mathbf{R}_1^5$ whose curvatures satisfy (5.3.3), then, there exists a null generalized helix

C in $\mathbf{R}_1^5$ *whose projection onto* Σ *is just exactly* $\bar{\alpha}$.

Theorem 3.4. *A null Cartan curve C in* $\mathbf{R}_1^5$ *is a generalized helix with null axis* v *if and only if it is a geodesic of a Lorentzian ruled surface whose directrix is a timelike generalized helix in* $\mathbf{R}_1^5$ *(with null axis and constant first curvature) and whose rulings have the direction of the axis.*

(c) *s***-degenerate curves in Lorentzian space forms.** Let E be a real vector space with a symmetric bilinear mapping $g : E \times E \to R$. We say that g is degenerate on E if there exists a vector $\xi \neq 0$, of E, such that

$$g(\xi, u) = 0, \quad \forall u \in E,$$

otherwise g is called non-degenerate. The radical or the null space of E, with respect to g, is a subspace *Rad E* of E defined by

$$Rad\, E = \{\xi \in E \,;\, g(\xi, u) = 0\,,\ u \in E\}.$$

The dimension of *Rad E*, denoted by *nullity E*, is called the nullity degree of g. Clearly, g is degenerate or non-degenerate on E if and only if *nullity E* > 0 or *nullity E* $= 0$, respectively. Let $(M_1^n, \bar{g})$ be an oriented Lorentzian manifold and C be a differentiable curve in M_1^n. For any vector field V along C, let V' be the covariant derivative of V along C. Suppose $E_i(t) = Span\{C'(t), C''(t), \ldots, C^i(t)\}$, where $t \in I$ and $i = 1, 2, \ldots, n$. Let d be the number defined by $d = \max\{i : E_i(t)i, \forall t\}$.

Definition 3.3 ([39]). *The curve C in* M_1^n *is said to be an s-degenerate curve if for all* $1 \leq i \leq d$, *dim Rad* $E_i(t)$ *is constant for all t, and there exists s,* $0 < s < d$, *such that Rad* $E_s \neq \{0\}$ *and Rad* $E_s = \{0\}$ *for all* $j < s$.

Note. Any 1-degenerate curve is the null curve discussed so far in this book. On the other hand, any s-degenerate curve ($s > 1$) is a spacelike curve which lives in a lightlike hypersurface of a Lorentzian space. Since the discussion on non-null curves is out of the scope of this book, in order to highlight the point that there are a variety curves (other than the null curves) in a null hypersurface, we refer [39] in which they have characterized the s-degenerate helices (i.e. s-degenerate curves with Cartan curvatures) in Lorentzian space forms and have obtained a complete classification of them in dimension four. Here we state their results and list some examples (see proofs in [39]) as follows:

Theorem 1. *Let* $\sigma_1, \ldots, \sigma_m : [-\varepsilon, \varepsilon] \to \mathbf{R}$ *be everywhere differentiable functions with* $\sigma_i > 0$ *for* $i \neq s, m$. *Let x be a point in a Lorentzian space form* $M_1^{m+2}(c)$ *and* $\{W_1^0, \ldots, W_{s-1}^0, \xi^0, W_s^0, N^0, W_{s+1}^0, \ldots, W_m^0\}$ *a positively oriented quasi-orthonormal basis of* $T_x M_1^{m+2}(c)$. *Then there exists a unique s-degenerate Cartan curve C in* $M_1^{m+2}(c)$, *with* $C(0) = x$, *whose null Cartan frame satisfies*

$$\xi(0) = \xi^0, \quad N(0) = N^0, \quad W_i(0) = W_i^0, \quad i \in \{1, \ldots, m\}.$$

Theorem 2. *If two s-degenerate Cartan curves C and $\bar{C}$ in $M_1^{m+2}(c)$ have same curvatures $\{\sigma_1, \ldots, \sigma_m\}$, where $\sigma_i : [-\varepsilon, \varepsilon] \to \mathbf{R}$ are differentiable functions, then there exists a Lorentzian transformation of $M_1^{m+2}(c)$ which maps C into $\bar{C}$.*

Example 1. Let $C_{(a, b)}$ be a null curve in $\mathbf{R}_1^4$ defined by

$$C_{(a, b)}(p) = \frac{1}{\sqrt{a^2 + b^2}} \left(\frac{b}{a} \cosh ap, \ \frac{b}{a} \sinh ap, \ \frac{a}{b} \sin bp, \ \frac{a}{b} \cos bp \right)$$

with $ab > 0$. Then $C_{(a, b)}$ is a helix with curvatures

$$\sigma_1 = ab, \qquad \sigma_2 = \frac{b^2 - a^2}{2ab}.$$

Theorem 3. *Let C be a 2-degenerate Cartan curve fully immersed in $\mathbf{R}_1^4$. Then C is a helix if and only if it is congruent to a helix of Example 1.*

There are exactly three types of helices in the De Sitter spacetime $\mathbf{S}_1^4 \subset \mathbf{R}_1^5$ as well as three types of helices in the anti-De Sitter spacetime $\mathbf{H}_1^4 \subset \mathbf{R}_1^5$ as follows:

Example 2 (type 1). Let $C_{(a, b)}$ be a null curve in $\mathbf{S}_1^4$ defined by

$$C_{(a, b)}(p) = (\ A_1, \ A_2 \sin ap, \ A_2 \cos ap, \ A_3 \sin bp, \ A_3 \cos bp\ ),$$

where $A_1 = \sqrt{\frac{(a^2-1)(1-b^2)}{a^2 b^2}}$, $A_2 = \frac{1}{a}\sqrt{\frac{1-b^2}{a^2-b^2}}$, $A_3 = \frac{1}{b}\sqrt{\frac{a^2-1}{a^2-b^2}}$. Then $C_{(a, b)}$ is a helix with curvatures

$$\sigma_1 = \sqrt{(a^2 - 1)(1 - b^2)}, \qquad \sigma_2 = \frac{a^2 + b^2 - 1}{2\sigma_1}.$$

Example 3 (type 2). Let $C_{(a, b)}$ be a null curve in $\mathbf{S}_1^4$ defined by

$$C_{(a, b)}(p) = \left(A_1 \cosh ap, \ A_1 \sinh ap, \ A_2 \sin bp, \ A_2 \cos bp, \ \frac{\sqrt{(a^2 + 1)(b^2 - 1)}}{ab} \right),$$

where $a \neq 0$ and $b^2 > 1$ and $A_1 = \frac{1}{a}\sqrt{\frac{b^2-1}{a^2+b^2}}$, $A_2 = \frac{1}{b}\sqrt{\frac{a^2+1}{a^2+b^2}}$. Then, $C_{(a, b)}$ is a helix with curvatures

$$\sigma_1 = \sqrt{(a^2 + 1)(b^2 - 1)}, \quad \sigma_2 = \frac{b^2 - a^2 - 1}{2\sigma_1}.$$

Example 4 (type 3). Let C_a, $a^2 > 1$, be a null curve in $\mathbf{S}_1^4$ defined by

$$C_a(p) = \left(A_1 p^2, \ A_2 p, \ A_3 p^2, \ \frac{1}{a^2} \sin ap, \ \frac{1}{a^2} \cos ap \right),$$

where $A_1 = \frac{\sqrt{a^4-1}}{2(a^2-1)}$, $A_2 = \sqrt{\frac{a^2-1}{a^2}}$, $A_3 = \frac{\sqrt{a^4-1}}{a^2} - \frac{\sqrt{a^4-1}}{2(a^2+1)}$. Then, C_a is a helix with curvatures

$$\sigma_1 = \sqrt{a^2 - 1}, \qquad \sigma_2 = \frac{1}{2}\sigma_1.$$

Theorem 4. *Let C be a 2-degenerate Cartan curve fully immersed in $\mathbf{S}_1^4$. Then C is a helix if and only if it is congruent to one in the Examples 2-4.*

Example 5 (type 1). Let $C_{(a,\,b)}$ be a null curve in $\mathbf{H}_1^4$ defined by

$$C_{(a,\,b)}(p) = (A_1 \cosh ap, \quad A_2 \cosh bp, \quad A_1 \sinh ap,$$

$$A_2 \sinh bp, \quad - \frac{\sqrt{(a^2+1)(b^2-1)}}{ab}),$$

where $A_1 = \frac{1}{a}\sqrt{\frac{1-b^2}{a^2-b^2}}$, $A_2 = \frac{1}{b}\sqrt{\frac{a^2-1}{a^2-b^2}}$ and $0 < b^2 < a^2$. Then, $C_{(a,\,b)}$ is a helix with curvatures

$$\sigma_1 = \sqrt{(a^2-1)(1-b^2)}, \qquad \sigma_2 = -\frac{a^2+b^2-1}{2\sigma_1}.$$

Example 6 (type 2). Let $C_{(a,\,b)}$, $a^2 > 1$, be a null curve in $\mathbf{H}_1^4$ defined by

$$C_{(a,\,b)}(p) = (A_1, \quad A_2 \cosh ap, \quad A_2 \sinh ap, \quad A_3 \sin bp, \quad A_3 \cos bp), \quad b \neq 0,$$

where $A_1 = \sqrt{\frac{(a^2-1)(1+b^2)}{a^2b^2}}$, $A_2 = \frac{1}{a}\sqrt{\frac{1+b^2}{a^2+b^2}}$, $A_3 = \frac{1}{b}\sqrt{\frac{a^2-1}{a^2+b^2}}$. Then $C_{(a,\,b)}$ is a helix with curvatures

$$\sigma_1 = \sqrt{(a^2-1)(1+b^2)}, \qquad \sigma_2 = \frac{b^2-a^2+1}{2\sigma_1}.$$

Example 7 (type 3). Let C_a, $a^2 > 1$, be a null curve in $\mathbf{H}_1^4$ defined by

$$C_a(p) = \left(A_1 p^2, \quad \frac{1}{a^2}\cosh ap, \quad \frac{1}{a^2}\sinh ap, \quad A_2 p, \quad A_3 p^2 \right),$$

where $A_1 = \frac{\sqrt{a^4-1}}{a^2} + \frac{\sqrt{a^4-1}}{2(a^2+1)}$, $A_2 = \sqrt{\frac{a^2-1}{a^2}}$, $A_3 = \frac{\sqrt{1-a^4}}{2(a^2+1)}$. Then, C_a is a helix with curvatures

$$\sigma_1 = \sqrt{a^2-1}, \qquad \sigma_2 = -\frac{1}{2}\sigma_1.$$

Theorem 5. *Let C be a 2-degenerate Cartan curve fully immersed in $\mathbf{H}_1^4$. Then C is a helix if and only if it is congruent to one of the Examples 5-7.*

Research problems

Null Cartan curves in $\mathbf{M}_q^{m+2}$. In previous chapters we have seen that, contrary to the non-degenerate case, the uniqueness of any type of general Frenet equations can not be assured even if one chooses a special parameter. Each type depends on the parameter of C and the choice of a screen distribution. However, for a null curve in a Lorentzian manifold, using the Natural Frenet equations (2.1.4) of chapter 2, we found a unique Cartan Frenet frame whose Frenet equations (5.1.1) have a minimum number of curvature functions which are invariant under Lorentzian transformations. This raises the following question: *Does there exist any unique Frenet frame for null curves in $\mathbf{M}_q^{m+2}$?* We, therefore, invite the readers to work on the following research problems.

(1) Find Cartan Frenet frame for a Type 1 non-geodesic null curve in $\mathbf{M}_2^{m+2}$.

(2) Find Cartan Frenet frame for a Type 2 non-geodesic null curve in $\mathbf{M}_2^{m+2}$.

(3) Find all possible Cartan Frenet frames of non-geodesic null curves in a semi-Riemannian manifold of index q, where $q > 2$.

Hint. One may use the Natural Frenet equations (3.4.1) for the type 1 and the Natural Frenet equations (3.4.8) - (3.4.9) for the type 2 in chapter 3. Alternatively, one may follow a paper by Ferrández-Giménez-Lucas [36] on *degenerate curves in pseudo-Euclidean spaces of index two.*

Chapter 6

Applications of null curves

In the literature there has been considerable work done on the geometric and/or physical use of null curves. In this chapter, we discuss a variety of specific problems followed by examples and cite some recent papers for further reading.

6.1 Soliton solutions in 3D Lorentzian space forms

To understand this application, we start with the classical case of a space-like curve $\gamma(s) \subset \mathbf{R}^3$ where s its arclength parameter. Let $\gamma(s, t)$ be the time variations of $\gamma(s)$ with $\{\gamma' = T, N, B, \kappa, \tau\}$ its moving Frenet frame along γ, where κ and τ are curvature and torsion functions. Denote the velocity vector by $\dot{\gamma} = \mathbf{v} = v_T T + v_N N + v_B B$, where overdots and primes are partial derivatives with respect to t and s respectively. The intrinsic equations are given in terms of $v'_N, \dot{\kappa}$ and $\dot{\tau}$.

The time variations of a spacelike curve have key roll in the study of physical systems such as vortex filaments in fluid mechanics, 1-dimensional classical continuum Heisenberg chains, elastic strings and soliton theory (or the theory of integrable systems), all of which has had enormous impact on mathematics and physics (for details see [79, 80, 96] and many more referred therein).

Da Rios and Levi-Civita were the first who had major contributions to 3-dimensional vortex filament dynamics, during the years 1906-1933 (see [96] for their related papers). The results of Da Rios are important in the study of integrable 1-dimensional systems and vortex filament motion, whereas, Levi-Civita's work was on asymptotic potential for slender tubes which is now the core of the mathematical formulation of potential theory and capacity theory (see references in [96]).

To identify the thin filament with the vortex line γ, free from self-interactions and smooth, the velocity at an extreme point of γ was obtained by Da Rios using the well-known *localized induction approximation*(**LIA**) for which

$$v_T = v_N = 0 \quad \text{and} \quad v_B = \kappa \quad \text{so that} \quad \mathbf{v} = \kappa B$$

is along the binormal direction. Thus the intrinsic equations (under **LIA**) are

$$v'_N = 0, \quad \dot{\kappa} = -\kappa\tau', \quad \dot{\tau} = (\kappa''/\kappa - \tau^2)' + \kappa\kappa'.$$

The motion of a thin vortex, under **LIA**, is governed by the equation

$$\frac{\partial\gamma}{\partial t} \wedge \nabla_{\frac{\partial}{\partial t}}\frac{\partial\gamma}{\partial t} = \frac{\partial\gamma}{\partial s}, \tag{6.1.1}$$

called the Betchov-Da Rios equation where ∇ is the Levi-Civita connection of the space. Note that, under **LIA**, total length, total curvature and total torsion are conserved quantities in time (see [96]).

The explicit classical solutions of the Betchov-Da Rios equation live in certain ruled surfaces in $\mathbf{R}^3$. Recently, Kim and Yoon [73] have done some work on the classification of ruled surfaces in Minkowski 3-space $\mathbf{R}^3_1$. Berros-Ferrández-Lucas-Meroño have found solutions of the Betchov-Da Rios living in certain ruled surfaces in 3-sphere and anti-De Sitter space $\mathbf{H}^3_1$ (see their two papers cited in [8]).

The purpose of our first application (in this chapter) is to present recent work of Berros-Ferrández-Lucas-Meroño [8] on soliton solutions of the Betchov-Da Rios equation (6.1.1) for a null curve. Let $C(p)$ be a null curve in a Lorentzian space form $(M = M^3_1(c), g)$ of constant curvature c, where p is a special parameter. Consider its associated Frenet frame $\{\xi = C', N, W\}$. To be consistent with usual notations used in physical applications, we set $-N = B$, i.e., $g(\xi, B) = -1$, $g(W, W) = 1$ and all other products are zero, satisfying the Frenet equations

$$\nabla_\xi\xi = \kappa W, \quad \nabla_\xi B = \tau W, \quad \nabla_\xi W = \tau\xi + \kappa B,$$

$\kappa = \kappa(p) \neq 0$ being a curvature function along C and τ a constant.

If the transversal vector field B moves along C, then, it generates a ruled surface given by the parameterization $((I \times \mathbf{R}), f)$ where $f : I \times \mathbf{R} \to M$ is defined by

$$(p, t) \to f(p, t) = C(p) + t\,B(p) = Exp_{C(p)}(tB(p)).$$

This ruled surface is known as a **B**-scroll over the null curve $C(s)$, first introduced by Graves [45], which we denote by $\mathcal{S}_c$. Some authors call this ruled surface by null scroll (see [5]). Then we have

$$\begin{aligned} f_p &= (d(Exp_c))_{tB}\,(\xi + t\tau W), \\ f_t &= (d(Exp_c))_{tB}\,B. \end{aligned}$$

For each fixed p, the curve $\gamma_p(t)$, defined by $t \to \gamma_p(t) = f(p, t)$, is the geodesic of M uniquely determined by the initial conditions $\gamma_p(0) = C(p)$ and $\gamma'_p(0) = B(p)$. Since f_p is a Jacobi field along $\gamma_p(t)$ and M is a space form, we can write

$$f_p = \xi_p(t) + t\tau W_p(t),$$

where $\xi_p(t)$ (resp. $W_p(t)$) are the parallel translations of $\xi(p)$ (resp. $W(p)$) along $\gamma_p(t)$. Let $\mathbf{n} = f_p \wedge f_t$ be a unit normal vector field to $\mathcal{S}_c$. Then, we have

$$\begin{aligned} \nabla_{f_p}f_p &= \tau^2 t f_p + \tau^4 t^3 f_t + (\kappa + \tau^3 t^2)\mathbf{n}, \\ \nabla_{f_p}f_t &= \nabla_{f_t}f_p = -\tau^2 t f_t + \tau\mathbf{n}. \end{aligned}$$

To look for reparametrizations of f which are solutions of the Betchov-Da Rios equation (6.1.1), we let $h \in F(\mathbf{R}^2)$ and write $h(u, v) = (p(u, v), t(u, v))$. Then, $\bar{f} = f \circ h$ is a solution of (6.1.1) if and only if

$$\bar{f}_u \wedge \nabla_{\bar{f}_u} \bar{f}_u = \bar{f}_v,$$

and $g(\bar{f}_u, \bar{f}_u) = \epsilon = \pm 1$. Using $\bar{f}_u = p_u f_p + t_u f_t$ and $\bar{f}_v = p_v f_p + t_v f_t$ and a simple computation yields

$$
\begin{aligned}
\nabla_{\bar{f}_u} \bar{f}_u &= p_{uu} f_p + t_{uu} f_t + 2p_u t_u \nabla_{f_p} f_t + p_u^2 \nabla_{f_u} f_u \\
&= (p_{uu} + \tau^2 t p_u^2) f_p + (t_{uu} + 2\tau^2 t p_u t_u + \tau^4 t^3 p_u^2) f_t \\
&\quad + (2\tau p_u t_u + (\kappa + \tau^3 t^2) p_u^2) \mathbf{n}.
\end{aligned}
$$

From $f_p \wedge \mathbf{n} = f_p + \tau^2 t^2 f_t$ and $f_t \wedge \mathbf{n} = -f_t$ we deduce that $\bar{f}$ is a solution of the Betchov-Da Rios equations if and only if the following system of partial differential equations hold:

$$\tau \left(\frac{t_u}{p_u} - \frac{t_v}{p_v} \right) = \frac{p_v}{p_u^3} + \kappa, \qquad \frac{t_u}{p_u} - \frac{t_v}{p_v} = \tau^2 t^2, \qquad \left(\frac{t_v}{p_v} \right)_u = \tau^2 t p_u \frac{t_v}{p_v}.$$

Now we let C be a generalized null cubic for which we know (see [61]) that $\tau = 0$. Then above equations reduce to

$$p_v = \kappa p_u^3, \qquad t_v = -\kappa p_u^2 t_u, \qquad 0 = p_{uu} t_u - p_u t_{uu}.$$

It follows that $t_u = b p_u$, for some function $b = b(v)$, which measures the slope of the u-curves (v constant). On the other hand, since $g(\bar{f}_u, \bar{f}_u) = -2p_u t_u = \epsilon$, we obtain $2bp_u^2 = -\epsilon$. Thus, p_u only depends on v, so $p(u, v) = h_1(v)u + h_2(v)$ for some differentiable functions h_1 and h_2. In particular, $p_{uu} = t_{uu} = 0$ and so

$$\nabla_{\bar{f}} \bar{f} = \kappa p_u^2 \mathbf{n}.$$

Now using above and the two compatibility conditions $p_{uv} = p_{vu}$ and $t_{uv} = t_{vu}$ it is easy to see that $p_u = a$ and $t_u = ab$, a and b are some constants related by $2ba^2 = -\epsilon$. Moreover, κ is also a constant and the u-curves are geodesics in the B-scroll S_c whose curvature in M is $\rho = \kappa a^2$. On the other hand, as the vector field $(1/\rho)\bar{f}_v$ is the binormal to u-curve, the torsion θ of the u-curve in M is given by

$$\theta = -bg \left(\nabla_{\bar{f}_u} \mathbf{n}, \frac{1}{\rho} \bar{f}_v \right) = \epsilon\rho.$$

Thus, we have proved the following result.

Theorem ([8]). *Let $C(p)$ be a generalized null cubic in a 3-dimensional Lorentzian space form M, where p is a special parameter. Let S_c be the B-scroll parameterized by $f(p, t)$. For any $h \in F(\mathbf{R}^2)$ consider $\bar{f} = f \circ h : \mathbf{R}^2 \to S_c$. Then, $\bar{f}$ is a solution of the Betchov-Da Rios soliton equation in M if and only if*

(1) *the curvature function κ along C is constant and*

(2) *$h(u, v) = (p(u, v), t(u, v))$ is given by*

$$p(u, v) = au + \kappa a^3 v + c_1, \quad t(u, v) = abu - kba^3 v + c_2$$

where $2ba^2 = -\epsilon = \pm 1$ is the causal character of the u-curve, a and b are non-zero constants and (c_1, c_2) is any couple of constants. Moreover, the u-curves are helices in M with curvature $\rho = \kappa a^2$ and torsion $\theta = \epsilon \kappa a^2$.

Example 1. Let $C(p)$ be a null curve in $\mathbf{R}_1^3$ defined by

$$C(p) = \kappa \left(\frac{p^3}{3} - \frac{p}{4}, \ \frac{p^2}{2}, \ \frac{p^3}{3} + \frac{p}{4} \right), \quad \kappa \neq 0.$$

It is straightforward to show that C is a generalized null cubic with constant curvature κ and Cartan frame $\{\xi, B, W\}$ given by

$$\xi = \kappa \left(p^2 - \frac{1}{4}, \ p, \ p^2 + \frac{1}{4} \right), \quad B = \frac{2}{\kappa}(1, \ 0, \ 1), \quad W = (2p, \ 1, \ 2p).$$

The B-scroll $\mathcal{S}_c$ associated to C is parameterized by

$$f(p, t) = \left(\kappa \left(\frac{p^3}{3} - \frac{p}{4} \right) + \frac{2t}{\kappa}, \ \kappa \frac{p^2}{2}, \ \kappa \left(\frac{p^3}{3} + \frac{p}{4} \right) + \frac{2t}{\kappa} \right).$$

Thus, as per above theorem, the soliton solutions of the Betchov-Da Rios equation in the B-scroll $\mathcal{S}_c$ of $\mathbf{R}_1^3$ are given by

$$\bar{f}(u, v) = \left(\bar{f}_1(u, v), \ \bar{f}_2(u, v), \ \bar{f}_3(u, v) \right),$$

where

$$\bar{f}_1 = \kappa \left(\frac{a^3(u + a^2 \kappa v)^3}{3} - \frac{a(u + a^2 \kappa v)^3}{4} \right) - \frac{\epsilon(u - a^2 \kappa v)^3}{a\kappa},$$

$$\bar{f}_2 = \frac{a^2 \kappa (u + a^2 \kappa v)^2}{2}.$$

$$\bar{f}_3 = \kappa \left(\frac{a^3(u + a^2 \kappa v)^3}{3} + \frac{a(u + a^2 \kappa v)^3}{4} \right) - \frac{\epsilon(u - a^2 \kappa v)^3}{a\kappa},$$

where a is a non-zero constant. The u-curves are helices in $\mathbf{R}_1^3$ with causal character $\epsilon = \pm 1$, curvature $\rho = \kappa a^2$ and torsion $\theta = \epsilon \kappa a^2$.

Example 2. Let $C(p)$ be a curve in a De Sitter spacetime $\mathbf{S}_1^3$ defined by

$$C(p) = \frac{\sqrt{2}}{2}(\cos A, \ \sin A, \ \cosh A, \ \sinh A),$$

where we set $A = \sqrt{\kappa} p$ and $\kappa > 0$. This curve is a generalized cubic in $\mathbf{S}_1^3$ with constant curvature κ and Cartan frame $\{\xi, B, W\}$ given by

$$\xi(p) = \frac{\sqrt{2\kappa}}{2}(-\sin A, \cos A, \sinh A, \cosh A),$$

$$B(p) = \frac{\sqrt{2\kappa}}{2\kappa}(\sin A, -\cos A, \sinh A, \cosh A),$$

$$W(p) = \frac{\sqrt{2}}{2}(-\cos A, -\sin A, \cosh A, \sinh A).$$

The B-scroll $\mathcal{S}_c$ associated to C is parameterized by

$$f(p, t) = \frac{\sqrt{2}}{2}\left(\cos A + \frac{t}{\sqrt{\kappa}}\sin A, \quad \sin A - \frac{t}{\sqrt{\kappa}}\cos A, \right.$$
$$\left. \cosh A + \frac{t}{\sqrt{\kappa}}\sinh A, \quad \sinh A + \frac{t}{\sqrt{\kappa}}\cosh A\right).$$

The soliton solutions of the Betchov-Da Rios equation in the B-scroll $\mathcal{S}_c$ of $\mathbf{S}_1^3$ are given by

$$\bar{f}(u, v) = (\sqrt{2}/2)\left(\bar{f}_1(u, v), \quad \bar{f}_2(u, v), \quad \bar{f}_3(u, v), \quad \bar{f}_4(u, v)\right), \quad \text{where}$$

$$\bar{f}_1 = \cos A_1 - A_2 \sin A_1, \qquad \bar{f}_2 = \sin A_1 + A_2 \cos A_1,$$
$$\bar{f}_3 = \cosh A_1 - A_2 \sinh A_1, \qquad \bar{f}_4 = \sinh A_1 + A_2 \cosh A_1$$

and $A_1 = \sqrt{\kappa}(au + \kappa a^3 v)$, $A_2 = \frac{\epsilon}{2\sqrt{\kappa}}(\frac{u}{a} - \kappa a^3 v)$ with a constant $a \neq 0$ and $\epsilon = \pm 1$.

6.2 Soliton solutions in Minkowski spaces

In section 3 of chapter 5 we used the classical Lancret theorem [106] to prove corresponding Lancret theorem for odd dimensional null curves and presented characterization theorems for 3-dimensional and 5-dimensional generalized null curves. We have seen, in previous section, that 3-dimensional generalized null helices (in particular, null cubics) have provided soliton solutions. To extend the results of previous section for any n-dimensional spaces, one must find a Lancret type theorem suitable for even as well as odd dimensional spaces to characterize null generalized helices of any dimension. For this purpose, as a first step, we start with the simple case of null curves in 4-dimensional Minkowski spaces $\mathbf{R}_1^4$ taken from [40].

Let $C(p)$ be a null Cartan curve in a 4-dimensional Minkowski space $(\mathbf{R}_1^4, g)$ with Cartan frame $\{\xi = C', W_1, N, W_2\}$ satisfying the Cartan equations

$$\xi' = W_1; \quad W_1' = -\sigma_1 \xi + N; \quad N' = \sigma_1 W_1 + \sigma_2 W_2; \quad W_2' = \sigma_2 \xi,$$

where the orientation is such that $g(\xi, N) = -1$ (instead of $+1$).

Definition 2.1. *A null Cartan curve $C(p)$ in $(\mathbf{R}_1^4, g)$ is called a generalized helix if there exists a constant vector $v \neq 0$ such that $g(\xi(p), v) = \mu \neq 0$ is constant.*

As before, the axis of v is the straight line generated by v and we assume unit length for a spacelike or a timelike axis of C. Following is the general characterization result:

Theorem 2.1. *A null Cartan curve $C(p)$ in $\mathbf{R}_1^4$ is a generalized helix if and only if its Cartan curvatures satisfy the following differential equation*

$$(\sigma_1')^2 = \sigma_2^2(2\sigma_1 + c), \quad \sigma_1' \neq 0, \quad c \text{ is a constant.} \tag{6.2.1}$$

Proof. Let $v = \mu_0(p)\xi + \mu_1(p)W_1 - \mu N + \mu_2(p)W_2$, where μ is a constant and other coefficients are differentiable functions of the special parameter p. Then, from $\frac{dv}{dp} = 0$ (since v is a constant vector) we deduce

$$\mu_1 = 0, \quad \mu_0 + \mu\sigma_1 = 0, \quad \mu_0' + \mu_2\sigma_2 = 0, \quad \mu_2' - \mu\sigma_2 = 0.$$

Since $\sigma_2 \neq 0$ and $\mu_0 = -\mu\sigma_1$, we deduce that $\mu_2 = \mu\frac{\sigma_1'}{\sigma_2}$. Thus, the axis of C is

$$v = -\mu\sigma_1\xi - \mu N + \mu\frac{\sigma_1'}{\sigma_2}W_2. \tag{6.2.2}$$

As $g(v, v) = \pm 1$, we obtain (6.2.1) with $c = \pm\mu^{-2}$. Finally, $\sigma_1' \neq 0$ as otherwise $\sigma_2 = 0$, and so $\mu\sigma_2 = 0$, which can not hold by definition of generalized helix.

Conversely, suppose the equation (6.2.1) holds. Consider a vector field v along C defined by (6.2.2), where $\mu = \sqrt{|c|}$. Then, one can easily show that v is constant and $g(\xi, v) = \mu$ which completes the proof.

Non-null axis v. Let Σ denote the hyperplane orthogonal to v, P the projection map onto Σ and $\bar\alpha = P(C)$ the non-null projection curve. Then one can write

$$C(p) = \bar\alpha(p) + \bar\mu v, \quad \bar\mu \text{ is a non-constant differentiable function.}$$

Clearly, the $\bar\alpha$ and v have same causal structure. Let $\alpha(s)$ represent the arc length parameterization of $\bar\alpha$ with curvatures k_1, k_2. Following is immediate.

Lemma 2.1. *Let $C(p)$ be a null Cartan curve in $\mathbf{R}_1^4$ and $\alpha(s)$ be the orthogonal projection of C onto a non-degenerate hyperplane Σ, where p and s are the special and the arc-length parameters of C and α respectively. Then p and s are linearly related if and only if $\bar\alpha$ is a curve in Σ with constant curvature k_1.*

Using above lemma, we have the following characterization result:

Theorem 2.2. *Let $C(p)$ be a null Cartan curve in $\mathbf{R}_1^4$, $v \neq 0$ a constant unit vector, Σ the hyperplane orthogonal to v in $\mathbf{R}_1^4$ and $\bar\alpha$ the projection of C onto Σ. Then, C is a generalized helix with axis v if and only if $\bar\alpha$ is a curve in Σ with* *constant curvature and non-constant torsion.*

Consequently, the following result is immediate:

Theorem 2.3. *A null Cartan curve C in $\mathbf{R}_1^4$ is a generalized helix with non-null axis if and only if it is a null geodesic of a Lorentzian cylinder constructed on a spacelike curve in $\mathbf{R}^3$ or $\mathbf{R}_1^3$ with constant curvature and non-constant torsion.*

Null axis v. We first show that there exists a close relation between timelike generalized helices and null Cartan generalized helices having both the same null axis. Indeed, consider a non-degenerate curve $\alpha(s)$ in $\mathbf{R}_1^4$ with Frenet frame $\{n_0, n_1, n_2, n_3\}$ such that $g(n_0, n_0) = \epsilon_0$ and $g(n_i, n_i) = \epsilon_i$, $1 \leq i \leq 3$. Suppose k_1, k_2, k_3 are the corresponding curvature functions. We say that α is a generalized helix if there exists a constant vector $v \neq 0$ such that $g(n_0(s), v) \neq 0$ is constant. The proof of the following theorem is common with the proof of Theorem 2.1.

Theorem 2.4. *Let α be a non-degenerate curve in $\mathbf{R}_1^4$. Then α is a generalized helix if and only if its curvature functions satisfy the following differential equation*

$$(A')^2 = \epsilon_0 \epsilon_1 k_3^2 (A^2 + c), \quad A' \neq 0, \quad A = k_1/k_2 \quad \text{and} \quad c \text{ is a constant.}$$

Now suppose α is a timelike generalized helix with null axis v. Choose v as

$$2v = n_0 + An_2 + \frac{1}{k_3}A'n_3,$$

satisfying $g(n_0, v) = -\frac{1}{2}$. Then the surface $\mathcal{S}_c$ locally parameterized by $f(s, w) = \alpha(s) + wv$ is a Lorentzian surface in $\mathbf{R}_1^4$. The null geodesics of $\mathcal{S}_c$ can be parameterized by $\bar{C}(s) = \alpha(s) - (s + \sigma)v$, where σ is a constant. Let p be a special parameter of $\bar{C}$ as a curve in $\mathbf{R}_1^4$ and write $C(p(s)) = \bar{C}(s) = \alpha(s) - (s + \sigma)v$. A straightforward computation leads to

$$\sqrt{k_1(s)}\, g(\xi, v) = -\frac{1}{2}.$$

Thus, C is a Cartan generalized helix if and only if k_1 is constant. Moreover, the Cartan curvatures of C and those of the timelike generalized helix α are related by

$$\sigma_1(p(s)) = \frac{k_2(s)^2 - k_1^2}{2k_1}, \quad \sigma_2(p(s))^2 = \frac{k_2(s)^2 k_3(s)^2 - k_2'(s)^2}{k_1^2}.$$

Based on above we have proved the following result (see [40]):

Theorem 2.5. *A null Cartan curve C in $\mathbf{R}_1^4$ is a generalized helix with null axis v if and only if it is a geodesic of a Lorentzian ruled surface whose directrix is a timelike generalized helix in $\mathbf{R}_1^4$ (with null axis and constant first curvature) and whose rulings have the direction of the axis.*

Soliton solutions. In [40] the authors have proposed a general equation, called *the null localized induction equation*, briefly denoted by NLIE, for finding soliton solutions of null Cartan curves in a n-dimensional Minkowski space as follows:

Definition [40]. *Let $C(p, t)$ be the evolution equations (also called the time variations) by null Cartan curves, in a n-dimensional Minkowski space $\mathbf{R}_1^n$, of a null curve $C(p) = C(p, 0)$, where p is a special parameter, Then, the equation*

$$\frac{\partial C}{\partial t} = \frac{\partial^2 C}{\partial p^2} \wedge \ldots \wedge \frac{\partial^n C}{\partial p^n} \qquad (6.2.3)$$

is called the null localized induction equation (NLIE).

Note that $C(p, t)$ is a parameterized surface and $V(p, t) = \frac{\partial C}{\partial t}$ is the *variation vector field* on this surface. The classical Betchov-Da Rios equation for $\mathbf{R}^3$ is usually written as $\frac{\partial \gamma}{\partial t} = kB$, where B and k are the binormal unit vector and the curvature function respectively of a spacelike curve $\gamma(p)$. In [40], the authors did the same for the (NLIE) null equation (6.2.3) in the three, four and five dimensional cases, respectively and obtained the following equations:

$$\mathbf{R}_1^3 \quad : \quad \frac{\partial C}{\partial t} = -(\sigma_1 \xi + N),$$

$$\mathbf{R}_1^4 \quad : \quad \frac{\partial C}{\partial t} = \sigma_2(\sigma_1 \xi + N) - \sigma_1' W_2,$$

$$\mathbf{R}_1^5 \quad : \quad \frac{\partial C}{\partial t} = -\sigma_2^2 \sigma_3(\sigma_1 \xi + N) - \sigma_1' \sigma_2 \sigma_3 W_2 - (\sigma_2^3 - \sigma_2 \sigma_1'' + \sigma_2' \sigma_1') W_3,$$

where the Cartan Frenet equations are taken from the Theorem 1.1 of chapter 5 with the orientation of their Frenet frames so that $g(\xi, N) = -1$ (instead of $+1$). Consequently, it is straightforward to prove the following result:

Proposition ([40]). *Let $C(p, t)$ be the evolution equation in null Cartan curves in $(\mathbf{R}_1^n, g)$, with the Cartan frame $\{\xi, W_1, N, W_2, \ldots, W_{n-2}\}$ associated to any curve of the evolution. Then the variation vector field V of $C(p, t)$ satisfies the following equations:*

$$(E_1) \ : \ g(\nabla_\xi V, \xi) = 0; \qquad (E_2) \ : \ g(\nabla_\xi^2 V, W_1) = 0.$$

Then, for three, four and five dimensional cases, the solutions of NLIE satisfy the following conditions:

3-dimensional case   : (E_1) : always; (E_2) : $\dfrac{\partial \sigma_1}{\partial p} = 0.$

4-dimensional case   : (E_1) : $\dfrac{\partial \sigma_2}{\partial p} = 0;$ (E_2) : always.

5-dimensional case   : (E_1) : $\dfrac{\partial}{\partial p}(\sigma_2^2 \sigma_3) = 0;$ (E_2) : $\dfrac{\partial \sigma_1}{\partial p} = 0.$

Example 3. Let $C(p)$ be a null Cartan helix in $\mathbf{R}_1^3$ with constant curvature σ_1 and Cartan frame $\{\xi, W, N\}$. Consider the evolution equation given by

$$C(p, t) = C(p) - t\left(\sigma_1 \xi(p) + N(p)\right),$$

which means that $C(p)$ evolves by translation along its axis. It is easy to show that $c(p, t)$ is a solution of the NLIE in $\mathbf{R}_1^3$.

Example 4. Let $C(p)$ be a null Cartan generalized helix in $\mathbf{R}_1^4$ with Cartan frame $\{\xi, W_1, N, W_2\}$ and the constant second curvature σ_2. Its axis (see (6.2.2)) is now written in the following form:

$$v = -\frac{\mu}{\sigma_2}\left(\sigma_2\sigma_1(p)\xi(p) + N - \sigma_1'(p)W_2(p)\right),$$

where $\mu = g(\xi, v)$ and $\sigma(p)$ satisfies (6.2.1). Then we get

$$\sigma(p) = \frac{1}{2}\left(\sigma_2^2(p+c)^2 - \frac{\epsilon}{\mu^2}\right),$$

where c is a constant and $C(p, t) = C(p) - (\sigma_2/\mu)$ is a solution of the NLIE in $\mathbf{R}_1^4$.

Example 5. Let $C(p)$ be a null Cartan helix in $\mathbf{R}_1^5$ with Cartan frame $\{\xi, W_1, N, W_2, W_3\}$. Its axis is given by

$$v = \sigma_3\sigma_2(\sigma_1\xi + N) + W_3,$$

where all three curvatures are constants. Then, $C(p, t) = C(p) - t\sigma_2^3 v$ is a simple solution of the NLIE in $\mathbf{R}_1^5$.

6.3 Mechanical systems and 3D null curves

It is well-known that, in a given mechanical problem, following are two essential steps: (a) obtain the differential equations of motion, and (b) solve those equations. The method of a French mathematician Lagrange (1736-1813) has been one of the most effective technique in obtaining the differential equations of the motion of a given mechanical system and, also, provide useful hints toward their integration. To understand how Lagrangian method works, consider the motion of a particle described by a differentiable (at least of class C^2) curve γ in an n-dimensional differentiable manifold M^n, represented by $x^i = x^i(t)$ where t is a general parameter. Then, we know from elementary differential geometry that a C^2 function $L(t, x^i, \dot{x}^i)$ of $2n+1$ independent variables is called the *Lagrangian* or *Lagrange function*, where we use the notation $\dot{x}^i = \frac{dx^i}{dt}$. Construct the integral of this Lagrangian L along the curve γ given by

$$I(\gamma) = \int_{t_1}^{t_2} L(t, x^i, \dot{x}^i)\, dt, \tag{6.3.1}$$

where the parameter values t_1 and t_2 correspond to two points P_1 and P_2 of a segment of γ. Clearly, the integral $I(\gamma)$ will in general depend on the choice of the curve γ joining P_1 and P_2.

Associate x^i as the general coordinates of a classical dynamical system with n degrees of freedom so that $\dot{x}^i$ are the components of the velocity of a particle for which t is the time parameter. If the dynamical system is conservative, then, a suitable Lagrangian may be defined by

$$L = T - V,$$

where $T = \frac{1}{2}a_{ij}\dot{x}^i\dot{x}^j$ is the kinetic energy of such a system, $V(x^i)$ a potential function and the coefficients a_{ij} are functions of the variables x^i. Above prescription for a Lagrangian gives rise to a problem in the calculus of variations as exemplified by the integral of (6.3.1). It is important to mention that the quantities defined by

$$E_i(L) = \frac{d}{dt}\left(\frac{\partial L}{\partial \dot{x}^i}\right) - \frac{\partial L}{\partial x^i} \qquad (6.3.2)$$

are the components of a covariant vector, called the *Euler-Lagrange vector of L* and the following holds: *In order that the curve γ has an extreme value to the integral (6.3.1) it is necessary that the Euler-Lagrange vector $E_i(L) = 0$ along γ.* This condition is known as the *Euler-Lagrange equations* defined by

$$\frac{d}{dt}\left(\frac{\partial L}{\partial \dot{x}^i}\right) - \frac{\partial L}{\partial x^i} = 0 \quad \text{or} \quad E_i(L) = 0, \qquad (6.3.3)$$

which are invariant under coordinate transformations. Any curve satisfying (6.3.3) is called an *extremal* (some authors call it *critical*) curve. We highlight the fact that the Euler-Lagrange equations are only necessary (but not sufficient) conditions for an extreme value of the integral (6.3.1).

However, although the application of this section involves use of Lagrangian in dealing with certain mechanical systems, the method of Lagrange has manifold uses in a variety of problems in mathematics and physics. In particular, use of Lagrangian in solving nonlinear differential equations of a given system of Einstein field equations is well-known.

Considerable work has been done on the search for Lagrangian describing spinning particles in massive and massless cases. In general, the classical models require some extra bosonic variables. Since the spinning particles are described by the geometry of its world trajectories, the most interesting and rather effective possibility (studied in the late 80's through 90's) was to supply those extra degrees of freedom by Lagrangian that depend on the first and the second curvatures of the world trajectories (see many papers in [7, 41]). In particular, recently Barros [7] considered a model of the spinning relativistic particles (both massive and massless) described by Lagrangian that depend linearly from the curvature of the world trajectories which gave the complete integration of the field equations in either a $2D$ background gravitational field or in any dimensional background with high rigidity, for example with constant curvature.

Related to the subject matter of this book, in Barros's paper [7] there is a discussion on null solutions which are precisely null geodesics of the Hopf tube on a horocycle. Following the approach in [7], Narsessian-Romas [84], Narsessian et al. [85] and Ferrández et al. [38, 41] have recently studied some mechanical systems along Cartan null curves whose Lagrangian are prescribed functions on the Cartan curvature of its world trajectories. In this section, we discuss the results of their papers [38, 41, 84, 85], some with details and others in brief.

Let $C(p)$ be a null Cartan curve in $(\mathbf{R}_1^3, g)$ where the metric g represents the background gravitational field and $\{\xi = C'(p), W(p), N(p)\}$ is its Cartan frame with respect to a special parameter p. Again, for physical reasons, the orientation is such that $g(\xi, N) = -1$ (instead of $+1$), $g(W, W) = 1$ and all other products vanish. The Cartan equations are

$$
\begin{aligned}
\nabla_\xi \xi &= W, \\
\nabla_\xi W &= -k\xi + N, \\
\nabla_\xi N &= -kW,
\end{aligned}
\tag{6.3.4}
$$

where ∇ is the Levi-Civita connection and k is the Cartan curvature function (also called torsion of the curve C). Recall from chapter 5 that k determines completely the null curve up to a Lorentzian transformation. Precisely, given a function k one can always construct a null curve parameterized by a special parameter, so that k is its curvature function. This allows us to say that any local geometrical scalar defined along C can always be expressed as a function of its curvature and derivatives. Let $\mathcal{C}$ be the set of all $3D$ null Cartan curves that have the same end points and the same Cartan frame at those endpoints. Suppose

$$
C_t = C(s, t) : [0, 1] \times (-\epsilon, \epsilon) \to \mathbf{R}_1^3
$$

denote the variation of Cartan null curves in $\mathcal{C}$ with $C(s, 0)$ the reparameterization of $C(p)$, with respect to another special parameter s. Then, $V = V(s, t) = \frac{\partial C_t}{\partial t}(s, t)$ is the variation vector field $V(s) = V(s, 0)$ on the parameterized surface $C(s, t)$. Also, we set $\frac{\partial C_t}{\partial s}(s, t) = \delta(s, t)\xi(s, t)$, where δ is a differentiable function. Throughout we consider special parameter for any member of $\mathcal{C}$, a function and a vector field etc.

Lemma. *Following results hold:*

(a) $g(\nabla_\xi V, \xi) = 0,$

(b) $\dfrac{\partial \delta}{\partial t} = V(\delta) = -\dfrac{1}{2}h\delta, \quad h = -g(\nabla_\xi^2 V, W),$ (6.3.5)

(c) $\dfrac{\partial k}{\partial t} = g(\nabla_\xi^3 V, N) + kg(\nabla_\xi V, N) + kh - \dfrac{1}{2}\xi(\xi(h)).$

Proof. (a) Set $\bar{\xi}(s, t) = \frac{\partial C_t}{\partial s}(s, t)$ so that $\bar{\xi}(s, t) = \delta(s, t)\xi(s, t)$, where δ is a differentiable function. Then,

$$
g\left(\frac{\partial C_t}{\partial p}(p, t), \ \frac{\partial C_t}{\partial p}(p, t)\right) = g(\bar{\xi}, \bar{\xi}) = 0.
$$

Thus, $Vg(\bar{\xi}, \bar{\xi}) = 2g(\nabla_V\bar{\xi}, \bar{\xi}) = 0$. Now $\bar{\xi} = \delta\xi$ and $[V, \bar{\xi}] = 0$ if and only if $\nabla_V\bar{\xi} = \nabla_{\bar{\xi}}V$. Hence,

$$g(\nabla_V\bar{\xi}, \bar{\xi}) = g(\nabla_{\bar{\xi}}V, \bar{\xi}) = \delta^2 g(\nabla_\xi V, \xi) = 0.$$

This implies that $g(\nabla_\xi V, \xi) = 0$ since $\delta \neq 0$ which proves (a).

(b) $\nabla_{\bar{\xi}}\bar{\xi} = \nabla_{\delta\xi}\delta\xi = \delta\xi(\delta)\xi + \delta^2 W \implies g(\nabla_{\bar{\xi}}\bar{\xi}, \nabla_{\bar{\xi}}\bar{\xi}) = \delta^4$. Also, $V(\delta^4) = 4\delta^3 V(\delta) = 2g(\nabla_V\nabla_{\bar{\xi}}\bar{\xi}, \nabla_{\bar{\xi}}\bar{\xi})$. On the other hand, since the curvature tensor $\mathcal{R}$ of $\mathbf{R}_1^3$ vanishes, we have

$$0 = \mathcal{R}(V, \bar{\xi})\bar{\xi} = \nabla_{[V, \bar{\xi}]}\bar{\xi} - \nabla_V\nabla_{\bar{\xi}}\bar{\xi} + \nabla_{\bar{\xi}}\nabla_V\bar{\xi}.$$

The first term on the righthand side vanishes and so we obtain

$$\nabla_V\nabla_{\bar{\xi}}\bar{\xi} = \nabla_{\bar{\xi}}\nabla_V\bar{\xi} = \nabla_{\bar{\xi}}^2 V = \delta\xi(\delta)\nabla_\xi V + \delta^2\nabla_\xi^2 V.$$

From these results, it is easy to deduce that $V(\delta) = \frac{1}{2}\delta g(\nabla_\xi^2 V, W)$ which proves (b)

(c) $[V, \bar{\xi}] = 0$ implies $[V, \xi] = \frac{1}{2}h\xi$ and $\nabla_V\xi = \nabla_\xi V + \frac{1}{2}h\xi$. On the other hand, it is easy to show that

$$\nabla_V W = \nabla_\xi^2 V + \frac{1}{2}\xi(h)\xi + hW.$$

Using the second Cartan equation (6.3.4) we obtain

$$\nabla_V\nabla_\xi W = -V(k)\xi - k\nabla_V\xi + \nabla_V N, \quad \text{where} \quad V(k) = \frac{\partial k}{\partial t}.$$

Above results and some computations prove (c), which completes the proof.

A vector field V along $C(p)$ in $\mathbf{R}_1^3$ is said to be a *Killing vector field* on $\mathbf{R}_1^3$ if it infinitesimally preserves the causal character, the special parameter p and the curvature of C. Using the three relations of (6.3.5) we say that the Killing vector fields along C in $\mathbf{R}_1^3$ are characterized by the equations

$$g(\nabla_\xi V, \xi) = V(\delta) = V(k) = 0. \tag{6.3.6}$$

Consider the action $I : \mathcal{C} \to \mathbf{R}$ given by

$$I(C_t) = \int_{C_t} L(k)dp, \tag{6.3.7}$$

where $L(k)$ is a differentiable Lagrangian function of the Cartan curvature k. By using Killing vector fields along curves as a key tool, we obtain and solve the motion equations for a few prescribed types of the Lagrangian by integrating the equation (6.3.7). In the sequel we say that an extremal curve is a *critical point* of the action I.

Definition 3.1. *A null curve $C(p)$ will be a critical point of the action I if*

$$\frac{d}{dt}\Big|_{t=0} I(C_t) = \frac{d}{dt}\Big|_{t=0} \int_{C_t} f(k_t)dp, \quad \forall\, C_t \in \mathcal{C}.$$

To compute the first-order variation of this action, along the fields space $\mathcal{C}$, we need to calculate $I'(0)$ where we set $I(t) = I(C_t)$ and $dp = \delta ds$. Then,

$$I(t) = \int_0^1 L(k_t)\delta ds \Longrightarrow I'(t) = \int_0^1 \{V(L(k_t))\delta + L(k_t)V(\delta)\}ds$$

$$= \int_0^1 \left(L'(k_t)V(k_t) - \frac{1}{2}L(k_t)h\right)\delta ds.$$

Now, if we take $C_0 = C$ and $k_0 = k$, then

$$I'(0) = \int_{a_0}^{a_1} \left(L'(k)V(k) - \frac{1}{2}L(k)h\right)dp.$$

In [41] the authors have done geometric study of actions I in $\mathbf{R}_1^3$ whose Lagrangian are arbitrary functions on the curvature of the particle path. We first discuss in details their simple prescription for L when the action I is linear in the curvature of the particle path, which has physical applications (see three papers [38, 84, 85]). This will be followed by a brief on their complicated mathematical case when L is a quadratic function. Let the action I be prescribed by

$$I(C_t) = \int_{C_t} (a + bk)dp, \tag{6.3.8}$$

where a and b are constants. Using the integration by parts and the Cartan equations (6.3.4) we obtain

$$I'(0) = [\Omega]_{a_0}^{a_1} - \frac{1}{2}\int_{a_0}^{a_1} g(V, E(C_t)\xi)dp, \quad \text{where}$$

$$\Omega = g(\nabla_\xi^3 V, bkW) + g(\nabla_\xi^2 V, bk(3N - k\xi))$$
$$\qquad + g(\nabla_\xi V, bkW) + g(V, P),$$
$$E(C_t) = bk''' + 3bkk' - ak'.$$

The *potential vector field*, called the *linear momentum* of the particle, is given by

$$P = (bk'' + bk^2 - ak)\xi - bk'W + (bk - a)N. \tag{6.3.9}$$

An easy computation shows that $E(C_t)\xi = \nabla_\xi P$. Thus, the *conserved linear momentum law* requires that P is constant along C_t if and only if

$$E(C_t) = bk''' + 3bkk' - ak' = 0 \tag{6.3.10}$$

which is the Euler Lagrange differential equation. Thus, $E(C_t) = 0$ if and only if C_t is a critical curve of I. Since our curves have the same endpoints and the same Cartan frames at those endpoints, $[\Omega]_{a_0}^{a_1}$ vanishes. Therefore, the first order variation is

$$I'(0) = -\frac{1}{2} \int_{a_0}^{a_1} g(V, E(C)\xi)dp. \qquad (6.3.11)$$

Consequently, we say that any curve $C_t \in \mathcal{C}$ is the null world line of a relativistic particle in $\mathbf{R}_1^3$ spacetime if and only if

(a) W, N and k are well defined on the entire world trajectory;

(b) The Euler Lagrange equation (6.3.10) is satisfied.

Solving the motion equations. Our goal is to integrate the motion equations of the Lagrangian giving models for relativistic particles that linearly involve the curvature of the null curves $C_t \in \mathcal{C}$. Note that the vector field P is Killing. To solve the motion equations we need another Killing vector field as follows:

Consider a vector field $X = (a + bk)\xi + 2bN + P \times C_t$. Since our local action is invariant under rotations, it is easy to see that X is constant along C_t. Therefore,

$$J = -P \times C_t + X = (a + bk)\xi + 2bN$$

is a Killing vector field along C_t that together with the Killing vector field P allows one to find non-trivial first order integrals of the Euler Lagrange equations. At this point we set $\phi - bk - a$. It follows that J satisfies

$$\nabla_\xi J = -bk'\xi + \phi W. \qquad (6.3.12)$$

First of all, we deal with the trivial case of constant Cartan curvature k. These curves are null helices (see chapter 5) and are always possible trajectories of the particles, for any choice of a and b. Next, we distinguish the following two types of solutions of the Euler Lagrange equation $E(C_t) = 0$.

Type 1: $\phi = 0$. This means that Cartan curvature $k = a/b = $ constant, that is, C_t is a helix. Using (6.3.9) and (6.3.12) we conclude that P vanishes and the Killing vector field $J = 2a\xi + 2bN$ is constant. Therefore, $g(J, J) = -8ab$ is a constant of the motion and a first integral of the equation $\phi = 0$.

Type 2: $\phi \neq 0$. For this case, one can show that

$$J = -p \times C_t + \epsilon dP^*, \quad \epsilon = \pm 1, \qquad (6.3.13)$$

where d is constant and P^* is a vector field with the same causal character as P and satisfies $g(P, P^*) = \epsilon$. Therefore, $g(P, J) = d$. The two Killing vector fields P and J can be interpreted as generators of the particle mass m and spin s, with following relation

$$g(P, P) = \epsilon e^2 = m^2, \quad g(P, J) = d = ms.$$

Case 1: $a \neq 0$; $b = 0$. This means that L is constant which represents the simplest non-trivial action I described by the motion of a particle, since I is proportional to the special parameter of the particle path. We have $P = -a(k\xi + N)$ and $J = a\xi$, so that $g(P, P) = -2a^2k = \epsilon e^2$ and $g(P, J) = a^2 = d$, and therefore $k = -(\epsilon e^2/2d)$, which is a constant. Thus, C_t is a Cartan helix with non-null axis given by the vector P. Observe that $d \neq 0$ otherwise $C_t = 0$ which is not possible.

Note 1. Case 1 was discussed (in details) by Narsessian and Ramos [84], where they have shown that the classical phase space of this system agrees with that of a massive spinning particle of spin $s = c^2/m$, where c is the coupling constant, and the particle states can be *anyonic*, i.e., their spin s can take any real value. They have also shown that the massive (tachyonic) solutions correspond to the null helices with negative (positive) curvature.

Note 2. In the case of tachyonic energy flow, the mass could be positive, negative or zero, according to the causal character of the vector field P. Timelike and lightlike trajectories are the natural ones in spacetime geometries, however, some recent experiments show the existence of super luminal particles (spacelike trajectories) without any breakdown of the principle of relativity. Moreover, neutrinos might be instances of tachyons as their square mass appears to be negative. Thus, to solve the motion equations when $\phi \neq 0$, we must consider all possible cases of P spacelike or timelike or null, which we deal separately as follows:

P is non-null. Consider a cylindrical coordinates system (r, θ, z) such that $P = e\partial_z$ and from equation (6.3.13) we get $J = \epsilon(d/e)\partial_z - e\partial_\theta$ and other non-zero products are $g(\partial_z, \partial_z) = \epsilon$, $g(\partial_\theta, \partial_\theta) = -\epsilon r^2$ and $g(\partial_r, \partial_r) = 1$. Consequently, a null world line of a relativistic particle in $\mathbf{R}_1^3$ can be described in cylindrical coordinates around a non-null P as follows:

$$r^2 = \frac{d^2}{e^4} - \frac{\epsilon}{e^2}g(J, J), \quad z' = \frac{\epsilon}{e}g(\xi, P), \quad \theta' = \frac{e(e^2 g(\xi, J) - \epsilon d g(\xi, P))}{\epsilon d^2 - e^2 g(J, J)},$$

where

$$g(J, J) = -4(a + k), \quad g(\xi, P) = a - k, \quad g(\xi, J) = -2.$$

P is null. Without any loss of generality, assume P is collinear with null vector $(1, 1, 0)$ and consider a coordinates system (r, θ, z) defined by

$$X(r, \theta, z) = \left(z - \frac{\epsilon r}{2}(\theta^2 + 1), \ z - \frac{\epsilon r}{2}(\theta^2 - 1), \ -\epsilon r\theta\right).$$

We call these coordinates the *null cylindrical coordinates* around P. The two surviving products are: $g(\partial_\theta, \partial_\theta) = r^2$ and $g(\partial_r, \partial_z) = \epsilon$ and $P = A\partial_z$, A is a non-zero constant. Choose $P^* = (\epsilon/A)(1, 0, 1)$. Then, we deduce that

$$p^* = -\frac{\epsilon}{2A}(\theta - 1)^2\partial_z - \frac{\theta - 1}{Ar}\partial_\theta + \frac{1}{A}\partial_r,$$

$$J = -\frac{d}{2A}(\theta - 1)^2\partial_z + \left(A - \frac{\epsilon d}{Ar}(\theta - 1)\right)\partial_\theta + \frac{\epsilon d}{A}\partial_r.$$

Consequently, a null world line of a relativistic particle in $\mathbf{R}_1^3$ can be described in cylindrical coordinates around a null P as follows:

$$d = 0, \quad r^2 = \frac{1}{A^2} g(J, J), \quad \theta' = A\frac{g(\xi, J)}{g(J, J)}, \quad z' = -\frac{ag(\xi, J)^2}{2g(J, J)g(\xi, P)}.$$

$$d \neq 0, \quad r' = \frac{\epsilon}{A} g(\xi, P), \quad \theta = \frac{A^2 r^2 - g(J, J)}{2\epsilon dr} + 1, \quad z' = \frac{Ar^2\theta^2}{2g(\xi, P)}.$$

where $g(J, J)$, $g(\xi, P)$ and $g(\xi, J)$ are same as given in the non-null case.

Case 2: $b \neq 0$. Without any loss of generality we normalize the constant b to be one. Two first integrals provided by the vector fields P and J are

$$(k')^2 - 2(k - a)(k'' + k^2 - ak) - \epsilon e^2 = 0,$$
$$-2k'' - 3k^2 + 2ak + a^2 - d = 0. \tag{6.3.14}$$

From above two differential equations we obtain

$$(k')^2 + k^3 - ak^2 + (d - a^2)k + a^3 - da - \epsilon e^2 = 0, \tag{6.3.15}$$

which can be written as $(k')^2 + Q(k) = 0$, Q is the polynomial

$$Q(X) = X^3 - aX^2 + (d - a^2)X + a^3 - da - \epsilon e^2.$$

By using standard techniques involving the elliptic functions, we obtain the solution of the equation $Q(X) = 0$ as follows: First assume that all the three roots a_1, a_2 and a_3 (such that $a_1 \leq a_2 \leq a_3$) are real. Then it is well-known that

$$a = a_1 + a_2 + a_3,$$
$$d - a^2 = a_1 a_2 + a_1 a_3 + a_2 a_3, \tag{6.3.16}$$
$$\epsilon e^2 + da - a^3 = a_1 a_2 a_3,$$

from which we deduce

$$a_1 \leq \frac{a}{3}, \quad a_2 \leq \frac{a - a_1}{2}.$$

Note 3. Since $Q(k) = -(k')^2$, therefore, k takes values only when Q is negative. Trivial solutions are $k(p) = a_i$, where a_i is a real root of Q, so that we find again the null Cartan helices. In this case $g(P, P) = -2k(k - a)^2$ and $g(P, J) = -3k^2 + 2ak + a^2$. Also, the massive and tachyonic sectors correspond with negative or positive curvature, respectively. In the following we analyze all possible cases of solutions.

I. Q has a real root of multiplicity 3: $\alpha = a_1 = a_2 = a_3$. For this case,

$$\alpha = a/3 \quad \text{and} \quad k(p) = \frac{a}{3} - \frac{4}{(p + A)^2}, \quad p \in (-\infty, a/3),$$

where A is a constant of integration which depends on the initial conditions satisfying that $(p + A)$ is always non-zero. From two equations of (6.3.14) or

(6.3.16) we find that the relation $8a^3 + 27\epsilon e^2 = 0$ and $4a^2 - 3d = 0$. Observe that the constant of motion ϵe^2 and d are completely determined by the constant a.

II. Q has two real roots, the lowest with multiplicity 2: $\alpha = a_1 = a_2 < a_3$. For this case, the root a_3 is given by $a - 2\alpha$ and following are two possibilities:

$$k(p) = a - 2\alpha + (3\alpha - a)\coth^2\left(\frac{1}{2}\sqrt{a - 3\alpha}\,(p + A)\right), \quad p \in (-\infty, \alpha).$$

$$k(p) = a - 2\alpha + (3\alpha - a)\tanh^2\left(\frac{1}{2}\sqrt{a - 3\alpha}\,(p + A)\right), \quad p \in (\alpha, a - 2\alpha].$$

The following relations hold: $-2\alpha(\alpha - a)^2 = \epsilon e^2$ and $-(\alpha - a)(3\alpha + a) = d$.

III. Q has two real roots, the greatest with multiplicity 2: $\alpha = a_1 < a_2 = a_3$. For this case, we have $a_2 = a_3 = (a - \alpha)/2$, with the solution

$$k(p) = \alpha + \frac{3\alpha - a}{2}\tan^2\left(\frac{1}{2}\sqrt{\frac{a - 3\alpha}{2}}\,(p + A)\right), \quad p \in (-\infty, \alpha].$$

Moreover, $(1/4)(\alpha - a)(\alpha + a)^2 = \epsilon e^2$ and $-(1/4)(\alpha + a)(3\alpha - 5a) = d$.

IV. Q has three distinct real roots: $a_1 < a_2 < a_3$. For this case, denote by $a_3 = a - a_1 - a_2$. Then, following are two possibilities:

$$k(p) = a_1 - (a_2 - a_1)\tan^2\left(\frac{1}{2}\sqrt{a - 2a_1 - a_2}\,(p + E),\ \sqrt{\frac{a - a_1 - 2a_2}{a - 2a_1 - a_2}}\right),$$

$$k(p) = a - a_1 - a_2$$
$$+ (a_1 + 2a_2 - a)\sin^2\left(\frac{1}{2}\sqrt{a - 2a_1 - a_2}(p + E),\ \sqrt{\frac{a - a_1 - 2a_2}{a - 2a_1 - a_2}}\right),$$

defined in the intervals $(-\infty, a_1]$ or $[a_2, a - a_1 - a_2]$, respectively. Moreover, $-(a_1 + a_2)(a_1 - a)(a_2 - a) = \epsilon e^2$ and $(a_1 + a)(a_2 + a) - (a_1 + a_2)^2 = d$.

V. Q has complex roots. Suppose a_1 and a_2 are complex so a_3 is real. Then,

$$k(p) = a_3 - (a_3 - a_2)\sin^2\left(\frac{1}{2}\sqrt{a_3 - a_1}(p + E),\ \sqrt{\frac{a_2 - a_3}{a_1 - a_3}}\right), \quad p \in (-\infty, a_3],$$

where if we set $a_1 = \alpha + i\beta$ and $a_2 = \alpha - i\beta$, then, the following relations hold:

$$-2\alpha\left((\alpha - a)^2 + \beta^2\right) = \epsilon e^2 \quad \text{and} \quad a^2 + 2\alpha a - 3\alpha^2 + \beta^2 = d.$$

Finally, using the cylindrical coordinates (see non-null and null cases of P) one can integrate the Cartan equations of the above obtained curves. The explicit integration of these equations is a difficult task, sometimes impossible even when the curvature is a nice function. However, it is clear from so far that null helices are always one of the trajectories of the relativistic particles.

Note 4. In [38], the authors have generalized above results for a $(2 + 1)$-dimensional spacetime manifold of constant curvature, with L linear in the curvature of the particle path. Their conclusions are the same as we have discussed for the case of $\mathbf{R}_1^3$ and include the work of Nersessian and Ramos [84] where they discussed a geometrical particle model for anyons for which the spin can take any real value. In addition to what we discussed so far, the authors of [38] have provided graphs of the world lines by using numerical integration for all the sub cases of the polynomial equation $Q(X) = 0$. Also, in [74], there is a discussion on $3D$ anyons with null world lines which may be constructed by reducing the model of spinning particles of a fixed mass to the null curves.

Moreover, in [85] the authors have considered a more complicated 3-dimensional system associated with null curves for which the action is a linear function in the curvature of the curve. They have shown that its mass and spin spectra are defined by 1-dimensional non-relativistic mechanics with cubic potential. Thus, their systems possess the typical properties of resonance-like particles.

L is a quadratic function on k**.** Finally, let the action I be described by

$$I(C_t) = \int_{C_t} (a + bk + ck^2)dp,$$

where a, b and $c \neq 0$ are constants. Without any loss of generality we set $c = 1$. Following exactly as in the linear case, the Euler Lagrange equation is given by

$$E(C_t) = 2k^{(5)} + (10k + b)k^{(3)} + 20k''k' + k'(15k^2 + 3bk - a) = 0.$$

We set $\phi = 2k'' + 3k' + bk - a$. For the case of $\phi = 0$, we obtain $P = 0$ and $J = -2(k^2 - a)\xi - 4k'W + 2(2k + b)N$ is a constant vector field with $g(J, J) = \epsilon j^2$. Then, the first family of solutions satisfy the following differential equation

$$(k')^2 + k^3 + \frac{b}{2}k^2 - ak - \left(\frac{8b + \epsilon j^2}{16}\right) = 0$$

which has the same nature as that of the equation (6.3.15) of the linear case 2, and the solutions are same as discussed for that family. If $\phi \neq 0$, then, we obtain

$$
\begin{aligned}
P &= \left(2k(4) + k''(8k + b) + 6(k')^2 + 3k^3 + bk^2 + ak\right)\xi \\
&\quad - \left(2k(3) + k'(6k + b)\right)W + (2k'' + 3k^2 + bk + a)N, \\
J &= (2k'' + k^2 + bk + a)\xi + 4k'W + 2(2k + b)N.
\end{aligned}
$$

From above equations we obtain the following three first integrals:

$$
\begin{aligned}
-2(2k^{(4)} &+ k''(8k + b) + 6(k')^2 + 3k^3 + bk^2 + ak)(2k'' + 3k^2 + bk - a) \\
&+ \left(2k(3) + k'(6k + b)\right)^2 + \epsilon e^2 = 0, \qquad\qquad (6.3.17)
\end{aligned}
$$

$$-(8k + 4b)k^{(4)} + 8k'k^{(3)} + 4(k'')^2 - (40k^2 + 24bk + 2b^2)k''$$
$$-8b(k')^2 - 15k^4 - 14bk^3 + (2a - 3b^2)k^2 + 2abk + a^2 - d = 0, \qquad (6.3.18)$$

$$2k^{(4)} + 10kk'' + bk'' + 5(k')^2 + 5k^3 + \frac{3}{2}bk^2 + ak + A = 0, \qquad (6.3.19)$$

where A is a constant. Combining these three equations (6.3.17) - (6.3.19) we obtain the following differential equation of degree two:

$$\frac{1}{16}(k')^2[-4(k'')^2 - 2(2k + b)(k')^2 + 5k^4 + 2bk^3 - 2ak^2 + 2A(2k + b)$$
$$+a^2 + d]^2 + (2k'' + 3k^2 + bk - a)[4kk'' - 2(k')^2 + 4k^3 + bk^2 + 2A]$$
$$- \epsilon e^2 = 0. \qquad (6.3.20)$$

The integration of this equation (6.3.20) is extremely difficult. In [41], the authors have discussed this case (in details) and by using computing methods they have drawn some graphs of their solutions.

Concluding remarks. We have so far seen that there are relativistic particle models based purely on the geometry of null curves in a 3-dimensional spacetime of constant curvature and these models are of independent interest. Thus, a priori there is no restriction to apply these ideas in other background gravitational fields of higher dimensions. We suggest considering actions in $(n + 1)$-dimensions ($n \geq 3$) whose Lagrangian depend on the curvature and study the trajectories of the relativistic particles in a prescribed model. Some limited work has been done on mechanical systems and $4D$ null curves. Here we brief on two papers [76, 83] in which the authors have discussed relativistic particle models associated with null paths in 4-dimensional Minkowski spacetimes. Since the details on these two papers involve higher knowledge on some concept in general theory of relativity, in this book we only provide an abstract of each of those two papers and leave up to the readers choice for detailed study through those papers.

Krüger ([76]). The author obtained two independent sets of Frenet - Darboux equations, with respect to two nonequivalent comoving Frenet frames of a null curve in a 4-dimensional Minkowski spacetime $\mathbf{R}_1^4$, whose orbits are on the group Spin(1, 3). The general solution of the Frenet-Darboux equations is derived by means of a quaternion decomposition of Spin(1, 3).

Based on these results, a universal form of Lagrangian dynamics for a null curve is constructed with the help of nonholonomic constraints. Noether's theorem is derived in order to express the equations of motion in terms of Lie derivatives with respect to the Poincaré group. This leads to an identification of momentum and angular momentum (spin). A new Lagrangian is postulated for a non-minimal coupling of an isotropic point charge to external fields. It implies a modification of Weyssenhoff's spin bivector and a generalization of his and Raabe's equations of motion (see reference [7] for their paper in [76]).

Nersessian and Ramos ([83]). They studied the simplest particle model associated with null paths in a 4-dimensional Minkowski spacetime. The action is given by the special (also called pseudo-arclength) parameter of the relativistic particle world line. They have shown that the reduced classical phase space of this system coincides with that of a massive spinning particle of spin $s = c^2/m$, where m is the particle mass and c is the coupling constant in front of the action. Consistency of the associated quantum theory requires the spin s to be an integer or a half integer number. If this holds then standard quantization techniques show that the corresponding Hilbert spaces are solution spaces of the standard relativistic massive wave equations. Therefore, this geometrical particle model provides with a unified description of Direc fermion ($s = 1/2$) and massive spin fields.

Chapter 7

Lightlike hypersurfaces

7.1 Introduction

Let $(\bar{M}, \bar{g})$ be a proper $(m+2)$-dimensional semi-Riemannian manifold of constant index $q \in \{1, \ldots, m+1\}$. We recall the following basic results. Details may be seen in any standard book on semi-Riemannian manifolds such as [82].

Recall from (2.1.1) that a linear connection $\bar{\nabla}$ on $(\bar{M}, \bar{g})$ is called a *metric* (*Levi-Civita*) connection if $\bar{g}$ is parallel with respect to $\bar{\nabla}$, i.e.,

$$(\bar{\nabla}_X \bar{g})(Y, Z) = X(\bar{g}(Y, Z)) - \bar{g}(\bar{\nabla}_X Y, Z) - \bar{g}(Y, \bar{\nabla}_X Z) = 0, \qquad (7.1.1)$$

for any $X, Y, Z \in \Gamma(T\bar{M})$. In terms of local coordinates system, we have

$$\bar{g}_{ij;\, k} = \partial_k \bar{g}_{ij} - \bar{g}_{ih} \Gamma^h_{jk} - \bar{g}_{jh} \Gamma^h_{ik} = 0 \qquad (7.1.2)$$

where

$$\Gamma^h_{ij} = \frac{1}{2} \bar{g}^{hk} \{\partial_j \bar{g}_{ki} + \partial_i \bar{g}_{kj} - \partial_k \bar{g}_{ij}\}, \quad \Gamma^h_{ij} = \Gamma^h_{ji}. \qquad (7.1.3)$$

Furthermore, if we set $\Gamma_{k|ij} = \bar{g}_{kh} \Gamma^h_{ij}$, then, (7.1.2) becomes

$$\bar{g}_{ij;\, k} = \partial_k \bar{g}_{ij} - \Gamma_{i|jk} - \Gamma_{j|ik} = 0.$$

The connection coefficients $\Gamma_{k|ij}$ and Γ^h_{ij} are called the *Christoffel symbols of first* and *second type* respectively. A result in semi-Riemannian geometry states (see O'Neill [82]) that their exists a metric connection $\bar{\nabla}$ which satisfies the following identity, so called *Koszul formula*

$$\begin{aligned}
2\bar{g}(\bar{\nabla}_X Y, Z) \;=\;& X(\bar{g}(Y, Z)) + Y(\bar{g}(X, Z)) - Z(\bar{g}(X, Y)) \\
+\;& \bar{g}([X, Y], Z) + \bar{g}([Z, X], Y) - \bar{g}([Y, Z], X), \qquad (7.1.4)
\end{aligned}$$

for any $X, Y, Z \in \Gamma(T\bar{M})$, where $[\,,]$ is Lie-bracket operator. In this book, we assume that $\bar{\nabla}$ is the Levi-Civita (metric) connection on $\bar{M}$. The *semi-Riemannian curvature tensor*, denoted by $\bar{R}$, of $\bar{M}$ is a $(1, 3)$ tensor field defined by

$$\begin{aligned}
\bar{R}(X, Y)Z \;=\;& \bar{\nabla}_X \bar{\nabla}_Y Z - \bar{\nabla}_Y \bar{\nabla}_X Z - \bar{\nabla}_{[X, Y]} Z, \text{ i.e.,} \qquad (7.1.5) \\
\bar{R}^t_{jhk} \;=\;& \partial_h \Gamma^t_{jk} - \partial_k \Gamma^t_{jh} + \Gamma^m_{jk} \Gamma^t_{mh} - \Gamma^m_{jh} \Gamma^t_{mk},
\end{aligned}$$

for any $X, Y, Z \in \Gamma(T\bar{M})$. The *torsion tensor*, denoted by $\bar{T}$, of $\bar{\nabla}$ is a $(1,2)$ tensor defined by

$$\bar{T}(X, Y) = \bar{\nabla}_X Y - \bar{\nabla}_Y X - [X, Y].$$

$\bar{R}$ is skew-symmetric in the first two slots. In case $\bar{T}$ vanishes on $\bar{M}$ we say that $\bar{\nabla}$ is *torsion-free* or *symmetric metric connection* on $\bar{M}$, which we assume in this book. The two *Bianchi's identities* are

$$\bar{R}(X, Y)Z + \bar{R}(Y, Z)X + \bar{R}(Z, X)Y = 0, \qquad (7.1.6)$$

$$(\bar{\nabla}_X \bar{R})(Y, Z, W) + (\bar{\nabla}_Y \bar{R})(Z, X, W) + (\bar{\nabla}_Z \bar{R})(X, Y, W) = 0, \text{ i.e.,} \qquad (7.1.7)$$

$$\bar{R}^i_{jkl} + \bar{R}^i_{klj} + \bar{R}^i_{ljk} = 0,$$

$$\bar{R}^i_{jkl;\, m} + \bar{R}^i_{jlm;\, k} + \bar{R}^i_{jmk;\, l} = 0.$$

The semi-Riemannian curvature tensor of type $(0, 4)$ is defined by

$$\begin{aligned}
\bar{R}(X, Y, Z, U) &= \bar{g}(\bar{R}(X, Y)Z, U), \quad \forall X, Y, Z, U \text{ on } \bar{M}, \text{ i.e.,} \\
\bar{R}_{ijhk} &= \bar{R}(\partial_h, \partial_k, \partial_j, \partial_i) = \bar{g}_{it}\bar{R}^t_{jhk}.
\end{aligned} \qquad (7.1.8)$$

Then by direct calculations we get

$$\begin{aligned}
\bar{R}(X, Y, Z, U) + \bar{R}(Y, X, Z, U) &= 0, \\
\bar{R}(X, Y, Z, U) + \bar{R}(X, Y, U, Z) &= 0, \\
\bar{R}(X, Y, Z, U) - \bar{R}(Z, U, X, Y) &= 0, \text{ i.e.,}
\end{aligned} \qquad (7.1.9)$$

$$\bar{R}_{ijkh} + \bar{R}_{jikh} = 0, \quad \bar{R}_{ijkh} + \bar{R}_{ijhk} = 0, \quad \bar{R}_{ijhk} - \bar{R}_{hkij} = 0.$$

Let $\{E_1, \ldots, E_{m+2}\}_x$ be a local orthonormal basis of $T_x\bar{M}$. Then,

$$g(E_i, E_j) = \epsilon_i \, \delta_{ij} \text{ (no summation in } i), \quad X = \sum_{i=1}^{m+2} \epsilon_i \, g(X, E_i) \, E_i,$$

where $\{\epsilon_i\}$ is the signature of $\{E_i\}$. Thus, we obtain

$$\bar{g}(X, Y) = \sum_{i=1}^{n} \epsilon_i \, \bar{g}(X, E_i) \, \bar{g}(Y, E_i).$$

The *Ricci tensor*, denoted by $\bar{R}ic$, is defined by

$$\bar{R}ic(X, Y) = trace\{Z \to \bar{R}(X, Z)Y\}, \qquad (7.1.10)$$

for any $X, Y \in \Gamma(T\bar{M})$. Locally, $\bar{R}ic$ and its *Ricci operator* $\bar{Q}$ are given by

$$\bar{R}ic(X, Y) = \sum_{i=1}^{m+2} \epsilon_i \, \bar{g}(\bar{R}(E_i, X)Y, E_i), \quad \bar{g}(\bar{Q}X, Y) = \bar{R}ic(X, Y), \text{ i.e.,} \qquad (7.1.11)$$

$$\bar{R}_{ij} = \bar{R}^t_{\ itj}, \qquad \bar{Q}^i_j = \bar{R}_{kj}g^{ki}.$$

$\bar{M}$ is *Ricci flat* if its Ricci tensor vanishes on $\bar{M}$. If $\dim(\bar{M}) > 2$ and

$$\bar{Ric} = kg, \quad k \text{ is a constant}, \tag{7.1.12}$$

then $\bar{M}$ is an *Einstein manifold*. For $\dim(\bar{M}) = 2$, any $\bar{M}$ is Einstein but k in (7.1.12) is not necessarily constant. The *scalar curvature* $\bar{r}$ is defined by

$$\bar{r} = \sum_{i=1}^{m+2} \epsilon_i \, \bar{Ric}(E_i, E_i) = \bar{g}^{ij} \bar{R}_{ij}. \tag{7.1.13}$$

Using (7.1.12) in (7.1.13) implies that $\bar{M}$ is Einstein if and only if $\bar{r}$ is constant and

$$\bar{Ric} = \frac{\bar{r}}{m+2} \, \bar{g}.$$

The *Weyl conformal curvature tensor* C of type $(1,3)$ is defined by

$$\begin{aligned}
\bar{C}(X,Y)Z &= \bar{R}(X,Y)Z + \frac{1}{m}\{\bar{Ric}(X,Z)Y - \bar{Ric}(Y,Z)X + \bar{g}(X,Z)\bar{Q}Y \\
&\quad - \bar{g}(Y,Z)\bar{Q}X\} - r\{m(m+1)\}^{-1}\{\bar{g}(X,Z)Y - \bar{g}(Y,Z)X\}, \text{ i.e.,}
\end{aligned}$$

$$\begin{aligned}
\bar{C}^h_{kij} &= \bar{R}^h_{kij} + \frac{1}{m}\left\{\delta^h_j \bar{R}_{ki} - \delta^h_i \bar{R}_{kj} + \bar{g}_{ki}\bar{R}^h_j - \bar{g}_{kj}\bar{R}^h_i\right\} \\
&\quad + r\{m(m+1)\}^{-1}\left\{\delta^h_i \bar{g}_{kj} - \delta^h_i \bar{g}_{ki}\right\}. \tag{7.1.14}
\end{aligned}$$

The tensor $\bar{C}$ vanishes for $\dim(\bar{M}) = 3$. Let $g' = \Omega^2 \bar{g}$ be a conformal transformation of $\bar{g}$ where Ω is smooth positive real function on $\bar{M}$. In particular, the conformal transformation is called *homothetic* if h is a non-zero constant. It is known that $\bar{C}$ is invariant under any such conformal transformation of the metric. If $\bar{g}$ is conformally related with a semi-Euclidean flat metric g' we say that $\bar{g}$ is *conformally flat* and $\bar{M}$ is then called a *conformally flat manifold*. It follows from a theorem of Weyl that $\bar{M}$ is conformally flat if and only if $\bar{C} \equiv 0$ for $\dim(\bar{M}) > 3$.

A vector field X on $\bar{M}$ is called a *Killing vector field* if

$$\pounds_X \bar{g} = 0, \tag{7.1.15}$$

where $\pounds_X$ is the *Lie derivative* with respect to the vector field X. See a note in section 7 for a brief on the concept of Lie derivatives.

Suppose π is a non-degenerate plane of $T_x\bar{M}$. Then, according to section 1.1 of chapter 1, the associated matrix G_x of $\bar{g}_x$, with respect to an arbitrary basis $B = \{u, v\}$, is of *rank* 2 and given by

$$G_p = \begin{pmatrix} g_{uu} & g_{uv} \\ g_{uv} & g_{vv} \end{pmatrix}, \quad \det(G_p) \neq 0. \tag{7.1.16}$$

Define a real number $K(\pi) = K_x(u, v) = \frac{\bar{R}(u,\, v,\, v,\, u)}{\det(G_x)}$, where $\bar{R}(u,\, v,\, v,\, u)$ is the 4-linear mapping on $T_x(M)$ by the curvature tensor as given in (7.1.8). The smooth function K which assigns to each non-degenerate tangent plane π the real number

$K(\pi)$ is called the *sectional curvature* of $\bar{M}$, which is independent of the basis $B = \{u, v\}$. If K is a constant c at every point of $\bar{M}$ we say that $\bar{M}$ is a *semi-Riemannian manifold of constant sectional curvature* c, denote by $\bar{M}(c)$, whose curvature tensor field $\bar{R}$ is given by [82, page 80]

$$
\begin{aligned}
\bar{R}(X, Y)Z &= c\{\bar{g}(Y, Z)X - \bar{g}(X, Z)Y\}, \quad \text{i.e.,} \quad &(7.1.17) \\
\bar{R}^h{}_{kij} &= c\{\delta^h_i \bar{g}_{jk} - \delta^h_j \bar{g}_{ki}\}.
\end{aligned}
$$

Suppose M is a $(m + 1)$-dimensional smooth manifold and $i : M \to \bar{M}$ is a smooth mapping such that each point $x \in M$ has an open neighborhood $\mathcal{U}$ for which i restricted to $\mathcal{U}$ is one to one and $i^{-1} : i(\mathcal{U}) \to M$ are smooth. Then, we say that $i(M)$ is an *immersed submanifold* of $\bar{M}$. If this condition globally holds then $i(M)$ is called an *embedded hypersurface* of $\bar{M}$, which we assume in this book. The embedded hypersurface has a natural manifold structure inherited from the manifold structure on $\bar{M}$ via the embedding mapping. At each point $i(x)$ of $i(M)$, the tangent space is naturally identified with an $(m + 1)$-dimensional subspace $T_{i(x)}M$ of the tangent space $T_{i(x)}\bar{M}$. The embedding i induces, in general, a symmetric tensor field, say g, on $i(M)$ such that

$$
g(X, Y)|_x = \bar{g}(i_*X, i_*Y)|_{i(x)}, \quad (7.1.18)
$$

for any $X, Y \in T_x(M)$. Here i_* is the differential map of i defined by $i_* : T_x \to T_{i(x)}$ and $(i_*X)(f) = X(f \circ i)$ for an arbitrary smooth function f in a neighborhood of $i(x)$ of $i(M)$. Henceforth, we write M and x instead of $i(M)$ and $i(x)$. As explained in chapter 1, due to the causal character of three categories (spacelike, timelike and null) of the vector fields of $\bar{M}$, there are three types of hypersurfaces M, namely, Riemannian, semi-Riemannian and lightlike(null) and g is a non-degenerate metric tensor field or a degenerate symmetric tensor field on M according as M is of the first two types and of the third type respectively.

We first assume that g is non-degenerate so that (M, g) is a semi-Riemannian hypersurface of $(\bar{M}, \bar{g})$. Define

$$
TM^\perp = \{V \in \Gamma(T\bar{M}) : g(V, W) = 0, \ \forall W \in \Gamma(T\bar{M})\} \quad (7.1.19)
$$

the normal bundle subspace of M in $\bar{M}$. Since M is a hypersurface of $\bar{M}$, $\dim(T_x M^\perp) = 1$. Following is the orthogonal complementary decomposition:

$$
T\bar{M} = TM \oplus_{orth} TM^\perp, \quad TM \cap TM^\perp = \{0\}. \quad (7.1.20)
$$

Here, both the tangent and the normal bundle subspaces are non-degenerate and any vector field of $T\bar{M}$ splits uniquely into a component tangent to M and a component perpendicular to M. Let $\bar{\nabla}$ and ∇ be the Levi-Civita connections on $\bar{M}$ and M respectively. Then, there exists a uniquely defined unit normal vector field, say $\mathbf{n} \in \Gamma(T\bar{M})$ and the Gauss-Weingarten formulas are

$$
\begin{aligned}
\bar{\nabla}_X Y &= \nabla_X Y + B(X, Y)\mathbf{n}, \\
\bar{\nabla}_X \mathbf{n} &= -\epsilon A_{\mathbf{n}} X, &(7.1.21)
\end{aligned}
$$

for any tangent vectors X and Y of M and $g(\mathbf{n}, \mathbf{n}) = \epsilon = \pm 1$ such that $\mathbf{n}$ belongs to $TM^{\perp}$ and $\nabla_X Y$, $A_{\mathbf{n}} X$ belong to the tangent space. Here $B(-, -)\mathbf{n}$ is *the second fundamental form tensor* and B is *the second fundamental form*, related with *the shape operator* $A_{\mathbf{n}}$ by

$$B(X, Y) = \bar{g}(A_{\mathbf{n}} X, Y), \quad \forall X, Y \in \Gamma(TM).$$

A point $x \in M$ is said to be *umbilical point* if $B(X, Y)_x = kg(X, Y)_x, \forall X, Y \in \Gamma(TM)$, where $k \in \mathbf{R}$ and depends on x. M is *totally umbilical* in $\bar{M}$ if every point of M is umbilical, i.e., if

$$B = \rho\, g, \tag{7.1.22}$$

where ρ is a smooth function. In particular, we say that M is *totally geodesic* in $\bar{M}$ if B vanishes identically. With respect to an orthonormal basis $\{E_1, \cdots, E_{m+1}\}$, the *mean curvature vector* μ of M is defined by

$$\mu = \frac{Trace(\epsilon B)}{m+1} = \frac{1}{m+1} \sum_{i=1}^{m+1} \epsilon_i B(E_i, E_i), \quad g(E_i, E_i) = \epsilon_i,$$

where $\epsilon = (\delta_{ij}\epsilon_j)$ is the signature matrix. M is *minimal* in $\bar{M}$ if $\mu = 0$. Any totally geodesic M is minimal in $\bar{M}$.

If g is degenerate on M, then, there exists a vector field $\xi \neq 0$ on M such that

$$g(\xi, X) = 0, \quad \forall X \in \Gamma(TM). \tag{7.1.23}$$

The *radical* (Artin [3, page 53]) or the *null space* (O'Neill [82, page 53]) of $T_x M$, at each point $x \in M$, is a subspace $Rad\, T_x M$ defined by

$$Rad\, T_x M = \{\xi \in T_x M : g_x(\xi, X) = 0, \forall X \in T_x M\}, \tag{7.1.24}$$

whose dimension is called the *nullity degree* of g and M is called a *lightlike hypersurface* of $\bar{M}$ [28]. Comparing (7.1.19) with (7.1.24), with respect to degenerate g, and any null vector being perpendicular to itself implies that $T_x M^{\perp}$ is also null and

$$Rad\, T_x M = T_x M \cap T_x M^{\perp}.$$

Since for a hypersurface M, $\dim(T_x M^{\perp}) = 1$, above relation implies that $\dim(Rad\, T_x M) = 1$ and $Rad\, T_x M = T_x M^{\perp}$. We call $Rad\, TM$ a *radical (null) distribution* (see section 7.4 for information on distributions) of M. Thus, for a lightlike hypersurface M, (7.1.20) does not hold because TM and $TM^{\perp}$ have a non-trivial intersection and their sum is not the whole of tangent bundle space $T\bar{M}$. In other words, a vector of $T_x \bar{M}$ cannot be decomposed uniquely into a component tangent to $T_x M$ and a component of $T_x M^{\perp}$. Therefore, the standard text-book definition of the second fundamental form and the Gauss-Wiengarten formulas do not work, in the usual way, for the lightlike case.

To deal with this anomaly, lightlike manifolds have been studied by several ways corresponding to their use in a given problem. Indeed, see Akivis-Goldberg [1, 2]; Bonnor [18]; Israel [65, 66]; Katsuno [72]; Nurowski-Robinson [88]; Penrose [89];

Rosca [98, 99] and more referred therein. In 1991, Bejancu-Duggal [13] introduced a general geometric technique to remove this anomaly by splitting the tangent bundle $T\bar{M}$ into three non-intersecting complementary (but not orthogonal) vector bundles (two of them null and one non-null). This result on lightlike hypersurfaces was presented in a 1996 Duggal-Bejancu's book [28]. In the same year Kupeli [78] published a book on singular semi-Riemannian geometry. Duggal-Bejancu's approach was basically extrinsic in contrast to the intrinsic one developed by Kupeli. Recently, there has been considerable amount of new material on lightlike hypersurfaces and their applications to some problems in general relativity, published by a large number of researchers. The objective of this chapter is to present the results obtained so far on the lightlike hypersurfaces and its applications in mathematics and physics. We also include a fresh and improved version of the material already appeared in [28, chapter 4] to make this volume a self contained book.

Our first task, in next section 7.2, is to introduce a non-degenerate screen distribution and a null transversal vector bundle so as to replace the equation (7.1.20) by another one which plays a key roll in finding the main induced geometric objects (such as second fundamental forms, shape operators, induced connection, curvature and Ricci tensors, scalar curvature etc) needed in the development of the lightlike geometry. In section 7.3, we collect all the basis results on two degenerate second fundamental forms. Since the screen distribution is not unique, we investigate its dependence on other induced objects discussed in this section. Section 7.4 includes results on integrable screen distributions and canonical screens [11]. Section 7.5 deals with curvature properties. In section 7.6, we make the first-ever attempt in mathematical literature to introducing the concept of induced *scalar curvature of genus zero* [27] for a specified class of lightlike hypersurfaces.

7.2   Screen and transversal bundles

Let (M, g) be a lightlike hypersurface of a proper $(m + 2)$-dimensional semi-Riemannian manifold $(\bar{M}, \bar{g})$ of constant index $q \in \{1, \ldots, m + 1\}$. As explained in section 7.1, for a lightlike hypersurface M, $TM^{\perp} = Rad\,TM$. Thus, in the sequel, $TM^{\perp}$ will be replaced by $Rad\,TM$ where ever it occurs. To develop the geometry of lightlike hypersurfaces (in line with the theory of non-null hypersurfaces) we show how the equation (7.1.20) can be changed by another one such that the tangent bundle $T\bar{M}$ splits into three non-intersecting complementary (but non-orthogonal) vector bundles. For this purpose we consider a complementary vector bundle $S(TM)$ of $Rad\,TM$ in TM. This means that

$$TM = Rad\,TM \oplus S(TM). \tag{7.2.1}$$

Following [28, chapter 4] we call $S(TM)$ a screen distribution on M. It is immediate from the decomposition equation (7.2.1) that $S(TM)$ is a non-degenerate distribution. Moreover, since we assume that M is paracompact, there always exists a screen $S(TM)$. Thus, along M we have the following decomposition

$$T\bar{M}_{|M} = S(TM) \oplus_{orth} S(TM)^{\perp}, \quad S(TM) \cap S(TM)^{\perp} \neq \{0\}, \tag{7.2.2}$$

that is, $S(TM)^\perp$ is orthogonal complement to $S(TM)$ in $T\bar{M}_{|M}$. Note that $S(TM)^\perp$ is also a non-degenerate vector bundle of rank 2. However, it includes $TM^\perp = Rad\,TM$ as its subbundle. Unfortunately, contrary to the non-null case, the equation (7.2.2) is not unique as it depends on the choice of a screen vector bundle which is not unique. This creates a problem with the change of $S(TM)$. In the literature, there are several specific approaches to deal with this problem. Kupeli [78] has shown that $S(TM)$ is canonically isometric to the factor vector bundle $TM^* = TM/TM^\perp$ and used canonical projection $\pi : TM \to TM^*$ in studying the intrinsic geometry of degenerate semi-Riemannian manifolds. In this book we will follow Duggal-Bejancu's [28] approach (which is primarily extrinsic and in line with the well-known differential geometry of Riemannian submanifolds). As the subject matter develops, we examine the dependence (or otherwise) of $S(TM)$ on the induced geometric objects of M when those objects are discussed. Moreover, we present several cases of lightlike hypersurfaces with either a canonical screen [11] or a good choice of screen distribution with interesting geometric and or physical properties [29]. We will also study those lightlike hypersurfaces which have a integrable (not necessarily canonical) screen distribution. In this section we prove that given a screen distribution $S(TM)$ with respect to a section ξ of $Rad\,TM$, there exists a unique null section of $T\bar{M}$ and not belonging to TM, called *lightlike transversal section*, which plays a roll similar to that of unique unit normal of a non-null hypersurface of a semi-Riemannian manifold. To show this, we state the following (proof is common with theorem 4.1 in chapter 1):

Theorem 2.1. *Let* $(M, g, S(TM))$ *be a lightlike hypersurface of a proper semi-Riemannian manifold* $(\bar{M}, \bar{g})$ *and* $\pi : ltr(TM) \to \bar{M}$ *be a subbundle of* $S(TC^\perp)^\perp$ *such that* $S(TM)^\perp = Rad\,TM \oplus ltr(TM)$. *Suppose* $V \in \Gamma^\infty(\mathcal{U}, ltr(TM))$ *is a locally defined nowhere zero section, defined on the open subset* $\mathcal{U} \subseteq \bar{M}$. *Then, for any non-zero section* ξ *of* $Rad\,TM$ *on a coordinate neighborhood* $\mathcal{U} \subset M$, *we have*

(i) $\bar{g}(\xi, V) \neq 0$ *everywhere on* $\mathcal{U} \subseteq \bar{M}$.

(ii) *If we consider* $N_V \in \Gamma^\infty(\mathcal{U}, S(TM^\perp)^\perp)$ *given by*

$$N_V = \frac{1}{\bar{g}(\xi, V)} \left\{ V - \frac{\bar{g}(V, V)}{2\bar{g}(\xi, V)} \xi \right\}, \qquad (7.2.3)$$

then $ltr(TM)$ *is a unique vector bundle over* M *of rank 1 such that on each* $\mathcal{U} \subset M$ *there is a unique vector field* $N_V \in \Gamma(ltr(TM)_{|\mathcal{U}})$ *satisfying*

$$\bar{g}(N_V, N_V) = 0, \quad \bar{g}(\xi, N_V) = 1. \qquad (7.2.4)$$

(iii) *The tangent bundle* $T\bar{M}$ *splits into the following three bundle spaces:*

$$T\bar{M}_{|M} = Rad\,TM \oplus S(TM) \oplus ltr(TM) = TM \oplus ltr(TM). \qquad (7.2.5)$$

Thus, given a screen distribution $S(TM)$, there exists a unique complementary vector bundle $ltr(TM)$ to TM in $T\bar{M}_{|M}$, where *ltr* is the abbreviation of the words (lightlike transversal) and $ltr(TM)$ is called the *lightlike transversal vector bundle* of M. At this point we write N instead of N_V for discussing the general properties of induced objects on M. Using (7.2.1), (7.2.5) and section 1.4 (Type I) of chapter 1, we say that there exists a local quasi-orthonormal basis of $\bar{M}$ along M, given by

$$F = \{\xi,\, N,\, W_a\}, \quad a \in \{1,\ldots,m\}, \tag{7.2.6}$$

where $\{\xi\}$, $\{N\}$ and $\{W_a\}$ are null basis of $\Gamma(Rad\,TM_{|\mathcal{U}})$, $\Gamma(ltr(TM)_{|\mathcal{U}})$ and orthonormal basis of $\Gamma(S(TM)_{|\mathcal{U}})$ respectively.

To investigate the dependence of induced objects on a change in the vector bundles $S(TM)$ and $ltr(TM)$, we consider two quasi-orthonormal frames fields $F = \{\xi,\, N,\, W_a\}$ and $F' = \{\xi',\, N',\, W'_a\}$ induced on $\mathcal{U} \subset M$ by $\{S(TM),\, ltr(TM)\}$ and $\{S'(TM),\, (ltr)'(TM)\}$ respectively. Using (7.2.4) and (7.2.5) we obtain

$$W'_a = \sum_{b=1}^{m} W_a^b(W_b - \epsilon_b\,\mathbf{f}_b\,\xi)$$

$$N' = N + \mathbf{f}\,\xi + \sum_{a=1}^{m} \mathbf{f}_a W_a,$$

where $\{\epsilon_a\}$ are signatures of orthonormal basis $\{W_a\}$ and W_a^b, $\mathbf{f}$ and $\mathbf{f}_a$ are smooth functions on $\mathcal{U}$ such that $\left[W_a^b\right]$ is $m \times m$ semi-orthogonal matrices. Also, computing $\bar{g}(N', N') = 0$ and using (7.2.4) and $\bar{g}(W_a, W_a) = 1$ we obtain

$$2\mathbf{f} + \sum_{a=1}^{m} \epsilon_a\,(\mathbf{f}_a)^2 = 0.$$

Using this in the second relation of above two equations, we get

$$W'_a = \sum_{b=1}^{m} W_a^b\,(W_b - \epsilon_b\,\mathbf{f}_b\,\xi), \tag{7.2.7}$$

$$N' = N - \frac{1}{2}\left\{\sum_{a=1}^{m} \epsilon_a\,(\mathbf{f}_a)^2\right\}\xi + \sum_{a=1}^{m} \mathbf{f}_a W_a. \tag{7.2.8}$$

These two relations will be used to investigate the transformation of the induced geometric objects (as those are discussed) when we change the pair $\{S(TM),\, ltr(TM)\}$ with respect to a change in the local basis, for which we need the following:

A *distribution* of rank r on M is a mapping D defined on M which assigns to each point x of M an r-dimensional linear subspace D_x of T_xM. Let $f : M' \to M$ be an immersion of M' in M. This means that the tangent mapping

$$(f_*)_x : T_x\,M' \to T_{f(x)}M\,,$$

is an injective mapping for any $x \in M'$. Suppose D is a distribution on M. Then M' is called an *integral manifold* of D if for any $x \in M'$ we have

$$(f_*)_x\,(T_x\,M') = D_{f(x)}\,.$$

If M' is a connected integral manifold of D and there exists no connected integral manifold $\bar{M}'$, with immersion $\bar{f} : \bar{M}' \to M$, such that $f(M') \subset \bar{f}(\bar{M}')$, we say that M' is a *maximal integrable manifold* or a *leaf* of D. The distribution D is said to be *integrable* if for any point $x \in M$ there exists an integral manifold of D containing x. Recall that the distribution D is involutive if for two vector fields X and Y belonging to D, the Lie-bracket $[X, Y]$ also belongs to D. We quote the following well-known theorem:

Theorem (Frobenius). *A distribution D on M is integrable, if and only if, it is involutive. Moreover, through every point $x \in M$ there passes a unique maximal integral manifold of D and every other integral manifold containing x is an open submanifold of the maximal one.*

From Frobenius theorem it follows that leaves of D determine a *foliation* on M of dimension r, that is, M is a disjoint union of connected subsets $\{L_t\}$ and each point x of M has a coordinate system $(\mathcal{U}; x^1, \ldots, x^{m+1})$ such that $L_t \cap \mathcal{U}$ is locally given by the equations

$$x^a = c^a, \qquad a \in \{r+1, \ldots, m+1\}, \tag{7.2.9}$$

where c^a are real constants, and (x^α), $\alpha \in \{1, \ldots, r\}$, are local coordinates on L_t.

Since $Rad\,TM = TM^\perp$ is a distribution of rank 1 on M, it is integrable and therefore, as per Frobenius theorem there exists an atlas of local charts $\{\mathcal{U}; u^o, \ldots, u^m\}$ such that $\partial_{u^o} \in \Gamma(Rad\,TM_{|\mathcal{U}})$. Thus, the matrix of the degenerate metric g on M with respect to the natural frames field $\{\partial_{u^o}, \ldots, \partial_{u^m}\}$ is as follows

$$[g] = \begin{bmatrix} 0 & 0 \\ 0 & g_{ab}(u^o, \ldots, u^m) \end{bmatrix}, \tag{7.2.10}$$

where $g_{ab} = g(\partial_{u^a}, \partial_{u^b})$, $a, b \in \{1, \ldots, m\}$, and $\det[g_{ab}] \neq 0$. The transformation of coordinates on M, with an integrable distribution $Rad\,TM$, has a special form. Precisely, considering another coordinate system $(\bar{\mathcal{U}}, \bar{u}^o; \bar{u}^a)$ on M and using (7.2.9), for both systems, we obtain

$$0 = d\bar{u}^a = \frac{\partial \bar{u}^a}{\partial u^b} d u^b + \frac{\partial \bar{u}^a}{\partial u^o} d u^o = \frac{\partial \bar{u}^a}{\partial u^o} d u^o,$$

which imply

$$\frac{\partial \bar{u}^a}{\partial u^o} = 0, \qquad \forall\, a \in \{1, \ldots, m\}.$$

Hence the transformation of coordinates on M is given by

$$\begin{aligned} \bar{u}^o &= \bar{u}^o(u^o, u^1, \ldots, u^m) \\ \bar{u}^i &= \bar{u}^i(u^1, \ldots, u^m). \end{aligned} \tag{7.2.11}$$

It follows that

$$\partial_{u^o} = F(u)\partial_{\bar{u}^o}, \qquad \partial_{u^a} = F_a^b(u)\partial_{\bar{u}^b} + F_a(u)\partial_{u^o},$$

where we put $F_a^b(u) = \frac{\partial \bar{u}^b}{\partial u^a}$, $F_a(u) = \frac{\partial \bar{u}^o}{\partial u^a}$ and $F(u) = \frac{\partial \bar{u}^o}{\partial u^o}$. As the screen distribution $S(TM)$ is transversal to the involutive distribution $Rad\,TM$, using the foliations theory we say that there exist m differentiable functions $S_a(u^o, \ldots, u^m)$ satisfying

$$S_a(u)F(u) = \bar{S}_b(u)F_a^b(u) + F_a(u), \tag{7.2.12}$$

with respect to the transformations (7.2.10), which provides a local basis of $\Gamma(S(TM))$ given by

$$\delta_{u^a} = \partial_{u^a} - S_a(u)\partial_{u^o}, \quad \delta_{u^a} = F_a^b(u)\delta_{\bar{u}^b}, \tag{7.2.13}$$

Hence, we obtain the local field of frames $\{\partial_{u^o}, N, \delta_{u^a}\}$ on $\bar{M}$, where $\{\partial_{u^o}, \delta_{u^a}\}$ is a local field of frames on M adapted to the decomposition (7.2.1), N is taken from theorem 2.1 and ξ is replaced by ∂_{u^o}.

Since, in general, the screen distribution of a lightlike hypersurface is not unique, one may look for a class of lightlike hypersurfaces with canonical screen or at least a screen with good properties. In this book we will show (at appropriate places) the existence of a large variety of semi-Riemannian manifolds having lightlike hypersurfaces with canonical screen or a screen with good properties. Based on the information we have so far, here we start with one of such class as follows:

Let $(\mathbf{R}_q^{m+2}, \bar{g})$ be a semi-Euclidean space with the metric $\bar{g}$ given by

$$\bar{g}(x, y) = -\sum_{i=0}^{q-1} x^i y^i + \sum_{a=q}^{m+1} x^a y^a. \tag{7.2.14}$$

Consider a hypersurface M of $\mathbf{R}_q^{m+2}$ locally given by the equations

$$x^A = \phi^A(u^0, \ldots, u^m), \quad \mathrm{rank}\left[\frac{\partial \phi^A}{\partial u^\alpha}\right] = m+1, \tag{7.2.15}$$

where $A \in \{0, \ldots, m+1\}$, $\alpha \in \{0, \ldots, m\}$ and $\{\phi^A\}$ are smooth functions on a coordinate neighborhood $\mathcal{U} \subset M$. Then we set

$$D^A = \begin{vmatrix} \frac{\partial \phi^0}{\partial u^0} & \cdots & \frac{\partial \phi^{A-1}}{\partial u^0} & \frac{\partial \phi^{A+1}}{\partial u^0} & \cdots & \frac{\partial \phi^{m+1}}{\partial u^0} \\ \cdot & & \cdot & \cdot & & \cdot \\ \cdot & & \cdot & \cdot & & \cdot \\ \frac{\partial \phi^0}{\partial u^m} & \cdots & \frac{\partial \phi^{A-1}}{\partial u^m} & \frac{\partial \phi^{A+1}}{\partial u^m} & \cdots & \frac{\partial \phi^{m+1}}{\partial u^m} \end{vmatrix}. \tag{7.2.16}$$

In this section, we use the following range of indices (unless otherwise stated):

$$A, B, C, \ldots \in \{0, \ldots, m+1\} \quad ; \quad \alpha, \beta, \gamma, \ldots \in \{0, \ldots, m\};$$
$$a, b, c, \ldots \in \{q, \ldots, m+1\} \quad ; \quad i, j, k, \ldots \in \{0, \ldots, q-1\}.$$

Theorem 2.2. Let (M, g) be a lightlike hypersurface of a semi-Euclidean space $(\mathbf{R}_q^{m+2}, \bar{g})$, given by (7.2.15). Then, for any non-zero section ξ of $Rad\,TM$ on a coordinate neighborhood $\mathcal{U} \in M$, there exists a canonical screen distribution *and a canonical lightlike transversal vector bundle on M.*

Proof. We first show that a hypersurface M of $\mathbf{R}_q^{m+2}$, given by (7.2.15), is lightlike, if and only if, on each $\mathcal{U}$, functions $\{\phi^A\}$ satisfy

$$\sum_{i=0}^{q-1}(D^i)^2 = \sum_{a=q}^{m+1}(D^a)^2, \tag{7.2.17}$$

and $Rad\,TM$ is locally spanned by

$$\xi = \sum_{i=0}^{q-1}(-1)^i\,D^i\,\partial_{x^i} + \sum_{a=q}^{m+1}(-1)^{a-1}\,D^a\,\partial_{x^a}. \tag{7.2.18}$$

Indeed, the natural frames field is given by $\partial_{u^\alpha} = \frac{\partial \phi^A}{\partial u^\alpha}\,\partial_{x^A}$, $\alpha \in \{0,\ldots,m\}$. Then one can check that ξ, given by (7.2.18), belongs to $\Gamma(Rad\,TM_{|\mathcal{U}})$. Hence, M is lightlike if and only if $\bar{g}(\xi,\xi) = 0$, which is equivalent to (7.2.18). Now consider a local section of $T\mathbf{R}_q^{m+2}$ defined on $\mathcal{U}$ by

$$V = \sum_{i=0}^{q-1}(-1)^{i-1}D^i\partial_{x^i}. \tag{7.2.19}$$

Then,

$$\bar{g}(V,\,\xi) = \sum_{i=0}^{q-1}(D^i)^2. \tag{7.2.20}$$

This section is nowhere tangent to M. Indeed, if there exists a point $x \in \mathcal{U}$ such that $\bar{g}_x(V_x,\,\xi_x) = 0$, then by (7.2.17) and (7.2.20) all determinants, of type as in (7.2.16), vanish at x. This is a contradiction as rank $\left[\frac{\partial \phi^A}{\partial u^\alpha}\right] = m+1$ at x.

Thus, as per theorem 2.1, we use V given by (7.2.19) and satisfying (7.2.20) to obtain the following null transversal vector bundle

$$N = \left(\sum_{i=0}^{q-1}(D^i)^2\right)^{-1}\left\{V + \frac{1}{2}\xi\right\}. \tag{7.2.21}$$

The sections given by (7.2.19) on each coordinate neighborhood determine a vector bundle E over M of rank 1. Suppose $\{\mathcal{U}';u'^0,\ldots,u'^m\}$ is another coordinate system on M such that $\mathcal{U}\cap\mathcal{U}' \neq \emptyset$. The immersion of $\mathcal{U}'$ in $\mathbf{R}_1^{m+2}$ is now given by

$$x^A = \phi'^A(u'^0,\ldots,u'^m), \quad A \in \{0,\ldots,m+1\}.$$

Then, by straightforward calculations we get on $\mathcal{U}\cap\mathcal{U}'$:

$$V = \frac{D(u'^0,\ldots,u'^m)}{D(u^0,\ldots,u^m)}V',$$

where V' is given by (7.2.19) but with respect to the lightlike immersion of $\mathcal{U}'$ in $\mathbf{R}_q^{m+2}$. Consider the vector bundle $H = Rad\,TM \oplus L$ over M, which is non-degenerate with respect to $\bar{g}$. Indeed, suppose there exists a point $x \in M$ and a non-zero vector $X_x \in H_x$ such that

$$\bar{g}_x(X_x, \xi_x) = 0 \quad \text{and} \quad \bar{g}(X_x, V_x) = 0.$$

Then, from the first equality it follows that $X_x \in T_xM$. But $T_xM \cap H_x = Rad\,T_xM$, and, therefore, X_x should be collinear with ξ_x. Thus $\bar{g}(X_x, V_x) \neq 0$, which contradicts the above second equality.

Take the orthogonal complimentary vector bundle $S(TM)$ to H in $T\mathbf{R}_q^{m+2}$ over M. Then, it is easy to see that $S(TM) \cap Rad\,TM = \{0\}$, and it is of rank m. Thus, it follows that $S(TM)$ is a distribution on M and $TM = Rad\,TM \oplus S(TM)$. Hence, $S(TM)$ is a screen distribution on M.

Finally, consider two quasi-orthonormal frames fields $F = \{\xi, N, W_a\}$ and $F' = \{\xi', N', W_a'\}$ induced on $\mathcal{U} \cap \mathcal{U}' \neq \emptyset \subset M$ by $\{S(TM), ltr(TM)\}$ and $\{S'(TM), (ltr)'(TM)\}$ respectively. Then, using (7.2.19), N and N' from (7.2.21) in (7.2.8) implies that all the functions $\mathbf{f_a}$ vanish. Therefore, we have

$$W_a' = \sum_{b=1}^{m} W_a^b\, W_b, \quad N' = N.$$

Consequently, $S(TM)$ is a canonical screen distribution and N, defined by (7.2.21), is a canonical lightlike vector bundle of M, which proves the theorem.

We close this section with the following examples of lightlike hypersurfaces:

Example 1. Let $(\mathbf{R}_2^4, \bar{g})$ be a 4-dimensional semi-Euclidean space of index 2 with signature $(-, -, +, +)$ of the canonical basis $(\partial_0, \ldots, \partial_3)$. Consider a Monge hypersurface M of $\mathbf{R}_2^4$ given by

$$x_3 = Ax_0 + Bx_1 + Cx_2, \quad A^2 + B^2 - C^2 = 1, \quad A, B, C \in \mathbf{R}.$$

Let $f(x_0, x_1, x_2, x_3) = x_3 - Ax_0 - Bx_1 - Cx_2$, then $M = f^{-1}(0)$ and since the gradient vector field $\nabla f = \sum \epsilon_i \frac{\partial f}{\partial x_i} \partial_i$ is orthogonal to all vectors tangent to the level surface M, it is easy to check that M is a lightlike hypersurface whose radical distribution $Rad\,TM$ is spanned by

$$\xi\,(= \nabla f) = A\partial_0 + B\partial_1 - C\partial_2 + \partial_3.$$

Let $V = -C\partial_2 + \partial_3$, then $g(V, V) = C^2 + 1$ and $g(\xi, V) = C^2 + 1$. Then, as per theorem 2.1, the lightlike transversal vector bundle is given by

$$ltr(TM) = Span\left\{N = -\frac{1}{2(C^2 + 1)}(A\partial_0 + B\partial_1 + C\partial_2 - \partial_3).\right\}$$

Since $u_0 = x_0$, $u_1 = x_1$, $u_2 = x_2$, $x_3 = Au_0 + Bu_1 + Cu_2$ and $\partial_{u_i} = \sum \frac{\partial x_A}{\partial u_i} \partial_A$, the tangent bundle $T\mathbf{R}_2^4$ is spanned by

$$\{\partial_{u_0} = \partial_0 + A\partial_3, \quad \partial_{u_1} = \partial_1 + B\partial_3, \quad \partial_{u_2} = \partial_2 + C\partial_3\}.$$

It follows that the corresponding screen distribution $S(TM)$ is spanned by

$$\{W_1 = B\partial_0 - A\partial_1, \quad W_2 = \partial_2 + C\partial_3\},$$

because $(A, B) \neq (0, 0)$, that is, $C^2 + 1 \neq 0$. Let $U_i = \frac{1}{A^2+B^2} W_i$, $i = 1, 2$, then $\{U_1, U_2\}$ is an orthonormal basis of $S(TM)$.

If we take $\bar{V} = \partial_3$, then since $g(\bar{V}, \bar{V}) = 1$ and $g(\xi, \bar{V}) = 1$, the lightlike transversal vector bundle is given by

$$l\bar{t}r(TM) = Span\left\{\bar{N} = -\frac{1}{2}(A\partial_0 + B\partial_1 - C\partial_2 - \partial_3)\right\}$$

and the corresponding screen distribution $\bar{S}(TM)$ is spanned by

$$\{\bar{W}_1 = B\partial_0 - A\partial_1, \quad \bar{W}_2 = AC\partial_0 + BC\partial_1 - (A^2 + B^2)\partial_2\}.$$

Example 2. Consider a Monge hypersurface M of $\mathbf{R}_2^4$ given by

$$x_0^2 + x_1^2 = x_2^2 + x_3^2.$$

It is easy to check that M is a lightlike hypersurface whose radical distribution $Rad\,TM$ is spanned by

$$\xi = x_0\partial_0 + x_1\partial_1 + x_2\partial_2 + x_3\partial_3.$$

Let $V = x_0\partial_0 + x_1\partial_1$, then $g(V, V) = g(\xi, V) = x_0^2 + x_1^2$. Then the lightlike transversal vector bundle

$$ltr(TM) = Span\left\{N = \frac{1}{2(x_0^2 + x_1^2)}(x_0\partial_0 + x_1\partial_1 - x_2\partial_2 - x_3\partial_3)\right\}$$

and the tangent bundle $T\mathbf{R}_2^4$ is spanned by

$$\{\partial_{u_0} = \mathbf{f}\,\partial_0 \pm x_0\,\partial_3, \quad \partial_{u_1} = \mathbf{f}\,\partial_1 \pm x_1\,\partial_3, \quad \partial_{u_2} = \mathbf{f}\,\partial_2 \mp x_2\,\partial_3\},$$

where $\mathbf{f} = \sqrt{x_0^2 + x_1^2 - x_2^2}$. It follows that the corresponding screen distribution $S(TM)$ is spanned by

$$\{W_1 = x_1\,\partial_0 - x_0\,\partial_1, \; W_2 = \mathbf{f}\,\partial_2 \mp x_2\,\partial_3\}.$$

Example 3. Let $(\mathbf{R}_2^4, \bar{g})$ be a 4-dimensional semi-Euclidean space of index 2 with signature $(-, -, +, +)$ of the canonical basis $(\partial_0, \ldots, \partial_3)$. Consider a hypersurface M of $\mathbf{R}_2^4$ given by

$$x_0 = x_1 + \sqrt{2}\sqrt{x_2^2 + x_3^2}.$$

For simplicity, we set $\mathbf{f} = \sqrt{x_2^2 + x_3^2}$. It is easy to check that M is a lightlike hypersurface whose radical distribution $Rad\,TM$ is spanned by

$$\xi = \mathbf{f}\,(\partial_0 - \partial_1) + \sqrt{2}\,(x_2\,\partial_2 + x_3\,\partial_3).$$

Then the lightlike transversal vector bundle is given by

$$ltr(TM) = Span\left\{N = \frac{1}{4\mathbf{f}^2}\left\{\mathbf{f}(-\partial_0 + \partial_1) + \sqrt{2}\,(x_2\,\partial_2 + x_3\,\partial_3)\right\}\right\}.$$

It follows that the corresponding screen distribution $S(TM)$ is spanned by

$$\{W_1 = \partial_0 + \partial_1,\ W_2 = -x_3\,\partial_2 + x_2\,\partial_3\}.$$

Example 4.   Let $(\mathbf{R}_1^4, \bar{g})$ be the Minkowski spacetime with signature $(-, +, +, +)$ of the canonical basis $(\partial_t, \partial_1, \partial_2, \partial_3)$. $(M, g = \bar{g}_{|M}, S(TM))$ is a lightlike hypersurface, given by an open subset of the lightlike cone

$$\left\{t(1, \cos u \cos v, \cos u \sin v, \sin u) \in \mathbf{R}_1^4 :\ t > 0,\ u \in (0, \pi/2),\ v \in [0, 2\pi]\right\}.$$

Then $Rad\,(TM)$ and $ltr(TM)$ are given by

$$Rad\,(TM) = Span\{\xi = \partial_t + \cos u \cos v\,\partial_1 + \cos u \sin v\,\partial_2 + \sin u\,\partial_3\},$$

$$ltr(TM) = Span\left\{N = \frac{1}{2}(-\partial_t + \cos u \cos v\,\partial_1 + \cos u \sin v\,\partial_2 + \sin u\,\partial_3)\right\},$$

respectively and the screen distribution $S(TM)$ is spanned by two orthonormal spacelike vectors

$$\{W_1 = -\sin u \cos v\,\partial_1 - \sin u \sin v\,\partial_2 + \cos u\,\partial_3,\quad W_2 = -\sin v\,\partial_1 + \cos v\,\partial_2\}.$$

7.3   Lightlike second fundamental forms

Let $(M, g, S(TM))$ be a lightlike hypersurface of a $(m + 2)$-dimensional semi-Riemannian manifold $(\bar{M}, \bar{g})$, $\bar{\nabla}$ the metric (Levi-Civita) connection on $\bar{M}$ with respect to $\bar{g}$ and $ltr(TM)$ the lightlike transversal vector bundle of M. We show that the existence of a unique null section N of $ltr(TM)$, for a null tangent section ξ of $Rad\,TM$, plays a key roll in setting up the *Gauss-Weingarten formulae* (for the lightlike case) from which we obtain the induced geometric objects such as linear connections, second fundamental forms, curvature tensor and Ricci tensor etc. In this section, we deal with the properties of two induced second fundamental forms. By using the second form of the decomposition in (7.2.5), we obtain

$$\begin{aligned}
\bar{\nabla}_X Y &= \nabla_X Y + h(X, Y), &\text{(7.3.1)}\\
\bar{\nabla}_X V &= -A_V X + \nabla_X^t V, &\text{(7.3.2)}
\end{aligned}$$

for any $X, Y \in \Gamma(TM)$ and $V \in \Gamma(ltr(TM))$, where $\nabla_X Y$ and $A_V X$ belong to $\Gamma(TM)$ while $h(X, Y)$ and $\nabla_X^t V$ belong to $\Gamma(ltr(TM))$. It is easy to check that ∇ is a torsion-free induced linear connection on M, h is a $\Gamma(ltr(TM))$-valued symmetric $F(M)$-bilinear form on $\Gamma(TM)$, A_V is a $F(M)$-linear operator on $\Gamma(TM)$. We call ∇^t an induced linear connection on $ltr(TM)$. Following [28, chapter 4], h and A_V are called the *second fundamental form* and the *shape operator* respectively, of M in

$\bar{M}$. Also, (7.3.1) and (7.3.2) are called the *global Gauss* and *Weingarten formulae*, respectively.

Locally, suppose $\{\xi, N\}$ is a pair of sections on $\mathcal{U} \subset M$ in theorem 2.1. Then, define a symmetric $F(\mathcal{U})$-bilinear form B and a 1-form τ on $\mathcal{U}$ by

$$B(X, Y) \;=\; \bar{g}(h(X, Y), \xi), \qquad\qquad (7.3.3)$$
$$\tau(X) \;=\; \bar{g}(\nabla^t_X N, \xi), \qquad\qquad (7.3.4)$$

for any $X, Y \in \Gamma(TM_{|\mathcal{U}})$. It follows that

$$h(X, Y) = B(X, Y)N, \quad \nabla^t_X N = \tau(X)\, N . \qquad\qquad (7.3.5)$$

Hence, on $\mathcal{U}$, (7.3.1) and (7.3.2) become

$$\bar{\nabla}_X Y \;=\; \nabla_X Y + B(X, Y)N , \qquad\qquad (7.3.6)$$
$$\bar{\nabla}_X N \;=\; -A_N X + \tau(X)N , \qquad\qquad (7.3.7)$$

respectively. Since B is the only component of h on $\mathcal{U}$ with respect to N, we call B the *local second fundamental form* of M, and the equations (7.3.6), (7.3.7) the *local Gauss* and *Weingarten formulae*.

Note. Comparing the Gauss and Weingarten formulae (7.1.21) of the semi-Riemannian hypersurfaces with the above Gauss and Weingarten formulae of the lightlike hypersurfaces, it is important to observe that the lightlike case involves an extra 1-form τ (defined by (7.3.4)). Because of this (and several other differences such as degenerate metric and non-uniqueness of a screen etc) the lightlike case is difficult and rather different than the semi-Riemannian case. In this chapter, the reader will see that the 1-form τ plays an important roll in the geometry of lightlike hypersurfaces.

Example 5. Let $(\mathbf{R}^4_2, \bar{g})$ be a 4-dimensional semi-Euclidean space of index 2 with signature $(-, -, +, +)$ of the canonical basis $(\partial_0, \ldots, \partial_3)$ having a lightlike hypersurface M as given in example 3. Then, by direct calculations we obtain

$$\bar{\nabla}_X W_1 \;=\; \bar{\nabla}_{W_1} X = 0,$$
$$\bar{\nabla}_{W_2} W_2 \;=\; -x_2\, \partial_2 - x_3\, \partial_3,$$
$$\bar{\nabla}_\xi \xi = \sqrt{2}\, \xi, \qquad \bar{\nabla}_{W_2}\xi = \bar{\nabla}_\xi W_2 = \sqrt{2}\, W_2,$$

for any $X \in \Gamma(TM)$. Then, by using the Gauss formula (7.3.6) we get

$$\nabla_X W_1 = 0, \quad \nabla_{W_2} W_2 = -\frac{1}{2\sqrt{2}}\, \xi, \quad \nabla_{W_1} W_1 = 0,$$

$$\nabla_\xi W_2 = \nabla_{W_2}\xi = \sqrt{2}\, W_2.$$

Thus,

$$B(W_1, W_1) = 0 = B(W_1, W_2), \quad B(W_2, W_2) = -\sqrt{2}\, (x_2^2 + x_3^2).$$

Example 6. Let $(\mathbf{R}_1^4, \bar{g})$ be the Minkowski spacetime with signature $(-, +, +, +)$ of the canonical basis $(\partial_t, \partial_1, \partial_2, \partial_3)$ having $(M, g = \bar{g}_{|M}, S(TM))$ a lightlike hypersurface as given in example 4. Then, following as in above example, we obtain

$$B(W_1, W_2) = -\left(\frac{1}{t \cos u}\right), \quad B(W_2, W_2) = -\left(\frac{1}{t \cos u}\right), \quad B(W_1, W_2) = 0.$$

Now we investigate the dependence (or otherwise) of the second fundamental form of M and the function τ on a chosen screen distribution and a chosen section $\xi \in \Gamma(Rad\,(TM)_{|\mathcal{U}})$ respectively, for which we have the following result:

Proposition 3.1. *Let $(M, g, S(TM))$ be a lightlike hypersurface of a semi-Riemannian manifold $(\bar{M}, \bar{g})$. Then*

(a) *The local second fundamental form B of M on $\mathcal{U}$ is independent of a screen distribution.*

(b) *B and the 1-form τ (in the Weingarten equation) depend on the choice of a section $\xi \in \Gamma(Rad\,(TM)_{|\mathcal{U}})$.*

(c) *$d\tau$ is independent of the section ξ.*

Proof. Let $S(TM)$ and $S(TM)'$ be two screens on M with the second fundamental forms h and h' corresponding to $ltr(TM)$ and $ltr(TM)'$, respectively. Using (7.3.1) and (7.3.3) for both screens we have

$$B(X, Y) = \bar{g}(\bar{\nabla}_X Y, \xi) = B'(X, Y), \tag{7.3.8}$$

for any $X, Y \in \Gamma(TM_{|\mathcal{U}})$. Therefore, $B = B'$ on $\mathcal{U}$, which proves (a). Take $\bar{\xi} = \alpha\xi$, for some function α. Then, it follows that $\bar{N} = (1/\alpha)N$ and from (7.3.6) and (7.3.7) we obtain

$$\bar{B} = \alpha B, \quad \tau(X) = \bar{\tau}(X) + X(\log \alpha), \tag{7.3.9}$$

for any $X \in \Gamma(TM_{|\mathcal{U}})$, which proves that B and τ depend on the section ξ on U, proving (b). Finally, taking the exterior derivative d on both sides of the second term of (7.3.9) we get $d\tau = d\bar{\tau}$ on $\mathcal{U}$, which completes the proof.

Since $\bar{\nabla}$ is a metric connection on $\bar{M}$, (7.1.1) and (7.3.8) imply

$$B(X, \xi) = 0, \tag{7.3.10}$$

for any $X \in \Gamma(TM_{|\mathcal{U}})$. The following result is immediate from (7.3.10):

Corollary 3.1. *The second fundamental form of a lightlike hypersurface $(M, g, S(TM))$, of a semi-Riemannian manifold $(\bar{M}, \bar{g})$, is degenerate.*

Define a local 1-form η by

$$\eta(X) = \bar{g}(X, N), \quad \forall X \in \Gamma(TM_{|\mathcal{U}}). \tag{7.3.11}$$

Using (7.3.6), (7.1.1) for a metric connection $\bar{\nabla}$ on $\bar{M}$ and (7.3.11) we get

$$\begin{aligned}
0 &= (\bar{\nabla}_X \bar{g})(Y, Z) \\
&= X(\bar{g}(Y, Z)) - \bar{g}(\bar{\nabla}_X Y, Z) - \bar{g}(Y, \bar{\nabla}_X Z) \\
&= X(g(Y, Z)) - g(\nabla_X Y, Z) - g(Y, \nabla_X Z) \\
&\quad - B(X, Y)\,\bar{g}(Z, N) - B(X, Z)\,\bar{g}(Y, N) \\
&= (\nabla_X g)(Y, Z) - B(X, Y)\,\eta(Z) - B(X, Z)\,\eta(Y).
\end{aligned}$$

Thus, in general, the connection ∇ on M is not a metric connection and satisfies

$$(\nabla_X g)(Y, Z) = B(X, Y)\,\eta(Z) + B(X, Z)\,\eta(Y), \tag{7.3.12}$$

for any $X, Y, Z \in \Gamma(TM_{|U})$. In section 7.5, we discuss the geometric conditions for the existence of a unique induced metric connection on M and its effect on the other induced objects.

In the lightlike case, we also have another second fundamental form and its corresponding shape operator which we now explain as follows:

Let P denote the projection morphism of $\Gamma(TM)$ on $\Gamma(S(TM))$ with respect to the decomposition (7.2.1). We obtain

$$\nabla_X PY = \nabla_X^* PY + h^*(X, PY), \tag{7.3.13}$$
$$\nabla_X U = -A_U^* X + \nabla_X^{*t} U, \tag{7.3.14}$$

for any $X, Y \in \Gamma(TM)$ and $U \in \Gamma(TM^\perp)$, where $\nabla_X^* Y$ and $A_U^* X$ belong to $\Gamma(S(TM))$, ∇^* and ∇^{*t} are linear connections on $\Gamma(S(TM))$ and $TM^\perp$ respectively, h^* is a $\Gamma(TM^\perp)$-valued $F(M)$-bilinear form on $\Gamma(TM) \times \Gamma(S(TM))$ and A_U^* is $\Gamma(S(TM))$-valued $F(M)$-linear operator on $\Gamma(TM)$. We call them the *screen second fundamental form* and *screen shape operator* of $S(TM)$ respectively. Define on $\mathcal{U}$

$$C(X, PY) = \bar{g}(h^*(X, PY), N), \tag{7.3.15}$$
$$\varepsilon(X) = \bar{g}(\nabla_X^{*t}\xi, N), \tag{7.3.16}$$

for any $X, Y \in \Gamma(TM_{|\mathcal{U}})$. One can show that $\varepsilon(X) = -\tau(X)$. Thus, locally we obtain

$$\nabla_X PY = \nabla_X^* PY + C(X, PY)\xi, \tag{7.3.17}$$
$$\nabla_X \xi = -A_\xi^* X - \tau(X)\xi, \tag{7.3.18}$$

for any $X, Y \in \Gamma(TM)$. Here $C(X, PY)$ is called the local screen second fundamental form of $S(TM)$. It is well known that the second fundamental form and the shape operator of a non-degenerate hypersurface (in general, submanifold) are related by means of the metric tensor field. Contrary to this we see from (7.3.3) and (7.3.15) that, in case of lightlike hypersurfaces, there are interrelations between these geometric objects and those of its screen distributions. More precisely, the two local second fundamental forms of M and $S(TM)$ are related to their shape operators by

$$B(X, Y) = g(A_\xi^* X, Y), \quad \bar{g}(A_\xi^* X, N) = 0, \qquad (7.3.19)$$
$$C(X, PY) = g(A_N X, PY), \quad \bar{g}(A_N Y, N) = 0. \qquad (7.3.20)$$

Proposition 3.2. Let $(M, g, S(TM))$ be a lightlike hypersurface of a semi-Riemannian manifold. Then

(a) The shape operator A_N of M has a zero eigenvalue.

(b) The screen second fundamental form of $S(TM)$ is also degenerate.

(c) The screen shape operator A_ξ^* of $S(TM)$ is symmetric with respect to the second fundamental form of M.

(d) The linear connection ∇^* from (7.3.13) is a metric connection on $S(TM)$.

(e) An integral curve of $\xi \in \Gamma(Rad\, TM_{|U})$ is a null geodesic of both M and $\bar{M}$ with respect to the connections ∇ and $\bar{\nabla}$ respectively.

Proof. The second equality in (7.3.20) implies that A_N is $\Gamma(S(TM))$-valued. Hence, rank $A_N \le m$ which implies the existence of a non-zero $X_0 \in \Gamma(TM_{|U})$ such that $A_N X_0 = 0$, that proves (a). Then, (b) is immediate from the first relation in (7.3.20). Next, from (7.3.19) we obtain

$$B(X, A_\xi^* Y) = B(A_\xi^* X, Y),$$

for any $X, Y \in \Gamma(TM)$, so (c) holds. (d) follows from (7.3.6) and (7.3.17). To prove (e) we note that (7.3.10) and (7.3.19) imply

$$A_\xi^* \xi = 0,$$

that is, ξ is an eigenvector field for A_ξ^* corresponding to the zero eigenvalue. Thus, by (7.3.6), (7.3.10), (7.3.18) and above we obtain

$$\bar{\nabla}_\xi \xi = \nabla_\xi \xi = -\tau(\xi)\, \xi.$$

Suppose $\xi = \sum_{\alpha=0}^m \xi^\alpha \frac{\partial}{\partial u^\alpha}$ and consider an integral curve $C : u^\alpha = u^\alpha(t)$, $\alpha \in \{0, \dots, m\}$, $t \in I \subset \mathbf{R}$, i.e., $\xi^\alpha = \frac{du^\alpha}{dt}$, or equivalently $\xi = \frac{d}{dt}$. In case $\tau(\xi) \ne 0$, choose a new parameter t^* on the null curve C such that

$$\frac{d^2 t^*}{dt^2} + \tau\left(\frac{d}{dt}\right)\frac{dt^*}{dt} = 0\,.$$

As explained in chapters 1 through 4, there always exists a distinguished parameter on C. Choosing t^* such a parameter we obtain $\nabla_{\frac{d}{dt^*}} \frac{d}{dt^*} = 0$ which proves (e).

Example 7. Let M be the lightlike hypersurface of the 4-dimensional semi-Euclidean space $(\mathbf{R}_2^4, \bar{g})$ of index 2, as given in example 1. Then, by direct calculations we obtain

$$\bar{\nabla}_\xi \xi = \nabla_\xi \xi = 0, \qquad \bar{\nabla}_X W_i = \bar{\nabla}_{W_i} X = 0, \quad i = 1, 2,$$

for any $X \in \Gamma(TM)$. Thus, consider an integral curve $\mathcal{C} : u^\alpha = u^\alpha(t)$, $\alpha \in \{0, \dots, 3\}$, $t \in I \subset \mathbf{R}$, i.e., $\xi^\alpha = \frac{du^\alpha}{dt}$, or equivalently $\xi = \frac{d}{dt}$. Then the parameter t is a distinguished parameter on $\mathcal{C}$.

Example 8. Let M be the lightlike hypersurface of the 4-dimensional semi-Euclidean space $(\mathbf{R}_2^4, \bar{g})$ of index 2, as given in example 2. Then, by direct calculations we obtain

$$\bar{\nabla}_\xi \xi = \nabla_\xi \xi = \xi, \qquad\qquad \bar{\nabla}_{W_1} W_2 = \bar{\nabla}_{W_2} W_1 = 0,$$
$$\bar{\nabla}_{W_1} W_1 = -x_0 \partial_0 - x_1 \partial_1, \quad \bar{\nabla}_{W_2} W_2 = -x_2 \partial_2 + \mathbf{f} \partial_3,$$
$$\bar{\nabla}_\xi W_1 = \bar{\nabla}_{W_1} \xi = W_1, \qquad\qquad \bar{\nabla}_\xi W_2 = \bar{\nabla}_{W_2} \xi = W_2.$$

Consider an integral curve $\mathcal{C} : u^\alpha = u^\alpha(t)$, $\alpha \in \{0, \dots, 3\}$, $t \in I \subset \mathbf{R}$, i.e., $\xi^\alpha = \frac{du^\alpha}{dt}$, or equivalently $\xi = \frac{d}{dt}$. Choose a new parameter t^* on the null curve C such that

$$\frac{d^2 t^*}{dt^2} - \frac{dt^*}{dt} = 0, \quad \text{i.e., } t^* = c e^t + d, \ c, d \in \mathbf{R}.$$

Then we obtain $\nabla_{\frac{d}{dt^*}} \frac{d}{dt^*} = 0$, that is, t^* is a distinguished parameter of $\mathcal{C}$.

To study the dependence of the induced objects $\{\nabla, \tau, A_N, A_\xi^*\}$ on the screen distribution $S(TM)$, we let $\{\nabla', \eta', A_{N'}', A_\xi^{*\prime}\}$ be another set of induced objects with respect to another screen distribution $S(TM)'$ and its transversal bundle $ltr(TM)'$. Consider two quasi-orthonormal frame fields $F = \{\xi, N, W_a\}$ and $F' = \{\xi, N', W_a'\}$ induced on the coordinate neighborhood $\mathcal{U} \subset M$ by $\{S(TM), ltr(TM)\}$ and $\{S'(TM), (ltr)'(TM)\}$ respectively. Using the transformation equations (7.2.7) and (7.2.8) we obtain the relationships between the geometrical objects induced by the Gauss and Weingarten equations with respect to $S(TM)$ and $S(TM)'$ as follows:

$$\nabla_X' Y = \nabla_X Y + B(X, Y) \left\{ \frac{1}{2} \left(\sum_{a=1}^m \epsilon_a (\mathbf{f}_a)^2 \right) \xi - \sum_{a=1}^m \mathbf{f}_a W_a \right\}, \qquad (7.3.21)$$

$$\tau'(X) = \tau(X) + B(X, N' - N), \qquad (7.3.22)$$

$$A_{N'}' X = A_N X + \sum_{a=1}^m \left\{ \epsilon_a \mathbf{f}_a X(\mathbf{f}_a) - \tau(X)\epsilon_a(\mathbf{f}_a)^2 \right.$$
$$\left. - \frac{1}{2}\epsilon_a(\mathbf{f}_a)^2 B(X, N - N') - \mathbf{f}_a C(X, W_a) \right\} \xi$$
$$+ \sum_{a=1}^m \left\{ \mathbf{f}_a \left(\tau(X) + B(X, N' - N)\right) - X(\mathbf{f}_a) \right\} W_a$$

$$- \sum_{a=1}^{m} \mathbf{f}_a \nabla_X^* W_a - \frac{1}{2} \sum_{a=1}^{m} \epsilon_a (\mathbf{f}_a)^2 A_\xi^* X, \tag{7.3.23}$$

$$A_\xi^{*\prime} X = A_\xi^* X + B(X, N - N')\xi, \quad \forall X, Y \in \Gamma(TM_{|\mathcal{U}}). \tag{7.3.24}$$

The following result is immediate from (7.3.21), (7.3.22) and (7.3.24).

Proposition 3.3. *Let* $(M, g, S(TM))$ *be a lightlike hypersurface of a semi-Riemannian manifold. Then, the induced connection* ∇ *on* M, *the 1-form* τ *(in the equation (7.3.7)) and the shape operator* A_ξ^* *(in the equation (7.3.18)) all three are independent of* $S(TM)$, *if and only if, the second fundamental form* h *of* M *vanishes identically on* M.

Note. It follows from (7.3.23) that the shape operator A_N depends on the choice of $S(TM)$ even if h vanishes identically on M. Also, as in the non-degenerate case (see O'Neill [82, page 104]), we say that $h \equiv 0$ on M which is equivalent to M *totally geodesic* in $\bar{M}$. In section 7.5, we have studied the totally geodesic case and the implications of proposition 3.3 on local geometry of those hypersurfaces.

Now, we ask the following question: Is the local screen second fundamental form C independent of the choice of a screen? The answer is negative. Indeed, we prove the following result with respect to a change in screen distribution.

Proposition 3.4. *The screen second fundamental forms* C *and* C' *of the screen distributions* $S(TM)$ *and* $S(TM)'$, *respectively, are related as follows:*

$$C'(X, PY) = C(X, PY) - \frac{1}{2}\| W \|^2 B(X, Y) + g(\nabla_X PY, W), \tag{7.3.25}$$

where $W = \sum_{a=1}^{m} \mathbf{f}_a W_a$ is the characteristic vector field of the screen change.

Proof. Using (7.2.8) and (7.3.21) we get

$$
\begin{aligned}
C'(X, PY) &= \bar{g}(\nabla_X' PY, N') \\
&= \bar{g}\left(\nabla_X PY + B(X, Y)\{\frac{1}{2}(\sum_{a=1}^{m} \varepsilon_a(\mathbf{f}_a)^2)\xi - \sum_{a=1}^{m} \mathbf{f}_a W_a\}, N' \right) \\
&= \bar{g}(\nabla_X PY, N) + \bar{g}(\nabla_X PY, \sum_{a=1}^{m} \mathbf{f}_a W_a) \\
&\quad + B(X, Y)\left\{ \frac{1}{2}(\sum_{a=1}^{m} \varepsilon_a(\mathbf{f}_a)^2) - \sum_{b=1}^{m}\sum_{a=1}^{m} g(\mathbf{f}_a W_a, \mathbf{f}_b W_b) \right\} \\
&= C(X, PY) + g(\nabla_X PY, W) - \frac{1}{2}\| W \|^2 B(X, Y)
\end{aligned}
$$

which is the desired formula.

Remark 1. In next section we show that there exists a large class of lightlike hypersurfaces for which the screen second fundamental form is independent of the choice of a screen distribution.

Finally, in this section, we express the Gauss-Weingarten equations in a coordinate system on M, find all the coefficients of the induced linear connection ∇ on M and the two local second fundamental forms. Consider the Levi-Civita connection $\bar{\nabla}$ on $\bar{M}$ and by using (7.3.6), (7.3.7) and (7.3.10) we obtain

$$
\begin{aligned}
\bar{\nabla}_{\delta_{u^b}} \delta_{u^a} &= \Gamma^o_{ab}\partial_{u^o} + \Gamma^c_{ab}\delta_{u^c} + B_{ab}N \\
\bar{\nabla}_{\delta_{u^b}} \partial_{u^o} &= \Gamma^o_{ob}\partial_{u^o} + \Gamma^k_{oj}\delta_{u^k} \\
\bar{\nabla}_{\partial_{u^o}} \delta_{u^a} &= \Gamma^o_{ao}\partial_{u^o} + \Gamma^c_{ao}\delta_{u^c} \\
\bar{\nabla}_{\partial_{u^o}} \partial_{u^o} &= \Gamma^o_{oo}\partial_{u^o}, \\
\bar{\nabla}_{\delta_{u^a}} N &= -A^c_a\delta_{u^c} + \tau_a N \\
\bar{\nabla}_{\partial_{u^o}} N &= -A^c_o\delta_{u^c} + \tau_o N,
\end{aligned}
\tag{7.3.26}
$$

where $\{\Gamma^c_{ab}, \Gamma^o_{ab}, \Gamma^c_{ob}, \Gamma^c_{io}, \Gamma^o_{ao}, \Gamma^o_{ob}, \Gamma^o_{oo}\}$ are the coefficients of the induced linear connection ∇ on M with respect to the frames field $\{\partial_{u^o}, \delta_{u^a}\}$; $\{A^c_a, A^c_o\}$ are the entries of the matrix of $A_N : \Gamma(TM_{|U}) \to \Gamma(S(TM)_{|\mathcal{U}})$ with respect to the basis $\{\delta_{u^a}, \partial_{u^o}\}$ and $\{\delta_{u^a}\}$ of $\Gamma(TM_{|\mathcal{U}})$ and $\Gamma(S(TM)_{|\mathcal{U}})$ respectively; $B_{ab} = B\left(\delta_{u^a}, \delta_{u^b}\right) = B\left(\partial_{u^a}, \partial_{u^b}\right)$; $\tau_a = \tau\left(\delta_{u^a}\right)$, $\tau_o = \tau\left(\partial_{u^o}\right)$.

Taking into account the Lie-bracket $[X, Y] = (X^a\partial_a Y^b - Y^a\partial_a X^b)\partial_b$ with respect to a natural frames field, (7.2.12) and (7.2.13) we obtain

$$
[\delta_{u^a}, \delta_{u^b}] = S_{ab}\partial_{u^o}, \quad S_{ab} = \frac{\delta S_a}{\delta u^b} - \frac{\delta S_b}{\delta u^a},
\tag{7.3.27}
$$

and

$$
[\delta_{u^a}, \partial_{u^o}] = \frac{\partial S_a}{\partial u^o}\partial_{u^o}.
\tag{7.3.28}
$$

As $\bar{\nabla}$ is torsion-free, using (7.1.6), (7.3.27) and (7.3.28) we obtain

$$
\Gamma^c_{ab} = \Gamma^c_{ba} \; ; \; \Gamma^o_{ab} = \Gamma^o_{ba} + S_{ba} \; ; \; \Gamma^c_{ob} = \Gamma^c_{bo} \; ; \; \Gamma^o_{ob} = \Gamma^o_{bo} + \frac{\partial S_b}{\partial u^o},
\tag{7.3.29}
$$

and

$$
B_{ab} = B_{ba}.
\tag{7.3.30}
$$

Further on, we decompose the following Lie brackets:

$$
\begin{aligned}
{[N, \delta_{u^c}]} &= N^o_c\partial_{u^o} + N^d_c\delta_{u^d} + N_c N, \\
{[N, \partial_{u^o}]} &= N^o\partial_{u^o} + N^c_o\delta_{u^c} + N_o N.
\end{aligned}
\tag{7.3.31}
$$

With respect to the non-holonomic frames field $\{\partial_{u^o}, N, \delta_{u^a}\}$ of $\bar{M}$, the semi-Riemannian metric $\bar{g}$ has the matrix

$$
[\bar{g}] = \begin{bmatrix} 0 & 1 & 0 \\ 1 & 0 & 0 \\ 0 & 0 & g_{ab}(u^o, \ldots, u^m) \end{bmatrix},
\tag{7.3.32}
$$

where $g_{ab} = \bar{g}\,(\delta_{u^a}, \delta_{u^b}) = g\,(\partial_{u^a}, \partial_{u^b})$ are the functions in (7.2.9). As the matrix $[g_{ab}(u)]$ is invertible, consider its inverse matrix $[g^{ab}(u)]$. Then, by using the general identity (7.1.4) and (7.3.26) - (7.3.32), we obtain

$$\Gamma^k_{ab} = \frac{1}{2}\,g^{cd}\left\{\frac{\delta g_{ad}}{\delta u^b} + \frac{\delta g_{bd}}{\delta u^a} - \frac{\delta g_{ab}}{\delta u^d}\right\}, \tag{7.3.33}$$

$$\Gamma^o_{ab} = \frac{1}{2}\,\{S_{ba} + N^c_b g_{ca} + N^c_a g_{cb} - N(g_{ab})\} = g_{ac}A^c_b, \tag{7.3.34}$$

$$\Gamma^o_{ob} = \frac{1}{2}\left\{\frac{\partial S_b}{\partial u^o} + N_b + N^c_o g_{ab}\right\} = g_{bc}A^c_o + \frac{\partial S_b}{\partial u^o} = -\tau_b, \tag{7.3.35}$$

$$\Gamma^c_{bo} = \Gamma^c_{ob} = \frac{1}{2}g^{ca}\frac{\partial g_{ab}}{\partial u^o}, \tag{7.3.36}$$

$$\Gamma^o_{oo} = N_o = -\tau_o, \tag{7.3.37}$$

$$B_{ab} = -\frac{1}{2}\frac{\partial g_{ab}}{\partial u^o}. \tag{7.3.38}$$

By using (7.3.15), (7.3.16), (7.3.26), (7.3.34) and (7.3.35), we obtain $\bullet$

$$\Gamma^{*c}_{ab} = \Gamma^c_{ab}\,; \quad \Gamma^{*c}_{bo} = \Gamma^c_{bo}, \tag{7.3.39}$$

$$\begin{aligned} C_{ab} &= C\,(\delta_{u^b}, \delta_{u^a}) = \Gamma^o_{ab} = g_{ac}A^c_b \\ C_a &= C\,(\partial_{u^o}, \delta_{u^a}) = \Gamma^o_{ao} = g_{ik}A^k_o, \end{aligned} \tag{7.3.40}$$

and

$$A^{*a}_b = g^{ac}B_{bc} = -\Gamma^a_{ob}, \tag{7.3.41}$$

where $\{\Gamma^{*c}_{ab}, \Gamma^{*c}_{ao}\}$ are the coefficients of the metric connection ∇^* on $S(TM)$ with respect to the frames field $\{\partial_{u^o}, \delta_{u^c}\}$ and A^{*a}_b are the entries of $A^*_{\partial_{u^o}}$ with respect to the basis $\{\frac{\partial}{\partial u^a}\}$. As we know from (7.3.12) that, in general, the induced connection ∇ on M is not a metric connection. However, we show here that some covariant derivatives of local components of the metric vanish. Indeed, replacing Y and Z from (7.3.12) by δ_{u^a} and δ_{u^b} respectively, and by using (7.3.11) we obtain

$$g_{ab;\,c} = \frac{\delta g_{ab}}{\delta u^c} - \Gamma^h_{ac}g_{db} - \Gamma^d_{bc}g_{ad} = 0, \tag{7.3.42}$$

$$g_{ab;\,o} = \frac{\partial g_{ab}}{\partial u^o} - \Gamma^h_{ao}g_{db} - \Gamma^d_{bo}g_{ad} = 0. \tag{7.3.43}$$

Example 9. Let $(M, g = \bar{g}_{|M}, S(TM))$ be the lightlike hypersurface of the Minkowski spacetime $(\mathbf{R}^4_1, \bar{g})$, as given in example 4. Since

$$\partial_{u_0} = \partial_t + \cos u \cos v\, \partial_1 + \cos u \sin v\, \partial_2 + \sin u\, \partial_3,$$

$$\partial_{u_1} = -t \sin u \cos v\, \partial_1 - t \sin u \sin v\, \partial_2 + t \cos u\, \partial_3,$$

$$\partial_{u_2} = -t \cos u \sin v\, \partial_1 + t \cos u \cos v\, \partial_2,$$

$$N = \frac{1}{2}(-\partial_t + \cos u \cos v\, \partial_1 + \cos u \sin v\, \partial_2 + \sin u\, \partial_3),$$

the coefficient of the degenerate metric g on M are given by $g_{oo} = g_{o1} = g_{o2} = g_{12} = 0$ and $g_{11} = t^2$, $g_{22} = t^2 \cos^2 u$ and the matrix of g with respect to the natural frame fields $\{\partial_{u_0}, \partial_{u_1}, \partial_{u_2}\}$ is as follows

$$[g] = \begin{bmatrix} 0 & 0 & 0 \\ 0 & t^2 & 0 \\ 0 & 0 & t^2 \cos^2 u \end{bmatrix}.$$

Let $\{\bar{\mathcal{U}}; \bar{u}_0, \bar{u}_1, \bar{u}_2\}$ be the another coordinate system satisfying the coordinate transformation (7.2.11) such that

$$\partial_{\bar{u}_0} = \partial_{u_0} = \partial_t + \cos u \cos v\, \partial_1 + \cos u \sin v\, \partial_2 + \sin u\, \partial_3,$$
$$\partial_{\bar{u}_1} = t \sin u\, \partial_t + t\partial_3,$$
$$\partial_{\bar{u}_2} = -t \sin v\, \partial_1 + t \cos v\, \partial_2,$$
$$\bar{N} = \frac{1}{2 \cos^2 u}(-\partial_t + \cos u \cos v\, \partial_1 + \cos u \sin v\, \partial_2 - \sin u\, \partial_3).$$

The matrix of g with respect to the natural frame fields $\{\partial_{\bar{u}_0}, \partial_{\bar{u}_1}, \partial_{\bar{u}_2}\}$ is as follows

$$[g] = \begin{bmatrix} 0 & 0 & 0 \\ 0 & t^2 \cos^2 u & 0 \\ 0 & 0 & t^2 \end{bmatrix}.$$

It follows that

$$\partial_{\bar{u}_0} = \partial_{u_0}, \quad \partial_{\bar{u}_1} = t \sin u\, \partial_{u_0} + \cos u\, \partial_{u_1}, \quad \partial_{\bar{u}_2} = t \cos u\, \partial_{u_2},$$
$$\partial_{u_0} = \partial_{\bar{u}_0}, \quad \partial_{u_1} = -t \tan u\, \partial_{\bar{u}_0} + \sec u\, \partial_{\bar{u}_1}, \quad \partial_{u_2} = \sec u\, \partial_{\bar{u}_2},$$

i.e., $\quad F(u) = 1, \qquad F_1(u) = -\sec u, \quad F_2(u) = 0,$
$\qquad F_1^1 = -t \tan u, \quad F_1^2 = F_2^1(u) = 0, \quad F_2^2(u) = \sec u.$

This imply that the functions $S_a(u_0, u_1, u_2)$ appear in (7.2.12) are given by

$$S_1(u) = -t \tan u\, \bar{S}_1(u) - \sec u, \quad S_2(u) = \sec u\, \bar{S}_2(u).$$

Since the local basis of $\Gamma(S(TM))$ is the vector fields

$$\delta_{u_1} = \partial_{u_1} - S_1(u)\, \partial_{u_0} = \partial_{u_1}, \quad \delta_{u_2} = \partial_{u_2} - S_2(u)\, \partial_{u_0} = \partial_{u_2},$$

we obtain the local field of frames $\{\partial_{u_0}, N, \delta_{u_1}, \delta_{u_2}\}$ on $(\mathbf{R}_1^4, \bar{g})$. With respect to the non-holonomic frames field $\{\partial_{u^o}, N, \delta_{u^1}, \delta_{u_2}\}$ of $\mathbf{R}_1^4$, the semi-Riemannian metric $\bar{g}$ has the matrix

$$[\bar{g}] = \begin{bmatrix} 0 & 1 & 0 & 0 \\ 1 & 0 & 0 & 0 \\ 0 & 0 & t^2 & 0 \\ 0 & 0 & 0 & t^2 \cos^2 u \end{bmatrix}.$$

So that the Christoffel symbols Γ_{ij}^k and the coefficients B_{ij} of the 2-forms B can be computed by (7.3.33), (7.3.36) and (7.3.38):

$$\Gamma_{11}^1 = 0, \qquad \Gamma_{12}^1 = 0, \qquad \Gamma_{22}^1 = \sin u \cos u,$$
$$\Gamma_{11}^2 = 0, \qquad \Gamma_{12}^2 = -\tan u, \qquad \Gamma_{22}^2 = 0 \, ;$$
$$\Gamma_{1o}^1 = \frac{1}{t}, \qquad \Gamma_{1o}^2 = \Gamma_{2o}^1 = 0, \qquad \Gamma_{2o}^2 = \frac{1}{t} \, ;$$
$$B_{11} = -t, \qquad B_{12} = B_{21} = 0, \qquad B_{22} = -t \cos^2 u.$$

By using (7.3.26) we obtain

$$\Gamma_{oo}^o = 0, \quad B_{o1} = B_{1o} = B_{o2} = B_{2o} = 0,$$
$$\Gamma_{o1}^o = \Gamma_{1o}^o = 0, \quad \Gamma_{o2}^o = \Gamma_{2o}^o = 0, \quad S_1 = S_2 = 0,$$
$$\Gamma_{11}^o = -\frac{t}{2}, \quad \Gamma_{12}^o = \Gamma_{21}^o = 0, \quad \Gamma_{22}^o = -\frac{1}{2}t \cos^2 u, \quad S_{12} = 0,$$
$$A_o^1 = A_o^2 = 0, \quad A_1^1 = -\frac{1}{2t}, \quad A_2^1 = A_1^2 = 0, \quad A_2^2 = -\frac{1}{2t}.$$

Example 10. Let $(\mathbf{R}_2^4, \bar{g})$ be a 4-dimensional semi-Euclidean space of index 2 with signature $(-, -, +, +)$ of the canonical basis $(\partial_0, \ldots, \partial_3)$. Consider a hypersurface M of $\mathbf{R}_2^4$ given by

$$x_0 = u \cosh v + \sinh v, \quad x_1 = w, \quad x_2 = u + v, \quad x_3 = u \sinh v + \cosh v.$$

Using the local coordinate chart $\{M \, ; \, u, v, w\}$ we have

$$\partial_u = \xi = \cosh v \, \partial_0 + \partial_2 + \sinh v \, \partial_3, \qquad \partial_w = \partial_1,$$
$$\partial_v = (u \sinh v + \cosh v) \, \partial_0 + \partial_2 + (u \cosh v + \sinh v) \, \partial_3,$$
$$N = \frac{-1}{2 \cosh^2 v} \, (\cosh v \, \partial_0 - \partial_2 - \sinh v \, \partial_3).$$

Let $u = u_0$, $v = u_1$, $w = u_2$. The coefficient of the degenerate metric g on M are given by $g_{oo} = g_{o1} = g_{o2} = g_{12} = 0$ and $g_{11} = u^2$, $g_{22} = -1$ and the matrix of g with respect to the natural frame fields $\{\partial_{u_0}, \partial_{u_1}, \partial_{u_2}\}$ is as follows

$$[g] = \begin{bmatrix} 0 & 0 & 0 \\ 0 & u^2 & 0 \\ 0 & 0 & -1 \end{bmatrix}.$$

Let $\{\bar{\mathcal{U}}; \bar{u}_0, \bar{u}_1, \bar{u}_2\}$ be the another coordinate system satisfying the coordinate transformation (7.2.11) such that

$$\partial_{\bar{u}_0} = \partial_{u_0} = \cosh v \, \partial_0 + \partial_2 + \sinh v \, \partial_3,$$
$$\partial_{\bar{u}_1} = u \sinh v \, \partial_0 + u \cosh v \, \partial_3, \qquad \partial_{\bar{u}_2} = \partial_1,$$
$$\bar{N} = -\frac{1}{2} (\cosh v \, \partial_0 - \partial_2 + \sinh v \, \partial_3).$$

The matrix of g with respect to the natural frame fields $\{\partial_{\bar{u}_0}, \partial_{\bar{u}_1}, \partial_{\bar{u}_2}\}$ is as follows

$$[g] = \begin{bmatrix} 0 & 0 & 0 \\ 0 & u^2 & 0 \\ 0 & 0 & -1 \end{bmatrix}.$$

It follows

$$\partial_{\bar{u}_0} = \partial_{u_0}, \quad \partial_{\bar{u}_1} = -\partial_{u_0} + \partial_{u_1}, \quad \partial_{\bar{u}_2} = \partial_{u_2},$$
$$\partial_{u_0} = \partial_{\bar{u}_0}, \quad \partial_{u_1} = -\partial_{\bar{u}_0} + \partial_{\bar{u}_1}, \quad \partial_{u_2} = \partial_{\bar{u}_2},$$

i.e., $\quad F(u) = 1, \quad F_1(u) = -1, \quad F_2(u) = 0,$
$$F_1^1(u) = 1, \quad F_1^2 = F_2^1(u) = 0, \quad F_2^2(u) = 1.$$

This imply that the functions $S_a(u_0, u_1, u_2)$ appear in (7.2.12) are given by

$$S_1(u) = \bar{S}_1(u) - 1 = 1, \quad S_2(u) = \bar{S}_2(u) = 0.$$

Thus the local basis of $\Gamma(S(TM))$ is the vector fields

$$\delta_{u_1} = \partial_{u_1} - S_1(u)\,\partial_{u_0} = \partial_{u_1} - \partial_{u_0} = \partial_{\bar{u}_0},$$
$$\delta_{u_2} = \partial_{u_2} - S_2(u)\,\partial_{u_0} = \partial_{u_0} = \partial_{\bar{u}_2}.$$

Then we obtain the local field of frames $\{\partial_{u_0}, N, \delta_{u_1}, \delta_{u_2}\}$ on $(\mathbf{R}_2^4, \bar{g})$. With respect to the non-holonomic frames field $\{\partial_{u^o}, N, \delta_{u^1}, \delta_{u_2}\}$ of $\mathbf{R}_2^4$, the semi-Riemannian metric $\bar{g}$ has the matrix

$$[\bar{g}] = \begin{bmatrix} 0 & 1 & 0 & 0 \\ 1 & 0 & 0 & 0 \\ 0 & 0 & u^2 & 0 \\ 0 & 0 & 0 & -1 \end{bmatrix}.$$

So that the Christoffel symbols Γ_{ij}^k and the functions B_{ij} are given by $\Gamma_{ij}^k = 0$ for any $i, j, k \in \{1, 2\}$; $B_{11} = -u$ and all other $B_{ij} = 0$ for any $i, j \in \{1, 2\}$; $\Gamma_{1o}^1 = \frac{1}{u}$ and $\Gamma_{1o}^2 = \Gamma_{2o}^1 = \Gamma_{2o}^2 = 0$. Also by using (7.3.26) we obtain

$$\Gamma_{oo}^o = 0, \quad B_{o1} = B_{1o} = B_{o2} = B_{2o} = 0,$$
$$\Gamma_{o1}^o = \Gamma_{1o}^o = 0, \quad \Gamma_{o2}^o = \Gamma_{2o}^o = 0, \quad S_1 = S_2 = 0,$$
$$\Gamma_{11}^o = \frac{u}{2}, \quad \Gamma_{12}^o = \Gamma_{21}^o = 0, \quad \Gamma_{22}^o = 0, \quad S_{12} = 0,$$
$$A_o^1 = A_o^2 = 0, \quad A_1^1 = \frac{1}{2u}, \quad A_2^1 = A_1^2 = 0, \quad A_2^2 = 0.$$

7.4   Integrable screen distribution

Although the 1-dimensional null distribution $Rad\,TM$ of a lightlike hypersurface $(M, g, S(TM))$ is obviously integrable, in general, any screen distribution is not necessarily integrable. Indeed, we have the following example:

Example 11. Let $(\mathbf{R}_2^4, \bar{g})$ be a 4-dimensional semi-Euclidean space of index 2 with signature $(-, -, +, +)$ of the canonical basis $(\partial_0, \ldots, \partial_3)$. Consider a Monge hypersurface M of $\mathbf{R}_2^4$ given by

$$x_3 = x_0 + \sin(x_1 + x_2), \quad x_1 + x_2 \neq n\pi, \ n \in \mathbf{Z}.$$

It is easy to check that M is a lightlike hypersurface whose radical distribution $Rad\,TM$ is spanned by

$$\xi = \partial_0 + \cos(x_1 + x_2)\partial_1 - \cos(x_1 + x_2)\partial_2 + \partial_3.$$

Let $V = \partial_0 + \cos(x_1 + x_2)\partial_1$, then $g(V, V) = g(\xi, V) = -(1 + \cos^2(x_1 + x_2))$. Thus, as per theorem 2.1, the lightlike transversal vector bundle is given by $ltr(TM) = Span\{N\}$, where

$$N = \frac{-1}{2(1 + \cos^2(x_1 + x_2))}\{\partial_0 + \cos(x_1 + x_2)\partial_1 + \cos(x_1 + x_2)\partial_2 - \partial_3\}.$$

The tangent bundle $T\mathbf{R}_2^4$ is spanned by

$$\left\{\frac{\partial}{\partial u_0} = \partial_0 + \partial_3, \quad \frac{\partial}{\partial u_1} = \partial_1 + \cos(x_1 + x_2)\partial_3, \quad \frac{\partial}{\partial u_2} = \partial_2 + \cos(x_1 + x_2)\partial_3\right\}.$$

It follows that the corresponding screen distribution $S(TM)$ is spanned by

$$\{W_1 = \cos(x_1 + x_2)\partial_0 - \partial_1, \quad W_2 = \partial_2 + \cos(x_1 + x_2)\partial_3\,\}.$$

In this case $[W_1, W_2] = \bar{\nabla}_{W_1} W_2 - \bar{\nabla}_{W_2} W_1 = \sin(x_1 + x_2)\{\partial_0 + \partial_3\}$. Thus $\bar{g}([W_1, W_2], N) = \frac{\sin(x_1 + x_2)}{1 + \cos^2(x_1 + x_2)}$. This imply that $S(TM)$ is not integrable.

The investigation of integrable screens has been an important and desirable topic of study. The objective of this section is to present an up-to-date account of work done on lightlike hypersurfaces with integrable screen distributions. We start with the following two example:

Example 12. Let $(\mathbf{R}_1^{m+2}, \bar{g})$ be a Minkowski space with the metric $\bar{g}$ given by

$$\bar{g}(x, y) = -x^0 y^0 + \sum_{a=1}^{m+1} x^a y^a. \tag{7.4.1}$$

Consider a hypersurface M of $\mathbf{R}_1^{m+2}$ locally given by the equations (7.2.15). Then, as per theorem 2.1 and (7.2.21), M admits a canonical screen distribution $S(TM)$ whose canonical lightlike transversal vector bundle is given by

$$N = (D^0)^{-2}\left\{V + \frac{1}{2}\xi\right\}. \tag{7.4.2}$$

It is easy to see that on $\mathbf{R}_1^{m+2}$ the Levi-Civita connection $\bar{\nabla}$ is defined by

$$\bar{\nabla}_X Y = \sum_{i=0}^{m+1} X(Y^i)\partial_{x^i}, \quad \forall X, Y \in \Gamma(S(TM)),$$

with respect to a coordinate basis $\{\partial_{x^i}\}$. Using this, (7.3.6) and (7.4.2) we obtain

$$\begin{aligned}
\bar{g}([X, Y], N) &= (D^0)^{-1} \bar{g}\left(\bar{\nabla}_X Y - \bar{\nabla}_Y X, \partial_{x^0}\right) \\
&= -(D^0)^{-1}\left\{\bar{g}\left(X, \bar{\nabla}_Y \partial_{x^0}\right) - \bar{g}\left(Y, \bar{\nabla}_X \partial_{x^0}\right)\right\} = 0.
\end{aligned}$$

Hence, $[X, Y] \in \Gamma(S(TM))$, that is, $S(TM)$ is integrable.

Example 13. Let $\mathbf{R}_q^{m+2}$ be a semi-Euclidean space with its semi-Euclidean metric given by (7.2.14). Consider its lightlike hypersurface M defined by (7.2.15). Then, in particular, M given by

$$\sum_{i=0}^{q-1} (x^i)^2 - \sum_{a=q}^{m+1} (x^a)^2 = 0; \quad x \neq 0, \tag{7.4.3}$$

is the light cone Λ_{q-1}^{m+1} of $\mathbf{R}_q^{m+2}$. In order to show this, we define

$$\mathcal{U}_1 = \left\{(u^0, \dots, u^m) \in \mathbf{R}^{m+1}; \sum_{i=0}^{q-1} (u^i)^2 - \sum_{a=q}^{m} (u^a)^2 > 0\right\},$$

and the local immersion $\varphi_1^+ : \mathcal{U}_1 \to \mathbf{R}_q^{m+2}$ of Λ_{q-1}^{m+1} by

$$x^0 = u^0, \dots, x^m = u^m, \quad x^{m+1} = \left\{\sum_{i=0}^{q-1} (u^i)^2 - \sum_{a=q}^{m} (u^a)^2\right\}^{1/2}. \tag{7.4.4}$$

Thus, the tangent bundle $T\Lambda_{q-1}^{m+1}$ on $\varphi_1^+(\mathcal{U}_1)$ is spanned by

$$\partial_{u^i} = \partial_{x^i} + \frac{x^i}{x^{m+1}} \partial_{x^{m+1}} \quad \text{and} \quad \partial_{u^a} = \partial_{x^a} - \frac{x^a}{x^{m+1}} \partial_{x^{m+1}},$$

$i \in \{0, \dots, q-1\}$, $a \in \{q, \dots, m\}$. Then it is easy to check that

$$\xi = \sum_{\alpha=0}^{m} x^\alpha \partial_{u^\alpha} = \sum_{A=0}^{m+1} x^A \partial_{x^A}, \tag{7.4.5}$$

and

$$N = \frac{1}{2 \sum_{i=0}^{q-1} (x^i)^2} \left\{-\sum_{i=0}^{q-1} x^i \partial_{x^i} + \sum_{a=q}^{m+1} x^a \partial_{x^a}\right\}. \tag{7.4.6}$$

Consider a coordinate neighborhood $\mathcal{U} \subset \Lambda_{q-1}^{m+1}$ such that $x^{m+1} > 0$ and $x^{q-1} \neq 0$ so that the local equations of Λ_{q-1}^{m+1} are given by (7.4.4). Then, by direct calculations using (7.2.19), the canonical screen distribution on Λ_{q-1}^{m+1} is locally spanned by $\{X_0, \ldots, X_{q-2}, Y_q, \ldots, Y_m\}$, where we set

$$X_p = x^{q-1}\partial_{x^p} - x^p \partial_{x^{q-1}} \quad ; \qquad Y_s = x^{m+1}\partial_{x^s} - x^s \partial_{x^{m+1}},$$

for any $p \in \{0, \ldots, q-2\}$, $s \in \{q, \ldots, m\}$. The Lie brackets of these vectors fields are expressed as follows:

$$[X_p, Y_s] = 0 \quad ; \quad [X_p, X_{\bar{p}}] = \frac{1}{x^{q-1}}\left\{x^{\bar{p}}X_p - x^p X_{\bar{p}}\right\} \quad ;$$

$$[Y_s, Y_{\bar{s}}] = \frac{1}{x^{m+1}}\left\{x^{\bar{s}}Y_s - x^s Y_{\bar{s}}\right\}.$$

Thus, we conclude the following:

Proposition 4.1. *The canonical screen vector bundle $S(TM)$ on the lightlike cone Λ_{q-1}^{m+1} of $\mathbf{R}_q^{m+2}$ is integrable.*

Remark 2. Example 11 shows (also see [28, page 81, example 1.2]) that, in general, the canonical screen $S(TM)$ of a lightlike hypersurface of $\mathbf{R}_q^{m+2}$, with $q > 1$, is not integrable. However, we now show that, under a geometric condition, there exists a large class of lightlike hypersurfaces of semi-Riemannian manifolds with integrable screen. For this, we recall the following definition:

Definition 4.1 ([4]). *A lightlike hypersurface $(M, g, S(TM))$ of a semi-Riemannian manifold is screen locally conformal if the shape operators A_N and A_ξ^* of M and its screen distribution $S(TM)$, respectively, are related by*

$$A_N = \varphi A_\xi^* \tag{7.4.7}$$

where φ is a non-vanishing smooth function on a neighborhood $\mathcal{U}$ in M.

Remark 3. In order to avoid trivial ambiguities, we consider $\mathcal{U}$ to be connected and maximal in the sense that there is no larger domain $\mathcal{U}' \supset \mathcal{U}$ on which the relation (7.4.7) holds. In case $\mathcal{U} = M$ the screen conformality is said to be global.

Example 14. Let $\mathbf{R}_1^{m+2}$ be a Minkowski space endowed with the metric (7.4.1). It follows from example 10 that the light cone Λ_0^{m+1} is given by

$$-(x^0)^2 + \sum_{a=1}^{m+1}(x^a)^2 = 0, \quad x = \sum_{A=0}^{m+1} x^A \partial_{x^A} \neq 0,$$

which is a lightlike hypersurface of $\mathbf{R}_1^{m+2}$ whose radical distribution is spanned by a global vector field

$$\xi = \sum_{A=0}^{n+1} x^A \partial_{x^A}. \tag{7.4.8}$$

The unique null transversal section, satisfying Theorem 2.1, is given by

$$N = \frac{1}{2(x^0)^2} \left\{ -x^0 \, \partial_{x^0} + \sum_{a=1}^{m+1} x^a \, \partial_{x^a} \right\} \tag{7.4.9}$$

and is also globally defined. As ξ is the position vector field we get

$$\bar{\nabla}_X \xi = \nabla_X X = X, \quad \forall X \in \Gamma(TM).$$

Then, $A_\xi^* X + \tau(X)\xi + X = 0$. As A_ξ^* is $\Gamma(S(TM))$−valued we obtain

$$A_\xi^* X = -PX, \quad \forall X \in \Gamma(TM). \tag{7.4.10}$$

Next, any $X \in \Gamma(S(T\Lambda_0^{m+1}))$ is expressed by $X = \sum_{a=1}^{m+1} X^a \partial_{x^a}$ where the components $(X^1, \cdots, X^{m+1})$ satisfy

$$\sum_{a=1}^{m+1} x^a X^a = 0. \tag{7.4.11}$$

Then,

$$\nabla_\xi X = \bar{\nabla}_\xi X = \sum_{A=0}^{m+1} \sum_{a=1}^{m+1} x^A \frac{\partial X^a}{\partial x^A} \frac{\partial}{\partial x^a},$$

$$\bar{g}(\nabla_\xi X, \xi) = \sum_{A=0}^{m+1} \sum_{a=1}^{m+1} x^a x^A \frac{\partial X^a}{\partial x^A} = -\sum_{a=1}^{m+1} x^a X^a = 0, \tag{7.4.12}$$

where (7.4.11) is differentiated with respect to each x^A. Using this and (7.4.12) we get $\nabla_\xi X \in \Gamma(S(T\Lambda_0^{m+1}))$, i.e., $A_N \xi = 0$. (7.4.9) and (7.4.11) implies

$$C(X, Y) = g(\nabla_X Y, N) = \bar{g}(\bar{\nabla}_X Y, N) = -\frac{1}{2(x^0)^2} g(X, Y),$$

that is,

$$g(A_N X, Y) = -\frac{1}{2(x^0)^2} g(X, Y), \quad X, Y \in \Gamma(S(T\Lambda_0^{m+1})).$$

Therefore, we have

$$A_N X = -\frac{1}{2(x^0)^2} PX, \quad \forall X \in \Gamma(T\Lambda_0^{m+1}). \tag{7.4.13}$$

Taking into account (7.4.10) and (7.4.13) we infer the following relation

$$A_N X = \frac{1}{2(x^0)^2} A_\xi^* X, \quad \forall X \in \Gamma(T\Lambda_0^{m+1}).$$

Thus, Λ_0^{m+1} is screen globally conformal lightlike hypersurface of $\mathbf{R}_1^{m+2}$ with positive conformal function $\varphi = \frac{1}{2(x^0)^2}$ globally defined on Λ_0^{m+1}.

Example 15. Consider a smooth function $F : \Omega \to \mathbf{R}$, where Ω is an open set of $\mathbf{R}^{m+1}$. Then

$$M = \{(x^0, \cdots, x^{m+1}) \in \mathbf{R}_q^{m+2} \; : \; x^0 = F(x^1, \cdots, x^{m+1})\}$$

is a *Monge hypersurface* [28]. The natural parameterization on M is given by

$$x^0 = F(v^0, \ldots, v^m) \; ; \quad x^{\alpha+1} = v^\alpha, \qquad \alpha \in \{0, \ldots, m\} \,.$$

Hence, the natural frames field on M is globally defined by

$$\partial_{v^\alpha} = F'_{x^{\alpha+1}} \partial_{x^0} + \partial_{x^{\alpha+1}}, \qquad \alpha \in \{0, \ldots, m\} \,.$$

Then

$$\xi = \partial_{x^0} - \sum_{s=1}^{q-1} F'_{x^s} \partial_{x^s} + \sum_{a=q}^{m+1} F'_{x^a} \partial_{x^a}$$

spans $TM^\perp$. Therefore, M is lightlike (i.e., $TM^\perp = Rad\,TM$), if and only if, the global vector field ξ span $Rad\,TM$ which means, if and only if, F is a solution of the partial differential equation

$$1 + \sum_{s=1}^{q-1} (F'_{x^s})^2 = \sum_{a=q}^{m+1} (F'_{x^a})^2 \,.$$

Along M consider the constant timelike section $V^* = \partial_{x^0}$ of $\Gamma(T\mathbf{R}_q^{m+2})$. Then $\bar{g}(V^*, \xi) = -1$ implies that V^* is not tangent to M. Therefore, the vector bundle $H^* = Span\{V^*, \xi\}$ is nondegenerate on M. The complementary orthogonal vector bundle $S^*(TM)$ to H^* in $T\mathbf{R}_q^{m+2}$ is a non-degenerate distribution on M and is complementary to $Rad\,TM$. Thus $S^*(TM)$ is a screen distribution on M. The transversal bundle $ltr^*(TM)$ is spanned by $N = -V^* + \frac{1}{2}\xi$ and $\tau(X) = 0$ for any $X \in \Gamma(TM)$. Indeed, $\tau(X) = \bar{g}(\bar{\nabla}_X N, \xi) = \frac{1}{2}\bar{g}(\bar{\nabla}_X \xi, \xi) = 0$. Therefore, Weingarten equations reduce to $\bar{\nabla}_X N = -A_N X$ and $\bar{\nabla}_X \xi = -A_\xi^* X$, which implies

$$A_N X = \frac{1}{2} A_\xi^* X, \quad \forall X \in \Gamma(TM).$$

Hence, any lightlike Monge hypersurface of $\mathbf{R}_q^{m+2}$ is screen globally conformal with constant positive conformal function $\varphi(x) = \frac{1}{2}$.

Remark 4. According to [28, page 120], we call $S^*(TM)$ the *natural screen distribution*. For the case $q = 1$ only, the natural and the canonical (see example 9) screen distributions coincide on lightlike Monge hypersurfaces. This is to say that, endowed with the canonical screen distribution, lightlike Monge hypersurfaces are screen globally conformal in Minkowski space.

To show the existence of a large class of lightlike hypersurfaces with an integrable screen, we first quote (with proof) the following general result:

Theorem 4.1 ([13]). *Let $(M, g, S(TM))$ be a lightlike hypersurface of a semi-Riemannian manifold $(\bar{M}, \bar{g})$. Then, the following assertions are equivalent:*

(i) $S(TM)$ is an integrable distribution.

(ii) $h^*(X, Y) = h^*(Y, X), \forall X, Y \in \Gamma(S(TM))$.

(iii) *The shape operator of M is symmetric with respect to g, i.e.,*

$$g(A_V X, Y) = g(X, A_V Y), \ \forall X, Y \in \Gamma(S(TM)), \ V \in \Gamma(tr(TM)).$$

Proof. (7.3.11) implies that a vector field X on M belongs to $S(TM)$, if and only if, on each $\mathcal{U} \subset M$ we have $\eta(X) = 0$. By using (7.3.17) and (7.3.11) we get

$$C(X, Y) - C(Y, X) = \eta([X, Y]), \quad \forall X, Y \in \Gamma(TM_{|\mathcal{U}}),$$

which together with (7.3.15) implies the equivalence of (i) and (ii). Finally, the equivalence of (ii) and (iii) follows from (7.3.20), which completes the proof.

Remark 5. Just as in the well-known case of locally product Riemannian or semi-Riemannian manifolds (see [82]), if $S(TM)$ is integrable then M is locally a product manifold $C \times M'$ where C is a null curve and M' is a leaf of $S(TM)$.

Examples. It is easy to see that the screen distributions of lightlike hypersurfaces of $(\mathbf{R}_2^4, \bar{g})$ (in example 5) and of $(\mathbf{R}_1^4, \bar{g})$ (in example 6) are integrable.

Now we prove the following main theorem, of this section, on integrable screens:

Theorem 4.2. *The screen distribution on a screen conformal lightlike hypersurface $(M, g, S(TM))$, of a semi-Riemannian manifold $(\bar{M}, \bar{g})$, is integrable. Moreover, M is screen locally (or globally) conformal if and only if the second fundamental forms B and C of M and $S(TM)$, respectively, satisfy*

$$C(X, PY) = \varphi B(X, Y), \quad \forall X, Y \in \Gamma(TM_{|\mathcal{U}}) \tag{7.4.14}$$

for some smooth φ on $\mathcal{U} \subseteq M$.

Proof. It follows from (7.3.19) that, for any lightlike hypersurface, the shape operator A_ξ^* of $S(TM)$ is symmetric with respect to g, i.e.,

$$g(A_\xi^* X, Y) = g(A_\xi^* Y, X), \quad \forall X, Y \in \Gamma(S(TM)), \quad \xi \in \Gamma(Rad\,TM).$$

This result and the relation (7.4.7), along with theorem 4.1(iii), implies that the screen distribution of a screen conformal lightlike hypersurface M is integrable, which proves the first part. For the second part, suppose M is screen conformal. Then, using (7.3.19), (7.3.20) and (7.4.7) we obtain

$$\begin{aligned} C(X, PY) &= g(A_N X, PY) = \varphi g(A_\xi^* X, PY) \\ &= \varphi B(X, PY) = \varphi B(X, Y), \quad \forall X, Y \in \Gamma(TM_{|\mathcal{U}}). \end{aligned}$$

The converse of the second part also holds, which completes the proof.

Applying theorem 4.2 to example 15, we conclude that (beside the special cases of Examples 10-12) their exists a large class of lightlike Monge hypersurfaces of $\mathbf{R}_q^{m+2}$ with integrable screen distribution. More precisely, we state the following:

Theorem 4.3 ([28]). *On a Monge hypersurface M of a semi-Euclidean space $\mathbf{R}_q^{m+2}$ there exists an integrable screen distribution.*

Remark 6. As the definition 4.1 allows $\mathcal{U} = M$, for which the screen conformality is global, it follows from theorem 4.2 that, besides being a local product manifold (see remark 3) there exists a screen globally conformal M which is a global product manifold $C \times M'$ where C is a global null curve and M' is a leaf of its integrable screen distribution $S(TM)$. In particular, as per example 12 and theorem 4.3, there exists a global product lightlike Monge hypersurface of a semi-Euclidean space $\mathbf{R}_q^{m+2}$.

Since, in general, the screen distribution is not unique and the same is true for an arbitrary integrable screen, it is reasonable to investigate the behavior of the screen conformality with respect to a change in screen. For this purpose, we denote by $\mathcal{S}^1$ the first derivative of $S(TM)$, of a screen conformal lightlike hypersurface $(M, g, S(TM))$, given by

$$\mathcal{S}^1(x) = Span\{[X, Y]_{|_x} : X_x, Y_x \in S(T_xM); \; x \in M\}. \tag{7.4.15}$$

As $S(TM)$ is integrable, $\mathcal{S}^1$ is its subbundle. Let $S(TM)$ and $S(TM)'$ denote two screen distributions on M, h and h' their second fundamental forms with respect to $ltr(TM)$ and $ltr(TM)'$ respectively, for the same $\xi \in \Gamma(TM^\perp|_\mathcal{U})$. Denote by ω the dual one form of the vector field $W = \sum_{a=1}^m \mathbf{f}_a W_a$ (see equation (7.3.25)) with respect to the metric tensor g, that is

$$\omega(X) = g(X, W), \quad \forall X \in \Gamma(TM). \tag{7.4.16}$$

Theorem 4.4 ([4]). *Let $(M, g, S(TM))$ be a screen conformal lightlike hypersurface of a semi-Riemannian manifold $(\bar{M}, \bar{g})$, with the first derivative $\mathcal{S}^1$ of $S(TM)$ given by (7.4.15). Then the one form ω in (7.4.16) vanishes identically on the $\mathcal{S}^1$. In particular, if $\mathcal{S}^1$ coincides with $S(TM)$, then, there is a unique screen conformal distribution, up to an orthogonal transformation and a unique lightlike transversal vector bundle. Moreover, for this class of hypersurfaces, the screen second fundamental form C is independent of the choice of a screen distribution.*

Proof. We know from (7.4.14) that for a screen conformal M

$$C(X, PY) = \varphi B(X, Y), \quad \forall \, X, Y \in \Gamma(TM_{|\mathcal{U}}).$$

Since the right hand side of above relation is symmetric in X and Y we obtain

$$g(\nabla_X PY - \nabla_Y PX, W) = 0, \quad \forall \, X, Y \in \Gamma(TM).$$

Therefore, we have $g(\nabla_X Y - \nabla_Y X, W) = 0$ for any $X, Y \in \Gamma(S(TM))$, that is, $\omega([X, Y]) = g([X, Y], W) = 0$ for any $X, Y \in \Gamma(S(TM))$, which proves the first part.

Now assume that $\mathcal{S}^1 = S(TM)$, that is, ω vanishes on $S(TM)$, which implies from (7.4.16) that $W = 0$. This further means that the functions f_a vanish. Thus, the transformation equation (7.2.7) and (7.2.8) become $W'_a = \sum_{b=1}^{m} W_a^b W_b$ $(1 \leq a \leq m)$ and $N' = N$ where (W_a^b) is an orthogonal matrix of $S(T_x M)$ at any $x \in M$. Finally, the independence of C follows by putting $W = 0$ in (7.3.25) which completes the proof.

Now we show that, besides lightlike Monge hypersurfaces, there are other types of lightlike hypersurfaces with integrable screens, some of them known in relativity. For this, we first present the following Mathematical Model:

Mathematical Model. Let $(\bar{M}, \bar{g})$ be a Lorentzian manifold with the metric

$$ds^2 = -dt^2 + e^\mu (dx^1)^2 \oplus f^2 \, d\Omega^2, \tag{7.4.17}$$

where we let $d\Omega^2 = \bar{g}_{ab} \, dx^a \, dx^b$ with respect to a local coordinate system $(t, x^1, \cdots, x^{m+1})$, $\dim(\bar{M}) = m + 2$, μ and f are functions of t and x^1 alone. Let $\mathbf{E} = \{e_0, e_1, \cdots, e_{m+1}\}$ be an orthonormal bases, such that e_0 is timelike and all others are spacelike unit vectors. Transform $\mathbf{E}$ into a quasi-orthonormal bases $\mathcal{E} = \{\partial_u, \partial_v, e_2, \ldots, e_{m+1}\}$ such that ∂_u and ∂_v are real null vectors satisfying $\bar{g}(\partial_u, \partial_v) = 1$ with respect to a new coordinate system $\{u, v, x^2, \ldots, x^{m+1}\}$. Then, ds^2 of $\bar{g}$ transforms into

$$ds^2 = -A^2(u, v) du \, dv + f^2 \, d\Omega^2$$

for some function $A(u, v)$ on M. The absence of du^2 and dv^2 in above line element implies that $\{v = \text{constant}\}$ and $\{u = \text{constant}\}$ are lightlike hypersurfaces and their intersection provides a leaf of their common screen distribution. Let $(M, g, S(TM), v = \text{constant})$ be a lightlike hypersurface of $\bar{M}$. Then, as per (7.2.10), the matrix of the metric g on M with respect to the frames field $\mathcal{E}$ is

$$[g] = \begin{bmatrix} 0 & 0 \\ 0 & f^2 \, d\Omega^2 \end{bmatrix}. \tag{7.4.18}$$

Let us call $\mathcal{G} = \{\mu, f, d\Omega^2\}$ the generating set of a family of lightlike hypersurfaces for prescribed values of its elements. For example, if $\mu = 0$, $f = 1$ and $d\Omega^2$ is a Euclidean metric, then, $\bar{M}$ is a Minkowski space.

Theorem 4.5. *Let $(M, g, S(TM), v = \text{constant})$ be a lightlike hypersurface of a Lorentzian manifold $(\bar{M}, \bar{g}, \mathcal{G})$, where $\bar{g}$ and g are given by (7.4.17) and (7.4.18) respectively and the prescription $\mathcal{G} = (\mu, f, d\Omega^2)$. Then, there exists a section ξ of $Rad\,TM$ with respect to which M is a screen conformal lightlike hypersurface of $\bar{M}$. Consequently, the screen distribution $S(TM)$ of M is integrable.*

Proof. Let $(t, x^1, \cdots, x^{m+1})$ be local coordinates on $\bar{M}$. Assume that $Rad\,T_x M$ is spanned by a null vector field ξ. Consider a natural basis

$\{\partial_t, \partial_{x^1}, \ldots, \partial_{x^m}\}$ such that ξ is given by

$$\xi = \xi_t \partial_t + \sum_{a=1}^{m} \xi^a \partial_{x^a}, \quad \sum_{a=1}^{m} (\xi^a)^2 = (\xi_t)^2. \tag{7.4.19}$$

It is easy to see that ∂_t is covariant constant vector field of $\bar{M}$, given by (7.4.17). This allows us to choose along M a timelike covariant constant vector field $V = -\partial_t$ which satisfies the condition $\bar{g}(V, \xi) = \xi_t \neq 0$ of theorem 2.1 (i). Using this and ξ, construct a non-degenerate vector bundle $\mathcal{B} = Span\{V, \xi\}$ on M. Take the complementary orthogonal vector bundle $S(TM)$ to $\mathcal{B}$ in $T\bar{M}$, which is a non-degenerate distribution on M complementary to $Rad\,TM$. This means that $S(TM)$ is a screen distribution on M. Using this, theorem 2.1 and (7.4.19) we obtain

$$N = (\xi_t)^{-1} \left(V + \frac{1}{2\xi_t} \xi \right), \tag{7.4.20}$$

the null transversal vector bundle of M. Using (7.3.7), (7.4.20) and (7.3.18) we get

$$\begin{aligned}
\tau(X) &= \bar{g}(\bar{\nabla}_X N, \xi) = X(\xi_t)^{-1} \bar{g}(V, \xi) + \frac{1}{2} (\xi_t)^{-2} \bar{g}(\bar{\nabla}_X \xi, \xi) \\
&= X(\xi_t)^{-1} \xi^0 = X(\ln \theta), \tag{7.4.21}
\end{aligned}$$

where we set $(\xi_t)^{-1} = \theta$. Using this value of τ, (7.4.20) and (7.3.18) we get

$$\begin{aligned}
\bar{\nabla}_X N &= X(\theta)V + \theta X(\theta)\xi + \frac{1}{2}\theta^2 \bar{\nabla}_X \xi \\
&= X(\theta)V + \frac{1}{2}\theta X(\theta)\xi - \frac{1}{2}\theta^2 A_\xi^* X. \tag{7.4.22}
\end{aligned}$$

On the other hand, substituting the value of τ in (7.3.7), we get

$$\bar{\nabla}_X N = -A_N X + X(\theta)V + \frac{1}{2}\theta X(\theta)\xi. \tag{7.4.23}$$

Equating (7.4.22) with (7.4.23) and putting back $\theta = (\xi^0)^{-1}$ we get

$$A_N = \frac{1}{2}(\xi_t)^{-2} A_\xi^*.$$

Thus, as per definition 4.1, $(M, g, S(TM), v = constant)$ is a screen conformal lightlike hypersurface of $(\bar{M}, \bar{g}, \mathcal{G} = (0, \mu, f, g'_{ab}))$ with conformal function $\varphi = \frac{1}{2}(\xi_t)^{-2}$. Consequently, it follows from theorem 4.2 that $S(TM)$ is integrable, which completes the proof.

The proof of the following result is immediate from theorems 4.4 and 4.5.

Corollary 4.1. *Under the hypothesis of theorem 4.5, if M admits the first derivative S^1 of $S(TM)$, given by (7.4.15), then, $S(TM)$ is unique up to an orthogonal transformation with a unique lightlike transversal vector bundle.*

Following results in [28, pages 117, 118] come from theorems 4.1 and 4.5.

(1) *The canonical screen distribution on any lightlike hypersurface M of a Minkowski space $\mathbf{R}_1^{m+2}$ is integrable.*

(2) *Let M be a lightlike hypersurface of $\mathbf{R}_1^{m+2}$ equipped with the canonical screen distribution $S(TM)$. Then the second fundamental form h^* of $S(TM)$ is symmetric on $S(TM)$ and the shape operator A_N of M is symmetric with respect to g.*

(3) *Any lightlike hypersurface M of $\mathbf{R}_1^{m+2}$ is locally a product $L \times M'$, where L and M' are an open set of a lightlike line and a Riemannian submanifold of $\mathbf{R}_1^{m+2}$, respectively.*

Physical example. Since the Einstein field equations are a complicated set of nonlinear partial differential equations, one often assume certain relevant symmetry conditions for a satisfactory representation of our universe. We refer a recent book [32] on *Symmetries of spacetimes and Riemannian manifolds*. Through extragalactic observations, it is reasonable to assume that our 4-dimensional universe is isotropic, that is, approximately spherical symmetric about each point in spacetime. This means (see Walker [110]) the universe is *spatially homogeneous*, that is, admits a 6-parameter group G_6 of isometries whose surfaces of transitivity are spacelike hypersurfaces of constant curvature. This further means that any point on one of these hypersurfaces is equivalent to any other point on the same hypersurface. Such a spacetime is called *Robertson-Walker spacetime* with metric

$$ds^2 = -dt^2 + S^2(t) d\Sigma^2, \tag{7.4.24}$$

where $d\Sigma^2$ is the metric of a spacelike hypersurface Σ with spherical symmetry and constant curvature $c = 1, -1$ or 0. With respect to a local spherical coordinate system (r, θ, ϕ), this metric is given by

$$d\Sigma^2 = dr^2 + f^2(r)(d\theta^2 + \sin^2\theta \, d\phi^2), \tag{7.4.25}$$

where $f(r) = \sin r$, $\sinh r$ or r according as $c = 1, -1$ or 0. The range of the coordinates is restricted from 0 to 2π or from 0 to ∞ for $c = 1$ or -1 respectively.

Thus, above described Robertson-Walker spacetime can be generated by

$$\mathcal{G} = \left\{0, \, f(r), \, d\Omega^2 = d\theta^2 + \sin^2\theta \, d\phi^2\right\}.$$

Take two null coordinates u and v such that $u = t + r$ and $v = t - r$. Thus, (7.4.24) with (7.4.25) transforms into a non-singular metric:

$$ds^2 = -du \, dv + f^2(r) \, d\Omega^2.$$

As explained in Mathematical Model, the absence of du^2 and dv^2 in this transformed metric implies that $\{v = \text{constant}\}$ and $\{u = \text{constant}\}$ are lightlike hypersurfaces of the Robertson-Walker spacetime, denoted by $\bar{M}$. Let $(M, g, v = \text{constant})$ be one of this lightlike pair and let D be the 1-dimensional distribution generated by the null vector $\{\partial_v\}$, in $\bar{M}$. Denote by L the 1-dimensional integral manifold of D. A leaf M' of the 3-dimensional screen of $\bar{M}$ is Riemannian with metric $d\Omega^2$.

Consequently, as per theorem 4.5, there exists a section ξ of $RadTM$ with respect to which M is a screen conformal lightlike hypersurface of the Robertson-Walker spacetime and, therefore, the screen distribution $S(TM)$ of M is integrable.

Observe that the Minkowski space, de-Sitter space and anti-de-Sitter space are all special cases of the general Robertson-Walker spacetimes. For details on these spacetimes and some more (which may satisfy Theorem 4.5) we refer [10, 56].

7.5   Induced curvature and Ricci tensors

Let $(M, g, S(TM))$ be a lightlike hypersurface of a semi-Riemannian manifold $(\bar{M}, \bar{g})$. Recall, from section 7.1, that we denote by $\bar{R}$ and R the curvature tensors of the Levi-Civita connection $\bar{\nabla}$ on $\bar{M}$ and the induced linear connection ∇ on M, respectively. Since $h(X, Y) \in \Gamma(ltr(TM))$ we set

$$(\nabla_X h)(Y, Z) = \nabla_X^t(h(Y, Z)) - h(\nabla_X Y, Z) - h(Y, \nabla_X Z).$$

Then, using (7.1.5), (7.3.1) and (7.3.2) we obtain

$$\begin{aligned}
\bar{R}(X, Y)Z &= R(X, Y)Z + A_{h(X,Z)}Y - A_{h(Y,Z)}X \\
&\quad + (\nabla_X h)(Y, Z) - (\nabla_Y h)(X, Z),
\end{aligned} \tag{7.5.1}$$

for any $X, Y, Z \in \Gamma(TM)$. Consider $\bar{R}$ of type $(0,4)$, defined by (7.1.8). Using (7.5.1) and the Gauss-Weingarten equations for M and $S(TM)$ (see section 7.3) we obtain:

$$\begin{aligned}
\bar{g}(\bar{R}(X, Y)Z, PW) &= g(R(X, Y)Z, PW) + \bar{g}(h(X, Z), h^*(Y, PW)) \\
&\quad - \bar{g}(h(Y, Z), h^*(X, PW)), \tag{7.5.2} \\
\bar{g}(\bar{R}(X, Y)Z, U) &= \bar{g}((\nabla_X h)(Y, Z) - (\nabla_Y h)(X, Z), U), \tag{7.5.3} \\
\bar{g}(\bar{R}(X, Y)Z, V) &= \bar{g}(R(X, Y)Z, V), \tag{7.5.4}
\end{aligned}$$

for any $X, Y, Z, W \in \Gamma(TM)$, $U \in \Gamma(RadTM)$, $V \in \Gamma(ltr(TM))$. We call (7.5.2)-(7.5.4) the global *Gauss-Codazzi type* equations for the lightlike hypersurface M. These equations depend on the screen distribution $S(TM)$. Finding of the transformation equations with respect to a change in screen is left as an exercise. The procedure is similar to the one presented in section 7.3. To find their local expressions, we use the local Gauss-Weingarten equations with respect to a null pair $\{\xi, N\}$ on $\mathcal{U} \subset M$ and obtain

$$\begin{aligned}
\bar{g}(\bar{R}(X,Y)Z, PW) &= g(R(X,Y)Z, PW) \\
&\quad + B(X, Z)C(Y, PW) - B(Y, Z)C(X, PW), \tag{7.5.5} \\
\bar{g}(\bar{R}(X, Y)Z, \xi) &= (\nabla_X B)(Y, Z) - (\nabla_Y B)(X, Z) \\
&\quad + B(Y, Z)\tau(X) - B(X, Z)\tau(Y), \tag{7.5.6} \\
\bar{g}(\bar{R}(X,Y)Z, N) &= \bar{g}(R(X,Y)Z, N), \tag{7.5.7}
\end{aligned}$$

for any $X, Y, Z, W \in \Gamma(TM_{|\mathcal{U}})$, where

$$(\nabla_X B)(Y, Z) = X(B(Y, Z)) - B(\nabla_X Y, Z) - B(Y, \nabla_X Z).$$

Using (7.3.17) and (7.3.18) in the right hand side of (7.5.7) we obtain

$$
\begin{aligned}
\bar{g}(\bar{R}(X, Y)PZ, N) &= (\nabla_X C)(Y, PZ) - (\nabla_Y C)(X, PZ) \\
&\quad + \tau(Y)C(X, PZ) - \tau(X)C(Y, PZ), \quad &(7.5.8) \\
\bar{g}(\bar{R}(X, Y)\xi, N) &= C(Y, A_\xi^* X) - C(X, A_\xi^* Y) - 2d\tau(X, Y), \quad &(7.5.9)
\end{aligned}
$$

for any $X, Y, Z \in \Gamma(TM_{|\mathcal{U}})$, where we set

$$(\nabla_X C)(Y, PZ) = X(C(Y, PZ)) - C(\nabla_X Y, PZ) - C(Y, \nabla_X^* PZ)$$

and in (7.5.9) we use the following well-known exterior derivative formula:

$$d\tau(X, Y) = \frac{1}{2}\{X(\tau(Y)) - Y(\tau(X)) - \tau([X, Y])\}.$$

To find the local expressions of Gauss-Codazzi equations in a coordinate system on M, we use an atlas of local charts and the transformation equations as given in section 7.2 (see the equations (7.2.9)-(7.2.13)) followed by the equations (7.3.26)-(7.3.40) of the section 7.3. Then, for $i, j, k, h \in \{0, \ldots, m\}$, using (7.1.8) we have the following local components of curvature tensors $\bar{R}$ and R:

$$
\begin{aligned}
\bar{R}_{ijkh} &= \bar{g}(\bar{R}(X_h, X_k)X_j, X_i), \quad R_{ijkh} = g(R(X_h, X_k)X_j, X_i), \\
\bar{R}_{ijk}^o &= \bar{g}(\bar{R}(X_k, X_j)X_i, N), \quad R_{ijk}^o = g(R(X_k, X_j)X_i, N).
\end{aligned}
$$

Thus, in terms of local coordinate, the local Gauss-Codazzi equations are:

$$
\begin{aligned}
\bar{R}_{abcd} &= R_{abcd} + C_{ac}B_{bd} - C_{ad}B_{bc}, &(7.5.10) \\
\bar{R}_{abod} &= R_{abod} + C_a B_{bd}, &(7.5.11) \\
\bar{R}_{aocd} &= R_{aocd} ; \bar{R}_{aood} = R_{aood}, &(7.5.12) \\
\bar{R}_{obcd} &= B_{bc;d} - B_{bd;c} + B_{bc}\tau_h - B_{bd}\tau_c = -R_{bocd}, &(7.5.13) \\
\bar{R}_{ojoh} &= -\partial_{u^o}(B_{bd}) + \Gamma_{bo}^c B_{cd} - \tau_o B_{bd} = -R_{bood}, &(7.5.14)
\end{aligned}
$$

where we put

$$B_{bc;d} = \delta_{u^d}(B_{bc}) - B_{ac}\Gamma_{bd}^a - B_{ba}\Gamma_{cd}^a.$$

$$
\begin{aligned}
\bar{R}_{abc}^o &= R_{abc}^o = C_{ab;c} - C_{ac;b} + \tau_b C_{ac} - \tau_c C_{ab} &(7.5.15) \\
\bar{R}_{abo}^o &= R_{abo}^o = C_{ab;o} - C_{a;b} + \tau_b C_a - \tau_o C_{ab}, &(7.5.16) \\
\bar{R}_{obc}^o &= R_{obc}^o = C_{db}(A^*)_c^d - C_{dc}(A^*)_b^d + 2d\tau(\delta_{u^b}, \delta_{u^c}), &(7.5.17) \\
\bar{R}_{ooc}^o &= R_{ooc}^o = C_d(A^*)_c^d + 2d\tau(\partial_{u^o}, \delta_{u^c}), &(7.5.18)
\end{aligned}
$$

where we set

$$
\begin{aligned}
C_{ab;c} &= \delta_{u^c}(C_{ab}) - \Gamma_{bc}^h C_{ad} - \Gamma_{bc}^o C_a - \Gamma_{ac}^d C_{db}, \\
C_{ab;o} &= \partial_{u^o}(C_{ab}) - \Gamma_{bo}^d C_{ad} - \Gamma_{bo}^o C_a - \Gamma_{ao}^d C_{db}, \quad &(7.5.19) \\
C_{a;b} &= \delta_{u^b}(C_a) - \Gamma_{ob}^d C_{ad} - \Gamma_{ob}^o C_a - \Gamma_{ab}^d C_d.
\end{aligned}
$$

Now we show that locally all the Gauss-Codazzi equations can be expressed in terms of the geometric objects induced by the global Gauss-Weingarten equations (7.3.6) and (7.3.7). First, by using (7.3.40) for C_{ab} and C_a in (7.5.10) and (7.5.11) we obtain

$$\begin{aligned}
\bar{R}_{abcd} &= R_{abcd} + g_{at}(A_c^t B_{bd} - A_d^t B_{bc}), \\
\bar{R}_{ijoh} &= R_{ijoh} + g_{it} A_o^t B_{jh},
\end{aligned} \qquad (7.5.20)$$

respectively. Then, from (7.5.19), (7.3.40), (7.3.42) and (7.3.43) we obtain the equivalence of (7.5.15) and (7.5.16) with

$$\bar{R}^o_{abc} = R^o_{abc} = g_{ad}\left\{ A^d_{b;\,c} - A^d_{c;\,b} + \tau_b A^d_c - \tau_c A^d_b \right\}, \qquad (7.5.21)$$

and

$$\bar{R}^o_{abo} = R^o_{abo} = g_{ad}\left\{ A^d_{b;\,o} - A^d_{o;\,b} + \tau_b A^d_o - \tau_o A^d_b \right\}, \qquad (7.5.22)$$

respectively, where we set

$$\begin{aligned}
A^d_{b;\,c} &= \frac{\delta A^d_b}{\delta u^c} - \Gamma^i_{bc} A^d_a - \Gamma^o_{bc} A^d_o + \Gamma^d_{ac} A^a_b, \\
A^d_{b;\,o} &= \frac{\partial A^d_b}{\partial u^o} - \Gamma^i_{bo} A^d_a - \Gamma^o_{bo} A^d_o + \Gamma^d_{ao} A^a_b, \\
A^d_{o;\,b} &= \frac{\delta A^d_o}{\delta u^b} - \Gamma^i_{ob} A^d_a - \Gamma^o_{ob} A^d_o + \Gamma^d_{ab} A^a_o.
\end{aligned}$$

Finally, using (7.3.40) and (7.3.41) we see that (7.5.17) and (7.5.18) become

$$\bar{R}^o_{obc} = R^o_{obc} = A^a_b B_{ac} - A^a_c B_{ab} + 2d\tau\,(\delta_{u^b}, \delta_{u^c}), \qquad (7.5.23)$$

and

$$\bar{R}^o_{ooc} = R^o_{ooc} = A^a_o B_{ac} + 2d\tau\,(\partial_{u^o}, \delta_{u^c}), \qquad (7.5.24)$$

respectively. In this way, all the Gauss-Codazzi equations, given by (7.5.12)-(7.5.14) and (7.5.20)-(7.5.24), are expressed locally by using the coefficients of ∇ and local components of h, A_N and τ.

Induced Ricci tensor. Using the equation (7.1.10) of the Ricci tensor $\bar{R}ic$ of $\bar{M}$, let $R^{(0,\,2)}$ denote the induced tensor of type $(0,2)$ on M given by

$$R^{(0,\,2)}(X,\,Y) = trace\{Z \to R(X,\,Z)Y\}, \quad \forall X, Y \in \Gamma(TM). \qquad (7.5.25)$$

Since, as per (7.3.12), the induced connection ∇ on M is not a metric connection, in general, $R^{(0,\,2)}$ is not symmetric. Therefore, in general, it is just a tensor quantity. Indeed, consider the induced quasi-orthonormal frame $\{\xi;\,W_a\}$ on M, where $RadTM = Span\{\xi\}$ and $S(TM) = Span\{W_a\}$ and let $E = \{\xi, N, W_a\}$ be the corresponding frames field on $\bar{M}$. Then, we obtain

$$R^{(0,\,2)}(X,\,Y) = \sum_{a=1}^m \epsilon_a\, g(R(X,\,W_a)Y,\,W_a) + \bar{g}(R(X,\,\xi)Y,\,N). \qquad (7.5.26)$$

where ϵ_a denotes the causal character (± 1) of respective vector field W_a. Using Gauss-Codazzi equations, we obtain

$$
\begin{aligned}
g(R(X, W_a)Y, W_a) \;=\;& \bar{g}(\bar{R}(X, W_a)Y, W_a) \\
& +\; B(X, Y)C(W_a, W_a) - B(W_a, Y)C(X, W_a).
\end{aligned}
$$

Substituting this in (7.5.26), then using the relations (7.3.19) and (7.3.20) we obtain

$$
\begin{aligned}
R^{(0,2)}(X, Y) \;=\;& \bar{R}ic(X, Y) + B(X, Y)tr A_N - g(A_N X, A_\xi^* Y) \\
& -\bar{g}(R(\xi, Y)X, N),
\end{aligned}
\tag{7.5.27}
$$

for all X, $Y \in \Gamma(TM)$ and $\bar{R}ic$ is the Ricci tensor of $\bar{M}$. This shows that $R^{(0,2)}$ is not symmetric. Therefore, in general, $R^{(0,2)}$ has no geometric or physical meaning similar to the symmetric Ricci tensor of $\bar{M}$. In 1996 book [28], $R^{(0,2)}$, given by (7.5.25), was called an induced Ricci tensor which is not correct. To highlight this correction, we introduce the following definition:

Definition 5.1. *Let $(M, g, S(TM))$ be a lightlike hypersurface of a semi-Riemannian manifold $(\bar{M}, \bar{g})$. A tensor field $R^{(0,2)}$ of M, given by (7.5.25), is called its induced Ricci tensor if it is symmetric.*

In the sequel, a symmetric $R^{(0,2)}$ tensor will be denoted by Ric. Now one may ask the following question: *Are there any lightlike hypersurfaces with symmetric Ricci tensor?* The answer is affirmative, for which we need the following:

Definition 5.2. *If any geodesic of a lightlike hypersurface M with respect to an induced connection ∇ is a geodesic of $\bar{M}$ with respect to $\bar{\nabla}$, we say that M is a totally geodesic lightlike hypersurface of $\bar{M}$.*

The following known general result shows that above definition is independent of the screen distribution.

Theorem 5.1 ([13]). *Let $(M, g, S(TM))$ be a lightlike hypersurface of a semi-Riemannian manifold $(\bar{M}, \bar{g})$. The following assertions are equivalent:*

(i) *M is totally geodesic.*

(ii) *h vanishes identically on M.*

(iii) *A_U^* vanishes identically on M, for any $U \in \Gamma(Rad\,TM)$.*

(iv) *There exists a unique torsion-free metric connection ∇ induced by $\bar{\nabla}$ on M.*

(v) *$Rad\,TM$ is a parallel distribution with respect to ∇.*

(vi) *$Rad\,TM$ is a Killing distribution on M.*

Suppose $(M, g, S(TM))$ is a totally geodesic lightlike hypersurface of a semi-Riemannian manifold $(\bar{M}(c), \bar{g})$ of constant sectional curvature c. Then, using (7.1.17) we have $\bar{R}(\xi, Y)X = \bar{g}(X, Y)\xi$. Since it follows from (7.5.7) that $\bar{g}(\bar{R}(\xi, Y)X, N) = \bar{g}(R(\xi, Y)X, N)$, using these both results in (7.5.27) and (ii) and (iii) of theorem 5.1, we obtain

$$Ric(X, Y) = \bar{R}ic(X, Y) - c\,\bar{g}(X, Y)$$

which is symmetric since $\bar{g}$ is symmetric. Consequently, we have

Proposition 5.1. *Any totally geodesic lightlike hypersurface of $\bar{M}(c)$ admits an induced symmetric Ricci tensor.*

Are there any others, with symmetric induced Ricci tensors, but not necessarily totally geodesic? Here is one such class.

Theorem 5.2 ([4]). *Let $(M, g, S(TM))$ be a locally (or globally) screen conformal lightlike hypersurface of a semi-Riemannian manifold $(\bar{M}(c), \bar{g})$ of constant sectional curvature c. Then, M admits an induced symmetric Ricci tensor.*

Proof. Proceeding as in the proof of Proposition 5.1. we obtain

$$R^{(0, 2)}(X, Y) = \bar{R}ic(X, Y) + B(X, Y)trA_N - g(A_N X, A_\xi^* Y) - c\bar{g}(X, Y).$$

g and B are symmetric and (7.4.7) of the definition of screen conformal M implies

$$\begin{aligned} R^{(0, 2)}(X, Y) - R^{(0, 2)}(Y, X) &= g(A_N X, A_\xi^* Y) - g(A_N Y, A_\xi^* X) \\ &= \varphi g([A_\xi^*, A_\xi^*]Y, X) = 0, \end{aligned}$$

which shows that $R^{(0, 2)} = Ric$. This completes the proof.

Note. Thus, as per section 7.4, a large variety of totally geodesic and screen conformal hypersurfaces of $\bar{M}(c)$ carry symmetric Ricci tensor. In general, we quote the following existence theorem (slightly reworded) on symmetric Ricci tensor.

Theorem 5.3 ([28]). *Let $(M, g, S(TM))$ be a lightlike hypersurface of a semi-Riemannian manifold $(\bar{M}, \bar{g})$. Then the tensor $R^{(0, 2)}$ of the induced connection ∇ is a symmetric Ricci tensor Ric, if and only if, each 1-form τ induced by $S(TM)$ is closed, i.e., $d\tau = 0$, on any $\mathcal{U} \subset M$.*

Suppose we assume that the tensor $R^{(0, 2)}$ is symmetric Ricci tensor Ric. By Theorem 5.3 and Poincaré lemma we obtain $\tau(X) = X(f)$, where f is a smooth function on $\mathcal{U}$. Setting $\alpha = \exp(f)$ in second equation of (7.3.9) we get $\bar{\tau}(X) = 0$ for any $X \in \Gamma(TM_{|\mathcal{U}})$. Thus,

Proposition 5.2 ([28]). *Let $(M, g, S(TM))$ be a lightlike hypersurface of $(\bar{M}, \bar{g})$. If the tensor $R^{(0,\, 2)}$ of ∇ is symmetric then there exists a pair $\{\xi, N\}$ on $\mathcal{U}$ such that the corresponding 1-form τ from (7.3.7) vanishes.*

Note. It is important to note that as per theorem 5.3 the existence of a symmetric induced Ricci tensor on M is equivalent to $d\tau = 0$, on any $\mathcal{U} \subset M$ and τ need not vanish since the converse of proposition 5.2 does not hold. Therefore, only vanishing of $d\tau$ is needed to get an induced symmetric Ricci tensor for M.

Example 16. Let $(\mathbf{R}_2^4, \bar{g})$ be a 4-dimensional semi-Euclidean space of index 2 with signature $(-, -, +, +)$ of the canonical basis $(\partial_0, \partial_1, \partial_2, \partial_3)$. Consider a hypersurface M of $\mathbf{R}_2^4$ given by

$$x_0 = u\cosh v + \sinh v, \quad x_1 = w, \quad x_2 = u + v, \quad x_3 = u\sinh v + \cosh v.$$

Then M is a lightlike hypersurface with the radical distribution

$$Rad\,TM = Span\,\{\xi = \cosh v\,\partial_0 + \partial_2 + \sinh v\,\partial_3\}.$$

The lightlike transversal vector bundle is given by

$$ltr(TM) = Span\left\{N = -\frac{1}{2}\left(\cosh v\,\partial_0 - \partial_2 + \sinh v\,\partial_2\right)\right\}.$$

Since the tangent bundle $T\mathbf{R}_2^4$ is spanned by

$$\{\partial_u = \xi,\ \partial_v = (u\sinh v + \cosh v)\,\partial_0 + \partial_2 + (u\cosh v + \sinh v)\,\partial_3,\ \partial_w = \partial_1\}.$$

It follows that the corresponding screen distribution $S(TM)$ is spanned by

$$\{W_1 = u\sinh v\,\partial_0 + u\cosh v\,\partial_3,\quad W_2 = \partial_1\,\}.$$

With respect to the local frame fields $\{W_0 = \xi,\ W_* = N,\ W_1, W_2\}$ of $\bar{M}$, the semi-Riemannian metric $\bar{g}$ has the matrix

$$[\bar{g}] = \begin{bmatrix} 0 & 1 & 0 & 0 \\ 1 & 0 & 0 & 0 \\ 0 & 0 & u^2 & 0 \\ 0 & 0 & 0 & -1 \end{bmatrix}.$$

By example 10, the Christoffel symbols Γ_{ij}^k and the functions B_{ij} are given by

$$\Gamma_{1o}^1 = \Gamma_{o1}^1 = \frac{1}{u}, \quad \Gamma_{11}^o = \frac{u}{2}, \quad A_1^1 = \frac{1}{2u}, \quad B_{11} = -u\,;$$

all other $\Gamma_{ij}^k = S_i = S_{ij} = A_j^i = 0, \quad \forall\, i,\, j,\, k \in \{0, 1, 2\}.$

Using the equation (7.5.10) and the fact $\bar{R}_{abcd} = 0$, we have

$$
\begin{aligned}
R_{ac} &= \frac{1}{u}\,C_{ac} + \frac{1}{u^2}\,C_{a1}B_{1c} - C_{a2}B_{2c} \\
&= -\frac{1}{2u}\,B_{ac} + \frac{1}{u^2}\,B_{a1}C_{1c} - B_{a2}C_{2c} = 0,
\end{aligned}
$$

because $C_{ab} = \Gamma^o_{ab}$. Similarly from (7.5.11) - (7.5.14), we have

$$R_{ao} = \frac{1}{u} C_a = 0, \qquad R_{oa} = R_{oo} = 0.$$

Thus this lightlike hypersurface M of $\mathbf{R}^4_2$ is Ricci flat.

Example 17. Let $(M, g = \bar{g}_{|M}, S(TM))$ be the lightlike hypersurface of the Minkowski spacetime $(\mathbf{R}^4_1, \bar{g})$, as given in example 4. By example 9, with respect to the non-holonomic frames field $\{\partial_{u^o}, N, \delta_{u^1}, \delta_{u_2}\}$ of $\bar{M}$, the semi-Riemannian metric $\bar{g}$ has the matrix

$$[\bar{g}] = \begin{bmatrix} 0 & 1 & 0 & 0 \\ 1 & 0 & 0 & 0 \\ 0 & 0 & t^2 & 0 \\ 0 & 0 & 0 & t^2 \cos^2 u \end{bmatrix}$$

and the Christoffel symbols Γ^k_{ij} and the coefficients B_{ij} and A^i_j of the 2-form B and $(1, 1)$-tensor A, for any $i, j, k \in \{0, 1, 2\}$, are given by

$$\Gamma^1_{22} = \sin u \cos u, \quad \Gamma^2_{12} = -\tan u, \quad \Gamma^1_{1o} = \frac{1}{t}, \quad \Gamma^2_{2o} = \frac{1}{t} ;$$

$$\Gamma^o_{11} = -\frac{t}{2}, \quad \Gamma^o_{22} = -\frac{1}{2}t \cos^2 u, \quad B_{11} = -t, \quad B_{22} = -t \cos^2 u,$$

$$A^1_1 = -\frac{1}{2t}, \quad A^2_2 = -\frac{1}{2t}$$

and all another forms are zero.

Using the equation (7.5.10) and the fact $\bar{R}_{abcd} = 0$, we have

$$\begin{aligned} R_{ac} &= \frac{2}{t} C_{ac} + \frac{1}{t^2} C_{a1} B_{1c} + \frac{1}{t^2 \cos^2 u} C_{a2} B_{2c} \\ &= \frac{1}{t} B_{ac} + \frac{1}{t^2} B_{a1} C_{1c} + \frac{1}{t^2 \cos^2 u} B_{a2} C_{2c}. \end{aligned}$$

Thus $R_{11} = -\frac{1}{2}$, $R_{12} = R_{21} = 0$ and $R_{22} = -\frac{1}{2}\cos^2 u$. Similarly from (7.5.11) - (7.5.14), we have

$$R_{ao} = \frac{2}{t} C_a = 0, \qquad R_{oa} = R_{oo} = 0.$$

Thus the Ricci curvature of this lightlike hypersurface is symmetric.

Based on discussion so far it would be appropriate to say that from the geometric point of view alone, the induced tensor $R^{(0,\,2)}$ on M must be symmetric, as without this property one only obtains tensorial relations. Physically, $R^{(0,\,2)}$ symmetric is essential. Consequently, as per above note, it is desirable to assume that only $d\tau$ (and not τ) vanishes locally (or globally) on M. Luckily, we have so far seen that there are large classes of hypersurfaces with symmetric Ricci tensor.

In particular, symmetric induced Ricci tensor has been useful in finding several good properties of lightlike hypersurfaces. Here is one such recent study. We need the following from [78]. Consider

$$\widetilde{TM} = TM/Rad(TM), \quad \Pi : \Gamma(TM) \to \Gamma(\widetilde{TM}) \quad \text{(canonical projection).}$$

Denote $\widetilde{X} = \Pi(X)$ and $\widetilde{g}(\widetilde{X}, \widetilde{Y}) = g(X, Y)$. It is easy to prove that the operator $\widetilde{A}_U : \Gamma(\widetilde{TM}) \to \Gamma(\widetilde{TM})$ defined by $\widetilde{A}_U(\widetilde{X}) = -(\Pi(\bar{\nabla}_X U))$, where $U \in \Gamma(Rad(TM))$ and $X \in \Gamma(TM)$ is a self-adjoint operator. Moreover, it is known that all Riemannian self-adjoint operators are diagonalizable. Let $\{k_1, \ldots, k_m\}$ be the eigenvalues. If $\widetilde{S}_{k_i}$, $1 \le i \le n$, is the eigenspace of k_i then

$$\widetilde{TM} = \widetilde{S}_{k_1} \perp \ldots \perp \widetilde{S}_{k_n},$$

where $\perp = \oplus_{orth}$ denotes a orthogonal direct sum. Choose a screen distribution $S(TM)$ and denote by $P_S : \Gamma(TM) \to \Gamma(S(TM))$ the corresponding projection. Let $V \in \Gamma(Rad\,TM)$ be a null tangent section on M and consider the shape operator $A_{V|S(TM)}$ defined by the equation (7.3.2), which is a self-adjoint and diagonalizable operator (since $S(TM)$ is Riemannian). Let $\{k_1^*, \ldots, k_n^*\}$ be the different eigenvalues on M and $S_{k_i^*}(TM)$, $1 \le i \le n$ the eigenspace of k_i^*, respectively. Then

$$S(TM) = S_{k_1^*}(TM) \perp \ldots \perp S_{k_n^*}(TM)$$

and we can find the following local adapted orthonormal basis of eigenvectors

$$\left\{ E_1^1, \ldots, E_{r_1}^1, E_1^2, \ldots, E_{r_2}^2, \ldots\ldots, E_1^n, \ldots, E_{r_n}^n \right\},$$

where $A_U(E_j^i) = k_i^* E_j^i$, with $1 \le i \le n$, $1 \le j \le r_i$ and r_i is the dimension of the eigenspace of k_i^*. Consider a map $\widetilde{P}_S : \Gamma(\widetilde{TM}) \to \Gamma(S(TM))$ defined by $\widetilde{P}_S(\widetilde{X}) = P_S X$. Then, $\widetilde{P}_S$ is a vector bundle isomorphism, and we have

$$\widetilde{P}_S(\widetilde{A}_U \widetilde{X}) = \widetilde{P}_S(-\Pi(\bar{\nabla}_X U)) = P_S(-\bar{\nabla}_X U) = P_S(-\bar{\nabla}_{P_S X} U) = A_U(P_S X).$$

Lemma 1. *Let $(M, g, S(TM))$ be a lightlike hypersurface of a Lorentzian manifold $(\bar{M}, \bar{g})$. With the above notations k is an eigenvalue of $\widetilde{A}_U$ if and only if k is an eigenvalue of A_U. Furthermore, $\widetilde{X}$ is an eigenvector of $\widetilde{A}_U$ associated with k if and only if $P_S X$ is an eigenvector of A_U associated with k.*

Proof. It is an immediate consequence of $\widetilde{P}_S$ being an isomorphism. $\blacksquare$

Therefore, we conclude that the eigenvalues associated with a null section are the same for all screen distributions. Thus we say that the eigenvalues $\{k_1, \ldots, k_n\}$ of A_U are the *principal curvatures* associated with the null tangent section U.

Lemma 2. *Let U and $\widehat{U} \in \Gamma(Rad\,TM)$ such that $\widehat{U} = \alpha U$. If k is an eigenvalue of A_U then αk is an eigenvalue of $A_{\widehat{U}}$ with the same multiplicity.*

Proof. Let $S(TM)$ be any screen distribution and we consider the shape operator $A_{\widehat{U}} : \Gamma(S(TM)) \to \Gamma(S(TM))$. From lemma 1 we know that the eigenvalues respect to $\widehat{U}$ do not depend on the screen, so if W is an eigenvector of A_U with respect to k then

$$
\begin{aligned}
A_{\widehat{U}}(W) &= P_S(-\bar{\nabla}_W \widehat{U}) = P_S(-\bar{\nabla}_W(\alpha U)) \\
&= P_S(-W(\alpha)U - \alpha \bar{\nabla}_W U) = \alpha k W.
\end{aligned}
$$

Thus, αk is an eigenvalue associated with $\widehat{U}$, which completes the proof.

Let $\{E_j^i \,;\, 1 \le i \le m, 1 \le j \le r_i\}$ be a local orthonormal basis of eigenvectors of $\Gamma(S(TM)|_{\mathcal{U}})$. In order to facilitate the notation and depending on the context we will also denote it by $\{E_a; 1 \le a \le m\}$ and so $A_U(E_a) = k_a E_a$, where k_a may be repeated. It is well-known that the *lightlike mean curvature* $H_U : M \to R$ with respect to a null tangent section U is given by

$$
H_U = -\sum_{a=1}^m B(E_a, E_a) = -\sum_{a=1}^m g(A_U(E_a), E_a).
$$

It is easy to show that H_U does not depend on both the screen distribution and the orthonormal basis, and so $H_U = -\sum_{a=1}^m k_a$, but, it depends on the local null section U. Now we look for a good choice of a structure $(S(TM), \xi)$.

A local null tangent section ξ is called *geodesic* if $\bar{\nabla}_\xi \xi = 0$ for which the integral curves of ξ are called the *null geodesic generators*. This condition has interesting geometric and physical meanings and also helps in simplifying the computations. If U is a null tangent section on M, then for all $x \in M$ we can scale U to be geodesic on a neighborhood $\mathcal{U}$ of x. Suppose ξ is geodesic on $\mathcal{U}$.

Definition 5.3. *Let $(M, g, S(TM))$ be a lightlike hypersurface of a Lorentzian manifold $(\bar{M}, \bar{g})$. A pair $(S(TM), \xi)$ is said to a global structure on M if and only if ξ is a non vanishing global null (GN) section on M.*

One can find a large class of interesting examples of lightlike hypersurfaces admitting a global structure. Suppose $(\bar{M}, \bar{g})$ is a time-oriented spacetime manifold. This means that there exists a smooth global timelike vector field, say V, on $\bar{M}$. Let M be a lightlike hypersurface in $\bar{M}$. Then V restricted to M is a global section of $Rad(TM) \oplus ltr(TM)$. Thus, the projection of $T\bar{M}|_M$ onto $Rad(TM)$ provides a non-vanishing GN section on M. In particular, recall the following mathematical model in section 4 of a Lorentzian manifold $\bar{M}$ with metric

$$
ds^2 = -dt^2 + e^\mu (dx^1)^2 \oplus f^2 \, d\Omega^2,
$$

where the form $d\Omega^2 = \bar{g}_{ab} \, dx^a \, dx^b$ with respect to a local coordinate system $(t, x^1, \cdots, x^{m+1})$, $\dim(\bar{M}) = m + 2$, μ and f are functions of t and x^1 alone. Suppose $\bar{M}$ is a orientable spacetime manifold, that is, $\bar{M}$ admits a global timelike vector field. Proceeding as exactly in that Mathematical Model, one can show that

$\bar{M}$ admits two lightlike hypersurfaces, each with a non-vanishing GN section.

Definition 5.4. *Let $(M, g, S(TM))$ be a lightlike hypersurface of a Lorentzian manifold $(\bar{M}, \bar{g})$ admitting a GN section ξ on M. A Riemannian distribution $D(TM)$ of TM is said to be ξ-distinguished if each section W of $D(TM)$ satisfies $\bar{\nabla}_W \xi \in \Gamma(D(TM))$. In particular, if a screen distribution $S(TM)$ is ξ-distinguished, then, the pair $(S(TM), \xi)$ is called a distinguished structure on M.*

Let $(S(TM), \xi)$ be a global structure of M. Consider the shape operator $A_{\xi}^{*} : \Gamma(S(TM)) \rightarrow \Gamma(S(TM))$ globally defined. From Lemma 1 we have that the eigenvalues respect to ξ do not depend on the screen. Thus, we say that the eigenvalues of A_{ξ}^{*} are the *principal curvatures* associated with the GN section ξ.

Proposition 5.3. *Let $(M, g, S(TM))$ be a lightlike hypersurface of a Lorentzian manifold $(\bar{M}, \bar{g})$ admitting a geodesic GN section ξ on M. Then, $S(TM)$ is ξ-distinguished if and only if the corresponding 1-form τ from (7.3.7) vanishes. Consequently, the Ricci tensor of the induced connection ∇ is symmetric.*

Proof. Consider a vector field X on M. Then, we have

$$\tau(X) = -\bar{g}(\bar{\nabla}_X \xi, N) = \bar{g}(\bar{\nabla}_{(P_S(X)+\lambda\xi)}\xi, N) = -\bar{g}(\bar{\nabla}_{P_S(X)}\xi, N).$$

Thus, $\tau = 0$ if and only if $\bar{\nabla}_{P_S(X)}\xi \in \Gamma(S(TM))$. Proposition 5.1 implies that the induced Ricci tensor is symmetric, which completes the proof.

In fact, the theorem 5.3 establishes that given a screen distribution $S(TM)$, then, the Ricci tensor is symmetric if and only if $d\tau = 0$ for any chosen GN section ξ. Moreover, the proposition 5.2 ensures that for this case there exists a pair $\{\xi, N\}$ on M such that the corresponding 1-form τ vanishes. Thus, the condition $d\tau = 0$ allows us to find distinguished structures $(S(TM), \xi)$ with a geodesic GN section ξ. On the existence of a unique ξ-distinguished Riemannian distribution $D_{\xi}(TM)$ we have the following result:

Theorem 5.4 ([29]). *Let $(M, g, S(TM))$ be a lightlike hypersurface of a Lorentzian manifold $(\bar{M}, \bar{g})$ admitting a GN section ξ. Suppose that $\bar{\nabla}_{\xi}\xi = -\rho\xi$ and $\{k_1, \cdots, k_m\}$ are the principal curvatures associated with ξ on M. Then there exist a unique ξ-distinguished Riemannian distribution $D_{\xi}(TM)$ such that:*

(i) *If $\rho \neq k_i$ for all $i \in \{1, \cdots, m\}$ we have the decomposition*

$$TM = Rad(TM) \perp D_{\xi}(TM)$$

and so $(S(TM) = D_{\xi}(TM), \xi)$ is a unique distinguished structure.

(ii) *If $\rho = k_{i_0}$ for some $i_0 \in \{1, \cdots, m\}$, we have the decomposition*

$$TM = Rad(TM) \perp S_{\rho}(TM) \perp D_{\xi}(TM),$$

where $S_{\rho}(TM)$ is any eigenspace associated with the eigenvalue $\rho = k_{i_0}$.

Proof. Suppose $S(TM)$ and $\widehat{S}(TM)$ are two different screens. Let $S_k(TM) \leq S(TM)$ and $\widehat{S}_k(TM) \leq S(\widehat{TM})$ be both vector subbundles associated with the same eigenvalue k. Let $\{E_1, \ldots, E_r\}$ be a orthonormal basis of $S_k(TM)$ and construct the set $\{\widehat{E}_1, \ldots, \widehat{E}_r\}$ where $\widehat{E}_j = P_{\widehat{S}}(E_j)$, $1 \leq j \leq r$. From Lemma 1 we have that $\{\widehat{E}_1, \ldots, \widehat{E}_r\}$ is a orthonormal basis of $\widehat{S}_k(TM)$ satisfying the formulas:

$$
\begin{aligned}
\bar{\nabla}_{E_j}\xi &= -kE_j - \tau(E_j)\xi, \\
\bar{\nabla}_{\widehat{E}_j}\xi &= -k\widehat{E}_j - \widehat{\tau}(\widehat{E}_j)\xi, \qquad 1 \leq j \leq r, \\
\widehat{E}_j &= E_j + \mu_j\xi.
\end{aligned}
$$

We are interested in finding μ_j such that $\widehat{\tau}(\widehat{E}_j)$ vanishes, so that $\widehat{S}_k(TM)$ is a ξ-distinguished Riemannian distribution. For this, we have

$$
\begin{aligned}
\bar{\nabla}_{\widehat{E}_j}\xi &= \bar{\nabla}_{(E_j+\mu_j\xi)}\xi = \bar{\nabla}_{E_j}\xi + \mu_j\bar{\nabla}_\xi\xi \\
&= -kE_j - \tau(E_j)\xi - \rho\mu_j\xi - k\widehat{E}_j = -k(E_j + \mu_j\xi)
\end{aligned}
$$

and, therefore, $\mu_j(\rho - k) = \tau(E_j)$. Accordingly, we obtain that if $\rho \neq k$ then it is enough to take $\mu_j = \tau(E_j)/(\rho - k)$ and trivially $\widehat{S}_k(TM)$ is unique with $\widehat{\tau} = 0$ on $\widehat{S}_k(TM)$. Thus, we have actually proved both statements by taking

$$
D(TM) = \bigoplus_{k \neq \rho} \widehat{S}_k(TM).
$$

If $k_i \neq \rho$ for any i, then $(\widehat{S}(TM), \xi)$ is a unique distinguished structure on M.

Example 18. Refer to example 13, in section 4, for a Monge hypersurface

$$
M = \{(x^0, \ldots, x^{n+1}) \in \mathbf{R}_1^{m+2} : x^0 = F(x^1, \ldots, x^{m+1})\}
$$

where $F : \Omega \to \mathbf{R}$ is a smooth function. M is lightlike if and only if F is a solution of the partial differential equation

$$
\sum_{a=1}^{m+1} \partial_a(F)^2 = 1. \tag{7.5.28}
$$

The radical and transversal vector bundles are globally spanned by

$$
\xi = \partial_0 + \sum_{a=1}^{m+1} \partial_a(F)\partial_a, \quad N = \frac{1}{2}\left\{-\partial_0 + \sum_{a=1}^{m+1} \partial_a(F)\partial_i\right\}.
$$

The corresponding screen distribution is given by $\{W_1, \ldots, W_m\}$ where

$$
W_a = \partial_{m+1}(F)\partial_a - \partial_a(F)\partial_{m+1}, \qquad 1 \leq a \leq m.
$$

If we derive the equation (7.5.28) with respect to x^b we obtain that

$$\sum_{a=1}^{m+1} \partial_a(F)\partial_{ab}(F) = 0$$

for all $1 \leq b \leq m+1$ and we get

$$
\begin{aligned}
\bar{\nabla}_\xi \xi &= \bar{\nabla}_{\left(\partial_0 + \sum_{a=1}^{m+1} \partial_a(F)\partial_a\right)} \left(\partial_0 + \sum_{a=1}^{m+1} \partial_a(F)\partial_a\right) \\
&= \sum_{b=1}^{m+1} \left(\sum_{a=1}^{m+1} \partial_a(F)\partial_{ab}(F)\right) \partial_b = 0,
\end{aligned}
$$

so ξ is a geodesic GN section. Now we prove that $S(TM)$ is ξ-distinguished.

$$
\begin{aligned}
\bar{\nabla}_{W_a} \xi &= \bar{\nabla}_{(\partial_{m+1}(F)\partial_a - \partial_a(F)\partial_{m+1})} \left(\partial_0 + \sum_{a=1}^{m+1} \partial_a(F)\partial_a\right) \\
&= \sum_{a=1}^{m+1} (\partial_{m+1}(F)\partial_{ba}(F) - \partial_b(F)\partial_{m+1,a}(F)) \partial_a.
\end{aligned}
$$

Bearing in mind that $\partial_b = \partial_{n+1}(F)^{-1}(W_b + \partial_b(F)\partial_{m+1})$ for all $1 \leq b \leq m$ we have

$$
\begin{aligned}
\bar{\nabla}_{W_a} \xi &= \sum_{b=1}^{m} (\partial_{ab}(F) - \partial_{m+1}(F)^{-1}\partial_a(F)\partial_{m+1,b}(F))W_b \\
&\quad + \sum_{b=1}^{m} (\partial_{ab}(F)\partial_b(F) - \partial_{m+1}(F)^{-1}\partial_b(F)\partial_a(F)\partial_{m+1,b}(F))\partial_{m+1} \\
&\quad + (\partial_{m+1}(F)\partial_{a,m+1}(F) - \partial_a(F)\partial_{m+1,m+1}(F))\partial_{m+1} \\
&= \sum_{b=1}^{m} (\partial_{ab}(F) - \partial_{m+1}(F)^{-1}\partial_a(F)\partial_{m+1,b}(F))W_b,
\end{aligned}
$$

where we have made use of $\sum_{b=1}^{m} \partial_{ab}(F)\partial_b(F) + \partial_{m+1}(F)\partial_{a,m+1}(F) = 0$. Therefore $\bar{\nabla}_{W_a}\xi \in \Gamma(S(TM))$ and so is distinguished. From above we can easily compute the null mean curvature H_ξ as follow

$$
\begin{aligned}
H_\xi &= \sum_{a=1}^{m} (\partial_{aa}(F) - \partial_{m+1}(F)^{-1}\partial_a(F)\partial_{m+1,a}(F)) \\
&= \sum_{a=1}^{m} \partial_{aa}(F) + \partial_{m+1}(F)^{-1}\partial_{m+1}(F)\partial_{m+1,m+1}(F) \\
&= \sum_{a=1}^{m+1} \partial_{aa}(F).
\end{aligned}
$$

7.6   Scalar curvature of genus zero

Since the induced symmetric tensor field g on M is degenerate, its inverse does not exist. Thus, it is not possible to find, from equation (7.5.26) of $R^{(0,2)}$ tensor, a scalar quantity for M, by the usual way of contraction. This raises the following question: Is there any other way to associate M with a scalar quantity which we may call an *induced scalar curvature* of M? Intuitively, the answer must be "YES", since M being a smooth paracompact manifold, several important concepts, related to the scalar curvature of the ambient manifold $\bar{M}$, can have no induced corresponding objects on M if the answer is "NO". Moreover, the semi-Riemannian geometry of an integral manifold of an integrable screen distribution remains incomplete if the answer is "NO".

We, therefore, look for a way to heal this missing gap. The non-uniqueness of screen distribution and its changing causal structure rules out the possibility of a single definition for an induced scalar curvature of any lightlike hypersurface M of a semi-Riemannian manifold. Although we have seen, in this chapter, the existence of many cases of a canonical screen (some of them integrable) and unique transversal vector bundle, but, the problem of scalar curvature must be classified subject to the causal structure of a screen, just as the case of null curves. We, therefore, start with the case of a hypersurface $(M, g, S(TM))$ of a Lorentzian manifold $(\bar{M}, \bar{g})$ for which we know that $S(TM)$ is Riemannian. This case is also physically important. Let $\{\xi; W_a\}$ be the quasi-orthonormal frame for TM induced from a frame $\{\xi; W_a, N\}$ for $T\bar{M}$ such that $Rad\,TM = Span\,\{\xi\}$ and $S(TM) = Span\,\{W_1, \ldots, W_m\}$. First we find the tensorial components $R^{(0,2)}(\xi, \xi)$ and $R^{(0,2)}(W_a, W_a)$ for each base vectors of T_xM. Setting $X = Y$ in (7.5.26) we get

$$R^{(0,2)}(X, X) = \sum_{b=1}^{m} g(R(X, W_b)X, W_b) + \bar{g}(R(X, \xi)X, N). \qquad (7.6.1)$$

Replacing X by ξ and using some of the Gauss-Codazzi equations we obtain

$$
\begin{aligned}
R^{(0,2)}(\xi, \xi) &= \sum_{a=1}^{m} g(R(\xi, W_a)\xi, W_a) - \bar{g}(R(\xi, \xi)\xi, N) \\
&= \sum_{a=1}^{m} g(R(\xi, W_a)\xi, W_a), \qquad (7.6.2)
\end{aligned}
$$

where the second term vanishes due to (7.5.9). Replacing each X successively by each base vector W_a of $S(TM)$ and then taking the sum of all these terms we get

$$
\begin{aligned}
\sum_{a=1}^{m} R^{(0,2)}(W_a, W_a) &= \sum_{a=1}^{m} \left\{ \sum_{b=1}^{m} g(R(W_a, W_b)W_a, W_b) \right\} \\
&\quad + \sum_{a=1}^{m} \bar{g}(R(W_a, \xi)W_a, N). \qquad (7.6.3)
\end{aligned}
$$

Finally, adding (7.6.3) and (7.6.2) we obtain a scalar r given by

$$
\begin{aligned}
r &= R^{(0,2)}(\xi, \xi) + \sum_{a=1}^{m} R^{(0,2)}(W_a, W_a) \\
&= \sum_{a=1}^{m} \left\{ \sum_{b=1}^{m} g(R(W_a, W_b)W_a, W_b) \right\} \\
&\quad + \sum_{a=1}^{m} \{ g(R(\xi, W_a)\xi, W_a) + \bar{g}(R(W_a, \xi)W_a, N) \}.
\end{aligned}
\tag{7.6.4}
$$

In general, r given by (7.6.4) can not be called a scalar curvature of M since it can not be calculated from a tensor quantity $R^{(0,2)}$. Moreover, it can only have a geometric meaning if $R^{(0,2)} = Ric$ and its value is independent of the screen, its corresponding transversal vector bundle and the null section ξ. We, therefore, need reasonable geometric conditions on M to get its scalar curvature. For this purpose, recently the first author Duggal [27] has introduced the following general notation (also applicable to M in a prescribed semi-Riemannian manifold):

We say that a lightlike hypersurface $(M, g, S(TM))$ of a semi-Riemannian manifold $\bar{M}$ and its any induced object Ω are of *genus* **s** if $g_{|S(TM)}$ is of constant signature **s**. Any induced object will then be labeled by Ω^s. For example, if $\bar{M}$ is Lorentzian, then, a screen of M will be denoted by $S(TM)^0$.

Denote by $\mathcal{C} = [M^0]$ a class of lightlike hypersurfaces of genus zero such that

(a) M^0 admits a canonical screen distribution $S(TM)^0$ that induces a canonical lightlike transversal vector bundle N^0

(b) M^0 admits a symmetric Ricci tensor Ric^0, that is, $d\tau$ (and not necessarily τ) vanishes on any $\mathcal{U} \subseteq M^0$.

For geometric and/or physical reasons, above two conditions are necessary to assign a well-defined scalar curvature to each member of $\mathcal{C}$. We know from sections 7.2 and 7.4 and Theorem 5.4 (also see Bejancu[11]) that there are plenty of such members of $\mathcal{C}$ with good properties. Now we introduce the following definition:

Definition 6.1 ([27]). *Let* $(M^0, g^0, S(TM)^0, \xi^0, N^0)$ *be a member of* $\mathcal{C}$. *Then, the scalar quantity* r, *given by (7.6.4) and denoted by* r^0, *is called the induced scalar curvature of genus zero of the lightlike hypersurface* M^0.

Since $S(TM)^0$ and N^0 are chosen canonical, we must assure the stability of r^0 with respect to a choice of the second fundamental form B^0 and the 1-form τ both of which (as per proposition 3.1) depend on the choice of a null section $\xi \in \Gamma(Rad\,TM_{|\mathcal{U}})$. It is easy to see that for a canonical vector bundle N^0, the function α in (7.3.9) will be non-zero constant which implies that, for this case, both B^0 and τ are independent of the choice of ξ, except for a non-zero constant factor. Consequently, r^0 is a well-defined induced scalar curvature of this class of lightlike hypersurfaces. To calculate r^0 we start with the following simple case:

Let $(M^0, g^0, S(TM)^0, \xi^0, N^0)$ be a hypersurface of a time orientable spacetime manifold $\bar{M}(c)$ of constant curvature c such that M^0 belongs to the class $\mathcal{C}$. Using (7.1.17) in (7.5.7) we have $\bar{g}(R(X, \xi)Y, N) = -c\bar{g}(X, Y)$. Finally, this result and using (7.1.17) in (7.5.5) reduces the equation (7.6.1) to

$$
\begin{aligned}
Ric^0(X, X) &= \sum_{b=1}^{m} g^0(R(X, W_b)X, W_b) - c\bar{g}(X, X) \\
&= \sum_{b=1}^{m} \{B^0(W_b, X)C^0(X, W_b) - B^0(X, X)C^0(W_b, W_b)\} \\
&\quad -cm\,\bar{g}(X, X).
\end{aligned}
\tag{7.6.5}
$$

Consequently, $Ric^0(\xi^0, \xi^0) = 0$ and the equations (7.6.3) and (7.6.4) reduce to

$$
\begin{aligned}
\sum_{a=1}^{m} Ric^0(W_a, W_a) &= \sum_{a=1}^{m} \left\{ \sum_{b1} [B^0(W_b, W_a)C^0(W_a, W_b) \right. \\
&\quad \left. - B^0(W_a, W_b)C^0(W_b, W_b)] \right\} - cm^2 = r^0.
\end{aligned}
\tag{7.6.6}
$$

It is immediate from above that, for this class of lightlike hypersurfaces M^0, the induced Ricci tensor and the corresponding scalar curvature of M^0 can be determined if one knows $(Ric^0)_{|S(TM)^0}$ and $(r^0)_{|S(TM)^0}$. More precisely,

$$
Ric^0 = (Ric^0)_{|S(TM)^0}, \quad r^0 = (r^0)_{|S(TM)^0} - cm^2.
\tag{7.6.7}
$$

Let M^0 be a screen conformal lightlike hypersurface of $\bar{M}(c)$, satisfying the Theorem 4.5 and its Corollary 4.1, so that M^0 belongs to $\mathcal{C}$. It is known [10] that a time orientable spacetime $\bar{M}(c)$ is globally hyperbolic (for a brief on globally hyperbolic spacetimes see section 1 of chapter 8) and admits a global timelike vector field. Take a fixed global timelike function and choose its gradient a timelike covariant constant vector field $V = -\partial_t$ in the proof of theorem 4.5. Thus, ξ^0 and N^0 (given by (7.4.19) and (7.4.20)), respectively, are globally defined on M^0. We recall from the proof of theorem 4.5 and (7.4.14) that following holds:

$$
\tau(X) = X(\ln \theta), \quad C^0(X, PW) = \frac{1}{2}\theta^2 B^0(X, W), \quad \theta = (\xi_t)^{-1},
$$

for any $X, W \in \Gamma(TM^0)$, where ξ_t is the first coefficient of ξ^0 in (7.4.19). Taking into account above relations we obtain the following value of r^0 from (7.6.6)

$$
r^0 = \frac{1}{2}\theta^2 \sum_{a=1}^{m} \left\{ \sum_{b=1}^{m} [(B^0)^2(W_a, W_b) - B^0(W_a, W_a)B^0(W_b, W_b)] \right\} - cm^2.
\tag{7.6.8}
$$

In particular, for 3 and 4 dimensional $\bar{M}(c)$ the values of r^0 are

$$
r^0 = -4c \quad \text{and} \quad r^0 = \theta^2 [(B^0)^2(W_1, W_2) - B^0(W_1, W_a)B^0(W_2, W_2)]
$$

respectively.

Example 19. Let M^0 be a lightlike hypersurface of a 4-dimensional Minkowski spacetime $(\mathbf{R}_1^4, \bar{g})$ as defined in Example 2. In this example, notice that

$$\xi = \partial_t + \cos u \cos v \, \partial_1 + \cos u \sin v \, \partial_2 + \sin u \, \partial_3$$

implies that its first coefficient $\xi_t = 1$. Thus, τ vanishes so $\theta = 1$. Also (see Example 6) the components of the second fundamental form are

$$B^0(W_1, W_2) = -\left(\frac{1}{t \cos u}\right), \quad B^0(W_2, W_2) = -\left(\frac{1}{t \cos u}\right), \quad B^0(W_1, W_2) = 0.$$

Thus, the induced scalar curvature of M^0 in $(\mathbf{R}_1^4, \bar{g})$ is given by

$$r^0 = -\frac{1}{(t \cos u)^2}.$$

7.7 Brief notes and exercises

Integral Curves and Lie Derivatives. Let V be a vector field on a real n-dimensional smooth manifold. The *integral curves* (*orbits*) of V are given by the following system of ordinary differential equations:

$$\frac{dx^i}{dt} = V^i(x(t)), \qquad i \in \{1, \ldots, n\},$$

where (x^i) is a local coordinate system on M and $t \in I \subset \mathbf{R}$. It follows from the well-known theorem on the existence and uniqueness of the solution of above equation that for any given point, with a local coordinate system, there is a unique integral curve defined over a part of the real line. Consider a mapping ϕ from $[-\delta, \delta] \times \mathcal{U}$ ($\delta > 0$ and $\mathcal{U}$ an open set of M) into M defined by $\phi : (t, x) \rightarrow \phi(t, x) = \phi_t(x) \in M$ and satisfying the following conditions:

(1) $\phi_t : x \in \mathcal{U} \rightarrow \phi_t(x) \in M$ is a diffeomorphism of $\mathcal{U}$ onto the open set $\phi_t(\mathcal{U})$ of M, for every $t \in [-\delta, \delta]$,

(2) $\phi_{t+s}(x) = \phi_t(\phi_s(x)), \, \forall \, t, \, s, \, t + s \in [-\delta, \delta]$ and $\phi_s(x) \in \mathcal{U}$.

In the above case, we say that the family ϕ_t is a *1-parameter group of local transformations* on M. The mapping ϕ is then called a *local flow* on M. If each integral curve of V is defined on the entire real line, we say that V is a *complete vector field* and it generates a *global flow* on M. A set of local (respectively, complete) integral curves is called a *local congruence* (respectively, *congruence*) *of curves* of V.

In the following we show how the flow ϕ is used to transform any object, say Ω, on M into another one of the same type as Ω, with respect to a point transformation $\phi_t : x^i \rightarrow x^i + tV^i$ along an integral curve through x^i. We denote by $\bar{\Omega}(x^i)$ the pull-back of $\Omega(x^i + tV^i)$ to the point x^i through the inverse mapping of ϕ_t. This defines a differentiable operator, denoted by $\pounds_V$, which assigns to an arbitrary Ω another object $\pounds_V \Omega$ of the same type as Ω and given by

$$(\pounds_V \Omega)(x^i) = \lim_{t \to 0} \frac{1}{t} \left[\bar{\Omega}(x^i) - \Omega(x^i) \right]. \tag{7.7.1}$$

The operator $\pounds_V$ is called the *Lie derivative* with respect to V. This definition holds for local as well as global flows, with the following basic properties:

(1) $\pounds_V(aX + bY) = aL_V X + bL_V Y$, for all $a, b \in \mathbf{R}$ and $X, Y \in \Gamma(TM)$: (linearity)

(2) $\pounds_V(T \otimes S) = (\pounds_V T) \otimes S + T \otimes \pounds_V S$, where $\otimes$ is the tensor product of any two objects T and S: (Leibnitz rule)

(3) $\pounds_V$ commutes with the contraction operator.

It follows from above properties that in order to compute the Lie derivative of any arbitrary Ω, it is sufficient to know the Lie derivatives of a function, a vector field and a 1-form. Indeed, the Lie derivatives of all other objects of higher order can be obtained by using the operations of tensor analysis and above three properties. We now compute these three Lie derivatives.

Functions. Let f be a scalar function on M. Then,

$$\begin{aligned} \pounds_V f &= \lim_{t \to 0} \frac{1}{t} \left(f(x^i + tV^i) - f(x^i) \right) \\ &- \lim_{t \to 0} \frac{1}{t} \left(f(x^i) + tV^j \partial_j f \quad f(x^i) \right) \\ &= V^j \partial_j f, \end{aligned}$$

where we neglect t^2 and higher powers. Index-free expression is

$$\pounds_V f = V(f). \tag{7.7.2}$$

Vector Fields. Let $X = X^i \partial_i$ be a vector field on M. Then, the pull-back of X, from $x^i + tV^i$ to x^i, is given by

$$\begin{aligned} \bar{X}^j(x) &= X^k(x + tV) \partial_k(x^j - tV^j) \\ &= \{ X^k(x) + tV^i \partial_i X^k + O(t^2) \} \{ \delta_k^j - t \partial_k V^j \} \\ &= X^j(x) + t \left(V^i \partial_i X^j - X^k(x) \partial_k V^j \right) + O(t^2). \end{aligned}$$

Hence, using above in (7.7.1) with $\Omega = X$, we get

$$L_V X^j = V^i \partial_i X^j - X^i \partial_i V^j.$$

As per the Lie-bracket (1.1.4), the index-free expression is

$$\pounds_V X = [V, X]. \tag{7.7.3}$$

1-forms. Let $\omega = \omega_i dx^i$ be a 1-form (linear differential form) on M. Take $X = X^i \partial_i$ an arbitrary vector field on M. Then, the contraction $\omega(X) = \omega_i X^i$ is a real scalar function, say f. Now, from (7.7.2) we obtain

$$L_V \omega(X) = L_V f = V(f) = V(\omega(X)).$$

Then, using the linear property of $\pounds_V$, we get

$$\pounds_V \omega(X) = (L_V \omega)(X) + \omega(L_V X) = V(\omega(X)).$$

Therefore, substituting the value of $\pounds_V X$ from (7.7.3), we obtain

$$
\begin{aligned}
(L_V \omega)(X) &= V(\omega(X)) - \omega[V, X] \\
(L_V \omega)_i &= V^j \partial_j(\omega_i) + \omega_j \partial_i(V^j).
\end{aligned}
\tag{7.7.4}
$$

It follows from above three Lie derivatives that if V is a vector field of class, say C^m, then the Lie derivative of a function, a vector field and a 1-form is of the same type but of class C^{m-1}. This is also true for higher geometric/tensor quantities.

We need the Lie derivative of the metric tensor g which is given by

$$
\begin{aligned}
(\pounds_V g)(X, Y) &= V(g(X, Y)) - g([V, X], Y) - g(X, [V, Y]) \\
&= g(\nabla_X V, Y) + g(\nabla_Y V, X)
\end{aligned}
\tag{7.7.5}
$$

for arbitrary vector fields X and Y on M and a metric connection ∇ of g. as defined in chapter 2. Locally, we have

$$
\begin{aligned}
\pounds_V g_{ij} &= \nabla_i v_j + \nabla_j v_i \\
&= v_{j;i} + v_{i;j} , \quad v_i = g_{ij} V^j.
\end{aligned}
\tag{7.7.6}
$$

The readers may find further information on Lie derivatives in standard books such as Kobayashi-Nomizu [75].

Totally umbilical lightlike hypersurfaces. Let $(M, g, S(TM))$ be a lightlike hypersurface of a $(m+2)$-dimensional semi-Riemannian manifold $(\bar{M}, \bar{g})$. A point $u \in M$ is said to be umbilical if

$$B_u(X_u, Y_u) = k g_u(X_u, Y_u), \quad \forall X_u, Y_u \in T_u M, \tag{7.7.7}$$

where B is the local second fundamental form of M defined around u and $k \in \mathbf{R}$. This definition is independent of any coordinate neighborhood $\mathcal{U}$ around u. Indeed, if we take $\mathcal{U}^*$ around u, then $B^* = \alpha B$ on $\mathcal{U} \cap \mathcal{U}^*$, where α is a non-zero smooth function. Thus B^* at u satisfies a relation as in (7.7.7). We say that M is *totally umbilical* if any point of M is umbilical. Then, it is easy to see that M is totally umbilical, if and only if on each $\mathcal{U}$ there exists a smooth function ρ such that

$$B(X, Y) = \rho g(X, Y), \quad \forall X, Y \in \Gamma(TM_{|\mathcal{U}}). \tag{7.7.8}$$

Above definition does not depend on the screen distribution of M. On the other hand, from ((7.3.10) and (7.3.19) we conclude that M is totally umbilical, if and only if, on each $\mathcal{U}$ there exists ρ such that

$$\overset{*}{A_\xi}(PX) = \rho PX, \quad \forall X \in \Gamma(TM_{|\mathcal{U}}). \tag{7.7.9}$$

Most of the work done on totally umbilical case is available in Duggal-Bejancu's book [28, Section 4.5, pages 106-114], other than the following new general result:

Theorem ([31]). *Let $(M, g, S(TM))$ be a proper totally umbilical r-lightlike or a co-isotropic submanifold of a semi-Riemannian manifold $(\bar{M}(\bar{c}), \bar{g})$ of a constant curvature $\bar{c}$. Then, the induced Ricci tensor on M is symmetric, if and only if its screen distribution $S(TM)$ is integrable.*

Finally, we refer [28, Section 4.4, pages 100-106]) for statement and the proof of the *fundamental theorem for lightlike hypersurfaces*.

Exercises

(1) Consider a hypersurface $M : x^3 = x^0 + \frac{1}{2}(x^1 + x^2)^2$ in $\mathbf{R}_2^4$. Prove that M is a lightlike hypersurface and

$$TM^\perp = Span\left\{\xi = \frac{\partial}{\partial x^0} + (x^1 + x^2)\frac{\partial}{\partial x^1} - (x^1 + x^2)\frac{\partial}{\partial x^2} + \frac{\partial}{\partial x^3}\right\}.$$

(2) Consider the unit pseudosphere S_1^3 of $\mathbf{R}_1^4$ given by

$$-(x^0) + (x^1)^2 + (x^2)^2 + (x^3)^2 = 1.$$

Cut S_1^3 by the hyperplane $x^0 - x^1 = 0$ and obtain a lightlike surface of S_1^3.

(3) Two vector fields U and V on a lightlike hypersurface M are said to be conjugate if $B(U, V) = 0$. Show that any null vector field $\xi \in \Gamma Rad\, TM_{|\mathcal{U}}$ is conjugate to any vector field X on M.

(4) Let $(M, g, S(TM))$ be a lightlike hypersurface of $(\bar{M}, \bar{g})$. Prove that the following assertions are equivalent:
(i) $S(TM)$ is parallel with respect to the induced connection ∇.
(ii) $\overset{*}{h}$ vanishes identically on M.
(iii) A_N vanishes identically on M.

(5) Let $(M, g, S(TM))$ be a lightlike hypersurface of $(\bar{M}, \bar{g})$. Prove that the restriction of the curvature form of ∇ on $TM^\perp$ is independent of $S(TM)$.

(6) Prove that the shape operator of any screen distribution of a lightlike hypersurface M of $\bar{M}$ is symmetric with respect to its second fundamental form.

(7) Let M be a lightlike Monge hypersurface of $\mathbf{R}_1^4$ and M^* be a leaf of the natural screen distribution (see remark 4 on page 214) $S^*(TM)$. Prove that the curvature tensors R and R^* of M and M^* respectively, are related by

$$R(X, Y)Z = \frac{1}{2} R^*(PX, PY)PZ, \quad \forall\, X, Y, Z \in \Gamma(TM).$$

(8) Prove that any lightlike surface of a 3-dimensional Lorentzian manifold is either totally umbilical or totally geodesic.

(9) Prove that the lightlike cone of a Euclidean space, of index $q \geq 1$, is its totally umbilical hypersurface.

Chapter 8

Geometry and physics of null geodesics

8.1 Introduction

In this final chapter we provide up-to-date information on the geometric and physical use of null geodesics, both in the study of null curves and lightlike hypersurfaces. For this purpose, we first review the following basic preliminaries needed to understand the subsequent sections:

Globally hyperbolic spacetime manifolds. A spacetime $(\bar{M}, \bar{g})$ is said to be globally hyperbolic if there exists a spacelike hypersurface Σ such that every endless causal curve intersects Σ once and only once. Such a hypersurface (if it exists) is called a *Cauchy surface*. If $\bar{M}$ is globally hyperbolic, then (a) $\bar{M}$ is homeomorphic to $\mathbf{R} \times \Sigma$, where Σ is a hypersurface of $\bar{M}$, and for each t, $\{t\} \times \Sigma$ is a Cauchy surface, (b) if Σ' is any compact hypersurface without boundary, of $\bar{M}$, then Σ' must be a Cauchy surface (see Beem-Ehrlich [10]). It is obvious from above discussion that Minkowski spacetime is globally hyperbolic. Now we highlight as to why globally hyperbolic spacetimes are physically important and also present a mathematical technique to construct an extension of this class to include a large class of time orientable Lorentzian manifolds. Recall the following theorem of Hopf-Rinow [59] on compact and complete Riemannian manifolds.

Hopf-Rinow Theorem. *For any connected Riemannian manifold M, the following are equivalent:*

(a) *M is metric complete, i.e., every Cauchy sequence converges.*

(b) *M is geodesic complete, i.e., the exponential map is defined on the entire tangent space $T_p(M)$ at each point $p \in M$.*

(c) *Every closed bounded subset of M is compact.*

Thus the Hopf-Rinow theorem maintains the equivalence of metric and geodesic completeness and, therefore, guarantees the completeness of all Riemannian metrics, for a compact smooth manifold, with the existence of minimal geodesics. Also, if any one of the **(a)** through **(c)** holds then the Riemannian function is obviously finite-valued and continuous. In the non-compact case, it is known through the work of Nomizu-Ozeki [86] that every non-compact Riemannian manifold admits a complete metric. Unfortunately, there is no analogue to the Hopf-Rinow theorem for a general Lorentzian manifold. In fact, we know now that the metric completeness and the geodesic completeness are unrelated for arbitrary Lorentz manifolds and their causal structure requires that a complete manifold must independently be spacelike, timelike and null complete. The singularity theorems (see Hawking-Ellis [56]) confirm that not all Lorentz manifolds are metric and/or geodesic complete. Also, the Lorentz distance function fails to be finite and/or continuous for an arbitrary spacetime (see Beem-Ehrlich [10]).

Based on above, it is natural to ask if there exists a class of spacetimes which shares some of the conditions of the Hopf-Rinow theorem. It has been shown in the works of Beem-Ehrlich [10] that the globally hyperbolic spacetimes turn out to be the most closely related physical model sharing some properties of Hopf-Rinow theorem. Indeed, timelike Cauchy completeness and finite compactness are equivalent and the Lorentz distance function is finite and continuous for this class (see Beem-Ehrlich [10]). Thus, the globally hyperbolic spacetimes are physically important. Although the Minkowski spacetime and the Einstein static universe are globally hyperbolic, to include some more physically important models one needs an extended case of the product spaces, called *warped product* which we now explain.

In 1969, Bishop and O'Neill [17] introduced a class of Riemannian manifolds, called warped product manifolds as follows:

Let (M_1, g_1) and (M_2, g_2) be two Riemannian manifolds, $h : M_1 \rightarrow (0, \infty)$ and $\pi : M_1 \times M_2 \rightarrow M_1, \bar{\pi} : M_1 \times M_2 \rightarrow M_2$ the projection maps given by $\pi(x, y) = x$ and $\bar{\pi}(x, y) = y$ for every $(x, y) \in M_1 \times M_2$. Denote the warped product manifold $(M = M_1 \times_h M_2, g)$, where

$$g(X, Y) = g_1(\pi_\star X, \pi_\star Y) + h(\pi(x, y)) g_2(\bar{\pi}_\star X, \bar{\pi}_\star Y), \quad \forall X, Y \in \Gamma(TM)$$

where $\star$ is symbol for the tangent map. They proved that M is a complete Riemannian manifold if and only if both M_1 and M_2 are complete Riemannian manifolds. In particular, they constructed a large variety of complete Riemannian manifolds of everywhere negative sectional curvature using warped products. Following this, Beem-Ehrlich [10] used above warped product and constructed a large extended class of globally hyperbolic manifolds as follows:

Let (M_1, g_1) and (M_2, g_2) be Lorentz and Riemannian manifolds respectively. Let $h : M_1 \rightarrow (0, \infty)$ be a C^∞ function and $\pi : M_1 \times M_2 \rightarrow M_1$, $\bar{\pi} : M_1 \times M_2 \rightarrow M_2$ the projection maps given by $\pi(x, y) = x$ and $\bar{\pi}(x, y) = y$ for every $(x, y) \in M_1 \times M_2$. Then, define the metric g given by

$$g(X, Y) = g_1(\pi_\star X, \pi_\star Y) + h(\pi(x, y)) g_2(\bar{\pi}_\star X, \bar{\pi}_\star Y), \quad \forall X, Y \in \Gamma(TM)$$

where $\pi_\star$ and $\bar{\pi}_\star$ are respectively tangent maps. They proved the following theorem:

Theorem (Beem-Ehrlich [10]). *Let (M_1, g_1) and (M_2, g_2) be Lorentzian and Riemannian manifolds respectively. Then, the Lorentzian warped product manifold $(M = M_1 \times_h M_2, g = g_1 \oplus_h g_2)$ is globally hyperbolic if and only if both the following conditions hold:*

(1) *(M_1, g_1) is globally hyperbolic.*

(2) *(M_2, g_2) is a complete Riemannian manifold.*

They did an extensive global study on causal and completeness properties of Lorentz manifolds as well as null cut loci, conjugate and focal points (also see Kupeli [77]) and Morse theory for non-null and null geodesics. Later on, Ehrlich et al. [33] (also see Anderson - Dahl - Howard and others cited therein) have done work on Jacobian and volume calculation for Lorentzian warped products. In this chapter we need information on the following two physically significant classes of warped product globally hyperbolic spacetimes:

I. Robertson-Walker spacetimes. Since the Einstein field equations are a complicated set of nonlinear partial differential equations, we often assume certain relevant symmetry conditions for a satisfactory representation of our universe. Through extragalactic observations we know that the universe is approximately spherically symmetric about an observer. In fact, it would be more reasonable to assume that the universe is isotropic, that is, approximately spherical symmetric about each point in spacetime. This means the universe is *spatially homogeneous* [110], that is, admits a 6-parameter group of isometries whose surfaces of transitivity are spacelike hypersurfaces of constant curvature. This implies that any point on one of these hypersurfaces is equivalent to any other point on the same hypersurface. Such a spacetime is called *Robertson-Walker spacetime* with metric

$$ds^2 = -dt^2 + S^2(t)d\Sigma^2, \tag{8.1.1}$$

where $d\Sigma^2$ is the metric of a spacelike hypersurface Σ with spherical symmetry and constant curvature $c = 1, -1$ or 0. With respect to a local spherical coordinate system (r, θ, ϕ), this metric is given by

$$d\Sigma^2 = dr^2 + f^2(r)(d\theta^2 + \sin^2\theta d\phi^2), \tag{8.1.2}$$

where $f(r) = \sin r$, $\sinh r$ or r according as $c = 1, -1$ or 0. The range of the coordinates is restricted from 0 to 2π or from 0 to ∞ for $c = 1$ or -1 respectively.

Using the framework of Lorentzian warped products, we now show that all Robertson-Walker spacetimes are globally hyperbolic. We know from (8.1.1) that $d\Sigma^2$ is a Riemannian metric of the spacelike hypersurface Σ. Set $M_1 = (a, b)$ for $(-\infty \leq a, b \leq \infty)$ as 1-dimensional space with negative definite metric $-dt^2$. Define $S^2(t) = h(t)$ where $h : (a, b) \to (0, \infty)$. Then, it follows from the metric (8.1.1) and the discussion on warped product that a Robertson-Walker spacetime (M, g) can be written as a Lorentzian warped product

$$\left(M = M_1 \times_h \Sigma, \ g = -dt^2 \oplus_h d\Sigma^2\right).$$

The map $\pi \,:\, M_0 \times_h \Sigma \to R$, given by $\pi(t, x) = t$, is a smooth timelike function on M whose each level surface $\pi^{-1}(t_0) = \{t_0\} \times \Sigma$ is a Cauchy surface. Consequently, it follows from above stated Beem-Ehrlich theorem that all Robertson-Walker spacetimes are globally hyperbolic.

See [56, pages 134-142] for more details on the Robertson-Walker spacetimes (also, see Duggal-Sharma [32] for information on spacetime symmetries).

II. Asymptotically flat spacetimes. One of the important areas of research in general relativity is the study of isolated systems, such as the sun and a host of stars in our universe. It is now well known that such isolated systems can best be understood by examining the local geometry of the spacetimes which are *asymptotically flat*, that is, their metric is flat at a large distance from a centrally located observer. It is the purpose of this sub-section to discuss such physical spacetimes.

First we define *stationary and static spacetimes*. A spacetime is stationary if it has a 1-parameter group of isometries with timelike orbits. Equivalently, a spacetime is stationary if it has a timelike Killing vector field, say V. A static spacetime is stationary with the additional condition that V is hypersurface orthogonal, that is, there exists a spacelike hypersurface Σ orthogonal to V. The general form of the metric of a static spacetime can be written as

$$ds^2 = -A^2(x^1, x^2, x^3)dt^2 + B_{\alpha\beta}(x^1, x^2, x^3)dx^\alpha dx^\beta \,,$$

where $A^2 = -V_a V^a$ and $\alpha, \beta = 1, 2, 3$. Static spacetimes have both the time translation symmetry ($t \to t+$ constant) and the time reflection symmetry ($t \to -t$). A spacetime is said to be *spherically symmetric* if its isometry group has a subgroup isometric to $SO(3)$ and its orbits are 2-spheres.

Case 1: Schwarzschild spacetimes. Let (M, g) be a 4-dimensional isolated system with a Lorentz metric g and 3-dimensional spherical symmetry. Choose a coordinate system (t, r, θ, ϕ) for which the most general form of g is given by

$$ds^2 = -e^{2\lambda}dt^2 + e^{2\nu}dr^2 + A\,dr\,dt + B\,r^2\left(d\theta^2 + \sin^2\theta d\phi^2\right),$$

where λ, ν, A and B are functions of t and r only due to the 3-dimensional symmetry. The inherent freedom in choosing some of the coefficients allows to consider a Lorentz transformation such that $A = 0$ and $B = 1$. Using this we get

$$ds^2 = -e^{2\lambda}dt^2 + e^{2\nu}dr^2 + r^2\left(d\theta^2 + \sin^2\theta d\phi^2\right), \quad \text{where} \tag{8.1.3}$$

$$g_{00} = -e^{2\lambda}, \qquad g_{11} = e^{2\nu}, \qquad g_{22} = r^2, \qquad g_{33} = r^2\sin^2\theta,$$

$$g_{ab} = 0, \,\forall\, a \neq b, \qquad |g| = -r^4\sin^2\theta e^{2(\lambda+\nu)}.$$

Assume that M is Ricci-flat, that is, $R_{ab} = 0$. Finding the Christoffel symbols of the second type, we calculate the four non-zero components of the Ricci tensor and then equating them to zero entails the following three independent equations:

$$\partial_r \lambda \;=\; \frac{e^{2\lambda}-1}{2r},$$

$$\partial_r \nu = \frac{1 - e^{2\lambda}}{2r},$$

$$\partial_t \nu = 0.$$

Adding first and second equations provide $\partial_r (\lambda + \nu) = 0$. Thus, $\lambda + \nu = f(t)$. Now integrating first equation and then using $\partial_r \lambda = -\partial_r \nu$, we get

$$e^{2\lambda} = e^{-2\nu} = \left(1 - \frac{2m}{r}\right),$$

where m is a positive constant. Thus, (8.1.3) takes the form

$$ds^2 = -\left(1 - \frac{2m}{r}\right) dt^2 + \left(1 - \frac{2m}{r}\right)^{-1} dr^2 + r^2 \left(d\theta^2 + \sin^2 \theta d\phi^2\right). \qquad (8.1.4)$$

This solution is due to Schwarzschild for which M is the *exterior Schwarzschild spacetime* $(r > 2m)$ with m and r as the mass and the radius of a spherical body. If we consider all values of r, then (8.1.4) is singular at $r = 0$ and $r = 2m$. It is well-known that $r = 0$ is an *essential singularity* and the singularity $r = 2m$ can be removed by extending (M, g) to another manifold say (M', g') as follows. Let

$$r' = \int \left(1 - \frac{2m}{r}\right)^{-1} dr = r + 2m \log (r - 2m),$$

be a transformation with a new coordinate system (u, r, θ, ϕ), where $u = t + r'$ is an advanced null coordinate. Then, (8.1.4) takes the following new form:

$$ds^2 = -\left(1 - \frac{2m}{r}\right) du^2 + 2du\, dr + r^2 \left(d\theta^2 + \sin^2 d\phi^2\right), \qquad (8.1.5)$$

which is non-singular for all values of r. Similarly, if we use a retarded null coordinate $v = t - r'$, then (8.1.4) takes the form

$$ds^2 = -\left(1 - \frac{2m}{r}\right) dv^2 - 2dv\, dr + r^2 \left(d\theta^2 + \sin^2 \theta d\phi^2\right). \qquad (8.1.6)$$

The exterior Schwarzchild spacetime, with metric (8.1.4) for $r > 2m$, can be regarded as a Lorentzian warped product manifold in the following way. Let $M_1 = \{(t, r) \in \mathbf{R}^2 : r > 2m\}$ be endowed with the Lorentzian metric

$$g_1 = -\left(1 - \frac{2m}{r}\right) dt^2 + \left(1 - \frac{2m}{r}\right) dr^2$$

and let M_2 be the unit 2-sphere S^2 with the usual Riemannian metric g_2 of constant sectional curvature 1 induced by the inclusion mapping $S^2 \to \mathbf{R}^3$. Then, $(M = M_1 \times_h M_2, g = g_1 \oplus_h g_2, r^2 = h)$ is the exterior Schwarzchild spacetime. Based on above Beem-Ehrlich theorem, (M, g) is globally hyperbolic since M_1 is globally hyperbolic and $M_2 = S^2$ can have a complete Riemannian metric.

Case 2: Reissner-Nordström spacetimes. Another solution of asymptotically flat category is due to Reissner-Nordström, which represents the spacetime (M, g) outside a spherically symmetric body having an electric charge e but no spin or magnetic dipole. Following a procedure similar to the case of Schwarzschild solution, the metric of this spacetime can be expressed by

$$
\begin{aligned}
ds^2 &= -\left(1 - \frac{2m}{r} + \frac{e^2}{r^2}\right) dt^2 + \left(1 - \frac{2m}{r} + \frac{e^2}{r^2}\right)^{-1} dr^2 \\
&+ \ r^2 \left(d\theta^2 + \sin^2\theta d\phi^2\right)
\end{aligned} \tag{8.1.7}
$$

for a local coordinate system (t, r, θ, ϕ). This metric is also asymptotically flat as for $r \to \infty$, it approaches the Minkowski metric and in particular if $e = 0$ then this is Schwarzschild metric. It is singular at $r = 0$ and $r = m \pm (m^2 - e^2)^{\frac{1}{2}}$ if $e^2 \le m^2$. While $r = 0$ is an essential singularity, the other two can be removed as follows. Consider a transformation

$$
r' = \int \left(1 - \frac{2m}{r} + \frac{e^2}{r^2}\right)^{-1} dr.
$$

Let (u, r, θ, ϕ) be a new coordinate system, with respect to (8.1.5), such that $u = t + r'$ the advanced null coordinate. Then, this metric transforms into

$$
ds^2 = -\left(1 - \frac{2m}{r} + \frac{e^2}{r^2}\right) du^2 + 2\, du\, dr + r^2 \left(d\theta^2 + \sin^2\theta\, d\phi^2\right)
$$

which is regular for values of r and represents an extended spacetime (M', g') such that M is embedded in M' and g' is g on the image of M on M'. Reissner-Nordström spacetimes also belong to globally hyperbolic spacetimes.

8.2   Globally null manifolds

Let (M, g) be a real n-dimensional smooth manifold where g is a symmetric tensor field of type $(0, 2)$. We assume that M is paracompact. The radical or the null space of $T_x(M)$ is a subspace, denoted by $RadT_x(M)$, of $T_x(M)$ defined by

$$
Rad\, T_x(M) = \{\xi_x \in Rad\, T_x(M) : g(\xi_x\, X) = 0,\ X \in T_x(M)\}. \tag{8.2.1}
$$

The dimension, say r, of $RadT_x(M)$ is called nullity degree of g. $Rad\, TM$ is called the radical distribution of rank r on M. Clearly, g is degenerate or non-degenerate on M if and only if $r > 0$ or $r = 0$ respectively. We say that (M, g) is a lightlike manifold if $0 < r \le n$. Consider a complementary screen distribution $S(TM)$ to $RadTM$ in TM whose existence is secured for paracompact M. It is easy to see that $S(TM)$ is semi-Riemannian. Therefore, we have

$$
TM = S(TM) \ \oplus \ Rad\, TM. \tag{8.2.2}
$$

The associated quadratic form of g is a map $h : T_x(M) \to \mathbf{R}$ given by $h(X) = g(X, X)$ for any $X \in T_x(M)$. In general, h is of type (p, q, r), where $p + q + r = n$, and q is the index of g. We use the following range of indices:

$$I, J, \in \{1, \ldots, q\}, \quad A, B, \in \{q+1, \ldots, q+p\}, \quad \alpha, \beta, \in \{1, \ldots, r\}$$

$$a, b, \in \{r+1, \ldots, n\}, \quad i, j, \in \{1, \ldots, n\}.$$

Using a well-known result from linear algebra, we have the following canonical form for h (with respect to a local basis of $T_x(M)$):

$$h = -\sum_{I=1}^{q}(\omega^I)^2 + \sum_{A=q+1}^{q+p}(\omega^A)^2, \tag{8.2.3}$$

where $\{\omega^1, \ldots, \omega^{p+q}\}$ are linearly independent local differential 1-forms on M. With respect to a local coordinates system (x^i), by replacing in (8.2.3) each $\omega^I = \omega_i^I \, dx^i$ and each $\omega^A = \omega_i^A \, dx^i$, we obtain

$$h = g_{ij} \, dx^i \, dx^j, \quad \operatorname{rank} |g_{ij}| = p + q < n,$$

$$g_{ij} = g(\partial_i, \partial_j) = -\sum_{I=1}^{q} \omega_i^I \omega_j^I + \sum_{A=q+1}^{q+p} \omega_i^A \omega_j^A. \tag{8.2.4}$$

Suppose $Rad\,TM$ is an integrable distribution. Then it follows from the Frobenius theorem that leaves of $Rad\,TM$ determine a foliation on M of dimension r, that is, M is a disjoint union of connected subsets $\{L_t\}$ and each point x of M has a coordinate system $(\mathcal{U}, x^i)$, where $i \in \{1, \ldots, n\}$ and $L_t \cap \mathcal{U}$ is locally given by the equations $x^a = c^a$, $a \in \{r+1, \ldots, n\}$ for real constants c^a, and (x^α), $\alpha \in \{1, \ldots, r\}$, are local coordinates of a leaf L of $Rad\,TM$. Consider another coordinate system $(\bar{\mathcal{U}}, \bar{x}^\alpha)$ on M. The transformation of coordinates on M, endowed with an integrable distribution, has the following special form.

$$0 = d\bar{x}^a = \frac{\partial \bar{x}^a}{\partial x^b} \, dx^b + \frac{\partial \bar{x}^a}{\partial x^\alpha} \, dx^\alpha = \frac{\partial \bar{x}^a}{\partial x^\alpha} \, dx^\alpha,$$

which imply $\frac{\partial \bar{x}^a}{\partial x^\alpha} = 0$, $\forall\, a \in \{r+1, \ldots, n\}$; $\alpha \in \{1, \ldots, r\}$. Hence the transformation of coordinates on M is given by

$$\bar{x}^\alpha = \bar{x}^\alpha(x^1, \ldots, x^n), \quad \bar{x}^a = \bar{x}^a(x^{r+1}, \ldots, x^n). \tag{8.2.5}$$

As g is degenerate on TM, by using (8.2.1) and (8.2.3) we obtain $g_{\alpha\beta} = g_{\alpha a} = g_{a\alpha} = 0$. Thus the matrix of g with respect to the natural frame $\{\partial_i\}$ becomes

$$(g_{ij}) = \begin{pmatrix} O_{r,\,r} & O_{r,\,n-r} \\ O_{n-r,\,r} & g_{ab}(x^1, \ldots, x^n) \end{pmatrix}.$$

By using the first group of equations in (8.2.5) one can show that

$$\partial_\alpha \, g_{ab} = 0, \quad \forall\, a, b \in \{r+1, \ldots, n\}, \quad \alpha \in \{1, \ldots, r\} \tag{8.2.6}$$

holds for any other system of coordinates adapted to the foliation induced by
$RadTM$. It is easy to see that the screen distribution $S(TM)$ is invariant with
respect the transformations (8.2.6).

In this book, we assume that $r = 1$. Thus, the 1-dimensional nullity distribution
$RadTM$ is integrable. Recall from equation (7.7.5) of chapter 7 that

$$(\pounds_X g)(Y, Z) = X(g(Y, Z)) - g([X, Y], Z) - g(Y, [X, Z]), \qquad (8.2.7)$$

for any $X, Y, Z \in \Gamma(TM)$ where $\pounds$ is the Lie-derivative operator. A vector field X
on M is called a *Killing vector field* if $\pounds_X g = 0$. A distribution D on M is called a
Killing distribution if each vector field of D is Killing.

Theorem 2.1. *Let* (M, g) *be an* n-*dimensional lightlike manifold, with* $RadTM$
of rank $r = 1$. *Then the following assertions are equivalent:*

(i) *There exists a local coordinate system on* M *satisfying* (8.2.6).

(ii) $RadTM$ *is a Killing distribution.*

(iii) *There exists a metric (Levi-Civita) connection* ∇ *on* M *with respect to the
degenerate metric tensor* g.

Proof. (i) $\Rightarrow$ (ii). Suppose (8.2.6) is satisfied on M. As 1-dimensional $Rad\,TM$
is obviously integrable, let $(\mathcal{U}; x^1, \ldots, x^n)$ be a coordinate system such that any
$X \in \Gamma(RadTM)$ is locally expressed by $X = X^1 \partial_1$. Then by using (8.2.7) and
(8.2.1) we see that $\pounds_X g = 0$ becomes

$$X^1 \{\partial_1(g(Y, Z)) - g([\partial_1, Y], Z) - g(Y, [\partial_1, Z])\} = 0, \qquad (8.2.8)$$

for any $Y, Z \in \Gamma(TM)$. By using (8.2.1) and taking into account that $RadTM$
is integrable, it is easy to check that in case at least one of vector fields Y and Z
belongs to $RadTM$, then, (8.2.8) is identically satisfied. Now if we take $Y = \partial_a$
and $Z = \partial_b$, $a, b \in \{2, \ldots, n\}$, then, using (8.2.6) we conclude that (8.2.8) holds.
Hence $RadTM$ is a Killing distribution, so (i) $\Rightarrow$ (ii).

(ii) $\Rightarrow$ (i). Suppose $RadTM$ is a Killing distribution, that is

$$(\pounds_X g)(Y, Z) = X(g(Y, Z)) - g([X, Y], Z) - g(Y, [X, Z]) = 0, \qquad (8.2.9)$$

for any $X \in \Gamma(RadTM)$ and $Y, Z \in \Gamma(TM)$. Considering $X = \partial_1 \in \Gamma(RadTM)$,
$Y = \partial_a$ and $Z = \partial_b$ in (8.2.9) we obtain (8.2.6), so (ii) $\Rightarrow$ (i).

(iii) $\Rightarrow$ (ii). Suppose there exists a metric connection ∇ on M with respect to
its degenerate metric g. Then using (8.2.7), (7.1.1) and (8.2.1) we get

$$
\begin{aligned}
(\pounds_X g)(Y, Z) &= \{X(g(Y, Z)) - g(\nabla_X Y, Z) - g(Y, \nabla_X Z)\} \\
&\quad + \{g(\nabla_Y X, Z) + g(Y, \nabla_Z X)\} \\
&= g(\nabla_Y X, Z) + g(Y, \nabla_Z X) \\
&= Y(g(X, Z)) + Z(g(X, Y)) - g(X, \nabla_Y Z) - g(X, \nabla_Z Y) \\
&= 0,
\end{aligned}
$$

for any $X \in \Gamma(Rad\,TM)$ and $Y, Z \in \Gamma(TM)$. Hence $Rad\,TM$ is a Killing distribution on M, so (iii) $\Rightarrow$ (ii).

(ii) $\Rightarrow$ (iii). As 1-dimensional $Rad\,TM$ is obviously integrable, using the fibre bundle theory, we construct an $(n + 1)$-dimensional manifold, denoted by $\bar{M}$, with local coordinates (x, x^a, y), where (x, x^a) are coordinates on M induced by the foliation determined by $Rad\,TM$ and (y) is a coordinate on 1-dimensional fiber of its vector bundle structure. From the proof of (ii) $\Rightarrow$ (i), the transformations equations (8.2.6) hold on $\bar{M}$ and $\bar{y} = \frac{\partial \bar{x}}{\partial y}$ which provides a 1-dimensional vector bundle NM over M, locally spanned by $\{\frac{\partial}{\partial y}\}$. Moreover, we have

$$T\bar{M} = TM \oplus NM. \tag{8.2.10}$$

Next, since M is paracompact we consider a Riemannian metric g^* on M and a screen distribution $S(TM)$ as the complementary orthogonal distribution to $Rad\,TM$ in TM with respect to g^*. Then using (8.2.2) in (8.2.10) we obtain

$$T\bar{M} = S(TM) \oplus Rad\,TM \oplus NM. \tag{8.2.11}$$

Now, denote by p, s and t the projection morphisms of $T\bar{M}$ on $S(TM)$, $Rad\,TM$ and NM respectively, and define a symmetric bilinear form $\bar{g}$ on $\bar{M}$ such that

$$\bar{g}(\bar{X}, \bar{Y}) = g(p\bar{X}, p\bar{Y}) + g^*(s\bar{X}, (t\bar{Y})^*) + g^*(s\bar{Y}, (t\bar{X})^*), \tag{8.2.12}$$

for any $\bar{X}, \bar{Y} \in \Gamma(T\bar{M})$. It is easy to see that $\bar{g}$ is a non-degenerate metric on $\bar{M}$ and the degenerate metric g is the restriction of $\bar{g}$ to $\Gamma(TM)$. Denote by $\bar{\nabla}$ the metric connection (i.e., $\bar{\nabla}\bar{g} = 0$) and set

$$\bar{\nabla}_X Y = \nabla_X Y + B(X, Y)\,\partial_y, \qquad \forall\, X, Y \in \Gamma(TM), \tag{8.2.13}$$

where $\nabla_X Y \in \Gamma(TM)$ and $B(X, Y)$ is a function on M. It follows that ∇ is a torsion-free linear connection on M and B is a symmetric form on $\Gamma(TM)$. Now $Rad\,TM$ integrable, $\bar{g}$ parallel with respect to $\bar{\nabla}$ and (8.2.11), (8.2.13) imply

$$0 = (\pounds_X g)(Y, Z) = -\bar{g}(X, \bar{\nabla}_Y Z + \bar{\nabla}_Z Y) = -B(Y, Z)\,g^*(X, \partial_x),$$

for $X \in \Gamma(Rad\,TM)$ and $Y, Z \in \Gamma(TM)$. Therefore, g^* being Riemannian, $B(Y, Z) = 0$, so g is parallel with respect to ∇ which completes the proof.

Definition 2.1 (Duggal [23]). *A lightlike manifold (M, g) is said to be a globally null manifold if it admits a single global null vector field and a complete Riemannian hypersurface.*

Example 1. Let $(\bar{M}, \bar{g})$ be an $(n+1)$-dimensional globally hyperbolic spacetime [10], with the line element of the metric $\bar{g}$ given by

$$ds^2 = -dt^2 + dx^1 + \bar{g}_{ab}\,dx^a\,dx^b, \quad (a, b = 2, \dots, n)$$

with respect to a coordinate system $(t, x^1, \dots, x^n)$ on $\bar{M}$. We choose the range $0 < x^1 < \infty$ so that above metric is non-singular. Take two null coordinates u

and v such that $u = t + x^1$ and $v = t - x^1$. Thus, above metric transforms into a non-singular metric:

$$ds^2 = - \, du \, dv + \bar{g}_{ab} \, dx^a \, dx^b.$$

The absence of du^2 and dv^2 in above metric implies that $\{v = \text{constant.}\}$ and $\{u = \text{constant.}\}$ are lightlike hypersurfaces of $\bar{M}$. Let $(M, g, r = 1, v = \text{constant.})$ be one of this lightlike pair and let D be the 1-dimensional distribution generated by the null vector $\{\partial_v\}$ in $\bar{M}$. Denote by L the 1-dimensional integral manifold of D. A leaf M' of the $(n-1)$-dimensional screen distribution of M is Riemannian with metric $d\Omega^2 = \bar{g}_{ab} \, x^a \, x^b$ and is the intersection of the two lightlike hypersurfaces. In particular, there will be many global timelike vector fields in globally hyperbolic spacetimes $\bar{M}$. If one is given a fixed global *time function* then its gradient is a global timelike vector field in a given $\bar{M}$. With this choice of a global timelike vector field in $\bar{M}$, we conclude that both its lightlike hypersurfaces admit a global null vector field. Now, using the celebrated Hopf-Rinow theorem one may choose a screen whose leaf M' is a complete Riemannian hypersurface of M. Thus, it is possible to construct a pair of globally null manifolds as hypersurfaces of a globally hyperbolic spacetime. In particular, a Minkowski spacetime can have a pair of hypersurfaces which are globally null manifolds.

Theorem 2.2. *Let $(M, g, S(TM))$ be an n-dimensional $(n \geq 3)$ globally null manifold, with a choice of screen distribution $S(TM)$. Then, $S(TM)$ is integrable if and only if $M = M' \times C'$ is a global product manifold, where M' is a leaf of $S(TM)$ and C' is a 1-dimensional integral manifold of a global null curve C in M.*

Proof. Choose an integrable screen $S(TM)$. Then, by Frobenius theorem, $S(TM)$ is also involutive. Let M' be the integral manifold of $S(TM)$. The global null curve C being 1-dimensional, it is both integrable and involutive. Therefore, (a) $\Rightarrow$ (b). Conversely, let (b) hold. Then, one can always choose an integrable screen $S(TM)$ for M with M' its integral manifold, which completes the proof.

Now we show that there is a large class of globally null product manifolds. Consider a class of globally null manifolds, denoted by $(M, g, S(TM), G)$, such that each member M of this class carries a smooth 1-parameter group G of isometries whose orbits are global null curves in M. This means that each $Rad\,TM$ is a Killing distribution and so, as per theorem 2.1, each M admits a metric connection ∇ with respect to its degenerate metric g.

Proposition 2.1. *Let $(M, g, S(TM), G)$ be an $(n+1)$-dimensional $(n > 1)$ globally null manifold, with a smooth 1-parameter group G of isometries whose orbits are global null curves in M. Suppose M' is the spacelike n-dimensional orbit space of the action G. Then, $(M, g, S(TM), G)$ is a global product manifold $M = M' \times C'$, where M' and C' are leafs of a chosen integrable screen distribution $S(TM)$ and the $Rad\,TM$ of M respectively.*

Proof. Let M' be the orbit space of the action $G \approx C'$, where C' is a

1-dimensional null leaf of $Rad\,TM$ in M. Then, M' is a smooth Riemannian hypersurface of M and the projection $\pi : M \to M'$ is a principle C'-bundle, with null fiber G. The global existence of null vector field, of M, implies that M' is Hausdorff and paracompact. The infinitesimal generator of G is a global null Killing vector field, say ξ, on M. Then, the metric g restricted to the screen distribution $S(TM)$, of M, induces a Riemannian metric, say g', on M'. Since ξ is non-vanishing on M, we can take $\xi = \partial_\theta$ a global null coordinate vector field for some global function θ on M. Thus, θ induces a diffeomorphism on M such that $(M = M' \times C', g = \pi^* g')$ is a global product manifold. Finally, the integrability of $S(TM)$ follows from the theorem 2.1, which completes the proof.

Physical model. Consider a 4-dimensional stationary spacetime $(\bar{M}, \bar{g})$ which is chronological, that is, $\bar{M}$ admits no closed timelike curves. It is well known [56] that a stationary $\bar{M}$ admits a smooth 1-parameter group, say $\bar{G}$, of isometries whose orbits are timelike curves in $\bar{M}$. A static spacetime is stationary with the additional condition that its timelike Killing vector field, say T, is hypersurface orthogonal, that is, there exists a spacelike hypersurface orthogonal to T. Our model will be applicable to both of these types. We denote by $\bar{M}'$ the Hausdorff and paracompact 3-dimensional Riemannian orbit space of the action $\bar{G}$. The projection $\bar{\pi} : \bar{M} \to \bar{M}'$ is a principal $\mathbf{R}$-bundle, with the timelike fiber $\bar{G}$. Let $T = \partial_t$ be the non-vanishing timelike Killing vector field, where t is a global time coordinate function on $\bar{M}'$. Then, the metric $\bar{g}$ induces a Riemannian metric $g_{\bar{M}'}$ on $\bar{M}'$ such that

$$\bar{M} = \mathbf{R} \times \bar{M}', \quad \bar{g} = -u^2 (dt + \eta)^2 + \bar{\pi}^* g_{\bar{M}'},$$

where η is a connection 1-form for the $\mathbf{R}$-bundle $\bar{\pi}$ and

$$u^2 = -\bar{g}(T, T) > 0.$$

It is known that a stationary spacetime $(\bar{M}, \bar{g})$ uniquely determines the orbit data $(\bar{M}', \bar{g}_{\bar{M}'}, u, \eta)$ as described above, and conversely. Suppose the orbit space $\bar{M}'$ has a non-empty metric boundary $\partial \bar{M}' \neq \emptyset$. Consider the maximal solution data in the sense that it is not extendible to a larger domain $(\mathcal{M}', g'_{\mathcal{M}'}, u', \eta') \supset (\bar{M}', g_{\bar{M}'}, u, \eta)$ with $u' > 0$ on an extended spacetime $\mathcal{M}'$. Under these conditions, it is known [56] that in any neighborhood of a point $x \in \partial \bar{M}'$, either the metric $g_{\bar{M}'}$ or the connection 1-form η degenerates, or $u \to 0$. The third case implies that the timelike Killing vector T becomes null and, there exists a Killing horizon $H = \{u \to 0\}$ of $\bar{M}$. This Killing horizon H is related to our class of globally null spaces $(M, g, S(TM), G)$ as follows:

As stationary spacetimes belong to the class of globally hyperbolic spacetimes [10], following example 1 we construct a globally null hypersurface $(M, g, S(TM), G)$ of the above considered stationary spacetime $(\bar{M}, \bar{g})$ such that $Rad\,TM$ is a Killing distribution. Thus, it follows from the theorem 5.1 of chapter 7 that our chosen globally null M is a totally geodesic hypersurface of $\bar{M}$. Now consider the case when the timelike vector field $T \in \bar{M}$ becoming null, that is,

$$Lim(T)_{u \to 0} = \xi,$$

where $\xi \in \Gamma(T(\bar{M}))$ is a null Killing vector field. Then the spacelike hypersurface $\bar{M}'$ of $\bar{M}$ degenerates to a 3-dimensional globally null hypersurface M of $\bar{M}$ such that a leaf M' of $S(TM)$ is identified with $\partial\,\bar{M}'$, that is,

$$M' = \partial\bar{M}' \subset \bar{M}$$

is a common 2-dimensional submanifold of both M and $\bar{M}$. Since M is totally geodesic in $\bar{M}$, we conclude that M admits a smooth 1-parameter group G of isometries (induced from the group $\bar{G}$ of $\bar{M}$) whose orbits are global null curves in M. Consequently, we have constructed a physical model of globally null product manifolds which are totally geodesic hypersurfaces of stationary spacetimes.

As an application, we suggest reading [26] on *harmonic maps, morphisms and globally null manifolds*.

Globally null warped product manifolds. Motivated by extensive and very effective use of warped products in Riemannian and Lorentzian geometry (see [17, 82, 10]), in 2001, first author of this book proposed the concept of *lightlike warped product manifolds* [23] which we recall as follows:

Let (N, g_N) and (F, g_F) be a lightlike and a Riemannian manifold of dimensions n_1 and n_2 respectively, where the $Rad\,TN$ is of rank r. Let $\pi : N \times F \to N$ and $\eta : N \times F \to F$ denote the projection maps given by $\pi(x, q) = x$ and $\eta(x, q) = q$ for $(x, q) \in N \times F$ respectively, where the projection π on N is with respect to the non-degenerate screen distribution $S(TN)$.

Definition 2.2 ([23]). *The product manifold $M = N \times F$ is said to be a lightlike warped product $N \times_f F$, with the degenerate metric g defined by*

$$g(X, Y) = g_N(\pi_\star X, \pi_\star Y) + (f \circ \pi)^2 g_F(\eta_\star X, \eta_\star Y), \qquad (8.2.14)$$

for every X, Y of M and $\star$ is the symbol for the tangent map.

It follows that $Rad\,TM$ of M still has rank r but $\dim(M) = n_1 + n_2$ and $\dim(S(TM)) = n_1 + n_2 - r$. Consistent with the theme of this book, we assume that (N, g_N) is a globally null manifold for which $r = 1$ and (F, g_F) is a complete Riemannian manifold. Thus, we have the following characterization theorem. The proof is straightforward and common with the proof of Lorentzian case [10]:

Theorem 2.3 ([23]). *Let (N, g_N) and (F, g_F) be lightlike and Riemannian manifolds respectively. Then the warped product $(M = N \times_f F, g)$ is globally null if and only if both the following conditions hold:*
  (1) (N, g_N) *is globally null.*
  (2) (F, g_F) *is a complete Riemannian manifold.*

8.3   Frenet frames along null geodesics

Let (M, g) be a n-dimensional globally null manifold having a metric (Levi-Civita) connection ∇ with respect to its degenerate metric tensor g. This means that, as per theorem 2.1 in this chapter, $RadTM$ is a Killing distribution. Let C be a smooth null curve in (M, g) given by

$$x^i = x^i(t), \ t \in I \subset \mathbf{R}, \ i \in \{1, \ldots, n\},$$

for a coordinate neighborhood $\mathcal{U}$ on C. Then, the tangent vector field

$$\frac{d}{dt} = \left(\frac{dx^1}{dt}, \ldots, \frac{dx^n}{dt} \right)$$

on $\mathcal{U}$ satisfies

$$g\left(\frac{d}{dt}, \frac{d}{dt} \right) = 0, \qquad i.e., \qquad g_{ij} \frac{dx^i}{dt} \frac{dx^j}{dt} = 0,$$

where $g_{ij} = g(\partial_i, \partial_j)$ and $i, j \in \{1, \ldots, n\}$. Denote by TC the tangent bundle of C which is a vector subbundle of TM, and is of rank 1. Then,

$$TC^\perp = \{V \in TM \ : \ g(V, \xi) = 0\} = TM$$

where ξ is null vector field tangent over C. We consider a class of null curves such that $RadTM = TC$ and both are generated by ξ. Then, we have

$$TM = RadTM \oplus S(TM) = TC \oplus S(TM). \tag{8.3.1}$$

Proposition 3.1. *Let* (M, g) *be an* n-*dimensional globally null manifold. Then, there exists a quasi-orthonormal frame*

$$F = \{\xi, W_1, \ldots, W_{n-1}\}, \ g(W_a, W_a) = \delta_{ab}, \ g(\xi, W_a) = 0, \tag{8.3.2}$$

for all $a \in \{1, \ldots, n-1\}$, *along a null curve* C, *generated by a null vector field* ξ, *on* M, *where* $\Gamma(S(TM))$ *is spanned by an orthonormal frame* $\{W_1, \ldots, W_{n-1}\}$.

Proof. Since g is a metric tensor on M, we have

$$(\nabla_X g)(Y, Z) = X(g(Y, Z)) - g([X, Y], Z) - g(Y, [X, Z]) = 0,$$

for all $X, Y, Z \in \Gamma(TM)$. Using this, $\frac{d}{dt} \equiv \xi$ null and (8.3.1), we obtain the following equations (procedure is similar to the one used in chapter 2)

$$
\begin{aligned}
\nabla_\xi \xi &= h\,\xi, \\
\nabla_\xi W_1 &= -k_1\,\xi + k_3\,W_2 + k_4\,W_3, \\
\nabla_\xi W_2 &= -k_2\,\xi - k_3\,W_1 + k_5\,W_3 + k_6\,W_4, \\
\nabla_\xi W_3 &= -k_4\,W_1 - k_5\,W_2 + k_7\,W_4 + k_8\,W_5, \\
&\quad \cdots\cdots\cdots\cdots\cdots\cdots\cdots \\
&\quad \cdots\cdots\cdots\cdots\cdots\cdots\cdots \\
\nabla_\xi W_{n-2} &= -k_{n-1}\,W_{n-4} - k_n\,W_{n-3} + k_{n+2}\,W_{n-1} + k_{n+3}\,W_n, \\
\nabla_\xi W_{n-1} &= -k_{2n-4}\,W_{n-3} - k_{2n-3}\,W_{n-2},
\end{aligned}
\tag{8.3.3}
$$

provided $n \geq 5$, where h and $\{k_1, \ldots, k_{2n-3}\}$ are smooth functions on $\mathcal{U}$ and $\{W_1, \ldots, W_{n-1}\}$ is an orthonormal basis of $\Gamma(S(TM)_{\mathcal{U}})$. For $n < 5$, above equations reduce to the following cases:

Case 1 $(n = 2)$.

$$
\begin{aligned}
\nabla_\xi \xi &= h\xi, \\
\nabla_\xi W_1 &= -k_1 \xi.
\end{aligned}
$$

Case 2 $(n = 3)$.

$$
\begin{aligned}
\nabla_\xi \xi &= h\xi, \\
\nabla_\xi W_1 &= -k_1 \xi + k_3 W_2, \\
\nabla_\xi W_2 &= -k_2 \xi - k_3 W_1.
\end{aligned}
$$

Case 3 $(n = 4)$.

$$
\begin{aligned}
\nabla_\xi \xi &= h\xi, \\
\nabla_\xi W_1 &= -k_1 \xi + k_3 W_2 + k_4 W_3, \\
\nabla_\xi W_2 &= -k_2 \xi - k_3 W_1 + k_5 W_3, \\
\nabla_\xi W_3 &= -k_4 W_1 - k_5 W_2.
\end{aligned}
$$

In general, for any $n > 1$ we call F (given by (8.3.2)) a Frenet frame on M along C with respect to a screen distribution $S(TM)$. The functions $\{k_1, \ldots, k_{2n-3}\}$ and the equations (8.3.3) (along with three cases for $n < 5$) are called curvature functions of C and Frenet equations for F, respectively. Thus, F, given by (8.3.2), is a quasi-orthonormal Frenet frame which completes the proof.

Example 2. Let $\mathbf{R}_1^4$ be a 4-dimensional Minkowski spacetime with a Lorentz metric of signature $(-++)$ and local coordinates (x, x^1, x^2, y). Let (M, g) be a globally null hypersurface of $\mathbf{R}_1^4$ such that $(x, x^1, x^2, y = \text{constant})$ are coordinates on M, induced by $Rad\, TM$. Consider a curve C in M defined by

$$
x = f(t) , \quad x^1 = -f(t) , \quad x^2 = a_1 , \quad y = a_2 , \quad t \in I \subset \mathbf{R},
$$

where a_1 and a_2 are suitable constants. Then,

$$
\frac{d}{dt} = (f'(t) , -f'(t) , 0 , 0) \quad \text{and} \quad g\left(\frac{d}{dt}, \frac{d}{dt}\right) = 0.
$$

Thus, C is a null curve of M, generated by a null vector field, say $\xi \equiv \frac{d}{dt}$. Choose a quasi-orthonormal set $\{\xi, W_1, W_2\}$ on M along C, where

$$
W_1 = (b\, f(t), -b\, f(t), 1, 0) \; ; \; W_2 = (c\, f(t), -c\, f(t), 0, 1)
$$

are orthonormal spacelike vectors which generate a screen distribution of M, b and c are suitable constants. Following are three Frenet equations:

$$
\nabla_\xi \xi = h\xi , \quad h \equiv \frac{f''(t)}{f(t)} ,
$$

$$\nabla_\xi W_1 = b\xi + 0 W_2 \ , \quad \nabla_\xi W_2 = c\xi + 0 W_1$$

such that, according to the case 2 $(n = 3)$, $k_1 = -b$, $k_2 = -c$, $k_3 = 0$.

Consider, with respect to a given screen distribution $S(TM)$, two Frenet frames F and F^* along two neighborhoods $\mathcal{U}$ and $\mathcal{U}^*$ respectively with non-null intersection. Then we have

$$\xi^* = \frac{dt}{dt^*} \xi,$$
$$W_a^* = A_a^b W_b, \quad a, b \in \{1, \ldots, n-1\},$$

where A_a^b are smooth functions on $\mathcal{U} \cap \mathcal{U}^*$ and the matrix $[A_a^b(x)]$ is an element of the orthogonal group $O(n)$ for any $x \in \mathcal{U} \cap \mathcal{U}^*$.

Proposition 3.2. *Let C be a null curve of an n-dimensional globally null manifold M and F, F^* be two Frenet frames on $\mathcal{U}$ and $\mathcal{U}^*$ with curvature functions $\{k_1, \ldots, k_{2n-3}\}$ and $\{k_1^*, \ldots, k_{2n-3}^*\}$ respectively, induced by the same screen vector bundle $S(TC^\perp)$. Suppose $\mathcal{U} \cap \mathcal{U}^* \neq \phi$ and $\Pi_{\alpha=1}^{2n-2} k_\alpha \neq 0$ on $\mathcal{U} \cap \mathcal{U}^*$. Then at any point of $\mathcal{U} \cap \mathcal{U}^*$ we have*

$$k_1^* = k_1 A_1, \quad k_2^* = k_2 A_2, \tag{8.3.4}$$

$$k_a^* = A_a k_a \frac{dt}{dt^*} , \quad a \in \{3, \ldots, 2n-3\}, \quad \text{where} \quad A_a = \pm 1. \tag{8.3.5}$$

Proof. Since $[A_a^b]$ is an orthogonal matrix we infer $A_a^a = A_a = \pm 1$ and $A_a^b = 0$ for all $a, b \in \{1, \ldots, n-1\}$. Then from the second and the third equations in (8.3.3) with respect to both F and F^* and taking into account that $k_a \neq 0$ implies $k_a^* \neq 0$ for $a = 4, 5$ we obtain the relations (8.3.4) and (8.3.5) for $a = 1, \ldots, 5$. Similarly, we obtain all the relations of (8.3.5).

Corollary 1. *Under the hypothesis of proposition 3.2, k_1 and k_2 are invariant functions up to a sign, with respect to any parameter transformations on C.*

Next, let $F = \{\frac{d}{dt}, N, W_1, \ldots, W_{n-1}\}$ and $\bar{F} = \{\frac{d}{d\bar{t}}, \bar{N}, \bar{W}_1, \ldots, \bar{W}_{n-1}\}$ be two Frenet frames with respect to $(t, S(TC^\perp), \mathcal{U})$ and $(\bar{t}, \bar{S}(TC^\perp), \bar{\mathcal{U}})$, respectively. Then the general transformations relating elements of F and $\bar{F}$ on $\mathcal{U} \cap \bar{\mathcal{U}} \neq \emptyset$, are

$$\frac{d}{d\bar{t}} = \frac{dt}{d\bar{t}} \frac{d}{dt} \tag{8.3.6}$$

$$\bar{W}_a = B_a^b \left(W_b - \frac{dt}{d\bar{t}} c_b \frac{d}{dt} \right), \tag{8.3.7}$$

where c_b and B_a^b are smooth functions on $\mathcal{U} \cap \bar{\mathcal{U}}$ and the $(n-1) \times (n-1)$ matrix $[B_a^b(x)]$ is an element of $O(m)$ for each $x \in \mathcal{U} \cap \bar{\mathcal{U}}$. Thus, by using (8.3.6) and the first equation in (8.3.3) for both F and $\bar{F}$ we obtain

$$\bar{h} = \frac{d^2 t}{d\bar{t}^2} \frac{d\bar{t}}{dt} + h \frac{dt}{d\bar{t}}.$$

Proposition 3.3. *Let C be a null curve of a globally null manifold M, with Frenet equations (8.3.3). Then, there exists a special parameter p on C such that the function h vanishes and, therefore, C is a null geodesic in M.*

Consider two Frenet frames F and $\bar{F}$ for two screen distributions $S(TM)$ and $\bar{S}(TM)$ respectively. Using the transformation equations (8.3.6) and (8.3.7) we conclude that proposition 3.3 will also hold for any screen distribution. Also, let p and $\bar{p}$ be two special parameters induced by t and $\bar{t}$ with respect to the same screen distribution. Then, for both p and $\bar{p}$, one can obtain a special parameter $\bar{p} = a\,p + b$, $a \neq 0$. Thus, we have the following result:

Corollary 2. *Let M be a globally null manifold. The existence of a null geodesic curve C, of M, is independent of both the parameter transformations on C and the screen distribution transformations.*

Example 3. Let C be the null curve as given in example 2. It follows that C is a null geodesic if $h \equiv \frac{f''(t)}{f(t)} = 0$. This implies that $f''(t) = 0$. Thus, $f(t) = c_1 t + c_2$ is a linear function. For this case, we may take $t = p$, a special parameter. Other two Frenet equations (from example 2) will be same.

Null 2-surfaces. Let (N, h) be a 2-surface of an $(n + 2)$-dimensional globally null manifold (M, g), $n > 0$, where h is the induced tensor field on N of g, i.e.,

$$h(X, Y) = g(X, Y), \quad \forall X, Y \in \Gamma(TN).$$

It follows that N is a null 2-surface of M if

$$Rad\,TM = Rad\,TN$$

which we assume. Since $S(TM)$ is Riemannian, it is always possible to decompose it such that

$$S(TM) = TH \oplus TV, \tag{8.3.8}$$

where TH and $TV = TH^{\perp}$ are horizontal and vertical distributions of $S(TM)$ respectively. In this section, we assume that $\dim(TH) = 1$ and, therefore, $\dim(TV) = n$. Thus, using (8.2.2) and (8.3.8), we have

$$TM_{|N} = (Rad\,TN \oplus TH) \oplus TV = TN \oplus TV. \tag{8.3.9}$$

Since both the distributions $Rad\,TN$ and TH are of rank 1 on N, they are integrable. Therefore, there exists an atlas of local charts

$$\{\mathcal{U}\,;\, u^0,\, u^1,\, u^2, \ldots,\, u^{n+1}\}$$

such that $\{\frac{\partial}{\partial u^0}, \frac{\partial}{\partial u^1}\} \in \Gamma(TN_{|\mathcal{U}})$. Thus, the matrix of the degenerate metric g on M with respect to the natural frames field $\{\frac{\partial}{\partial u^0}, \ldots, \frac{\partial}{\partial u^{n+1}}\}$ is as follows

$$[g] = \begin{bmatrix} 0 & 0 \\ 0 & g_{ij}(u^0, \ldots, u^{n+1}) \end{bmatrix}, \tag{8.3.10}$$

where

$$g_{ij} = g\left(\frac{\partial}{\partial u^i}, \frac{\partial}{\partial u^j}\right), \quad i, j \in \{1, \ldots, n+1\},$$

$\det[\,g_{ij}\,] \neq 0$ and $g_{a1} = g_{1a} = 0$ for all $a \in \{2, \ldots, n+1\}$. According to the general transformations on a foliated manifold, we have

$$\bar{u}^0 = \bar{u}^0(u^0, u^1, \ldots, u^{n+1}),$$
$$\bar{u}^i = \bar{u}^i(u^1, \ldots, u^{n+1}).$$

Using above transformations, a well-known procedure is available to obtain a local field of frames on M adapted to the decomposition (8.3.9). However, with respect to the tangent bundle space TM, we show that their exists a pseudo-orthonormal Frenet frame $F = \{\xi, W_1, W_2, \ldots, W_{n+1}\}$ on M along N, adapted to the decomposition (8.3.9) such that TN is spanned by $\{\xi, W_1\}$ and TV is spanned by $\{W_2, \ldots, W_{n+1}\}$. Consistent with the hypersurface theory in chapter 7 and also for physical reasons, we restrict to $n = 1$ so that M is a 3-dimensional globally null manifold. We assume that the degenerate metric g comes from a Lorentzian metric $\bar{g}$ of a 4-dimensional Minkowski spacetime $\mathbf{R}_1^4$. Suppose N is given by

$$x^A = x^A(u, v), \quad A \in \{0, 1, 2\}.$$

Then the tangent bundle of N is spanned by

$$\left\{\frac{\partial}{\partial u} = \frac{\partial x^A}{\partial u}\frac{\partial}{\partial x^A}; \quad \frac{\partial}{\partial v} = \frac{\partial x^A}{\partial v}\frac{\partial}{\partial x^A}\right\}.$$

By considering a vector field

$$\xi = \alpha\frac{\partial}{\partial u} + \beta\frac{\partial}{\partial v}, \tag{8.3.11}$$

we find that N is null, if and only if, the homogeneous linear system

$$\alpha\left(\sum_{a=1}^{2}\left(\frac{\partial x^a}{\partial u}\right)^2 - \left(\frac{\partial x^0}{\partial u}\right)^2\right) + \beta\left(\sum_{a=1}^{2}\frac{\partial x^a}{\partial u}\frac{\partial x^a}{\partial v} - \frac{\partial x^0}{\partial u}\frac{\partial x^0}{\partial v}\right) = 0,$$

$$\alpha\left(\sum_{a=1}^{2}\frac{\partial x^a}{\partial u}\frac{\partial x^a}{\partial v} - \frac{\partial x^0}{\partial u}\frac{\partial x^0}{\partial v}\right) + \beta\left(\sum_{a=1}^{2}\left(\frac{\partial x^a}{\partial v}\right)^2 - \left(\frac{\partial x^0}{\partial v}\right)^2\right) = 0,$$

has non-trivial solutions, where (α, β) are two variables. Denote

$$D^{AB} = \begin{vmatrix} \frac{\partial x^A}{\partial u} & \frac{\partial x^B}{\partial u} \\ \\ \frac{\partial x^A}{\partial v} & \frac{\partial x^B}{\partial v} \end{vmatrix}. \tag{8.3.12}$$

Proposition 3.4. *A 2-surface N of a 3-dimensional globally null manifold M is null, if and only if, on each coordinate neighborhood $\mathcal{U} \subset N$ we have*

$$\sum_{a=1}^{2}(D^{0\,a})^2 = \sum_{1\leq a<b\leq 2}(D^{a\,b})^2. \tag{8.3.13}$$

Next, we choose from the above homogeneous system

$$\alpha = \sum_{a=1}^{2} \frac{\partial x^a}{\partial u} \frac{\partial x^a}{\partial v} - \frac{\partial x^0}{\partial v} \frac{\partial x^0}{\partial v} ; \quad \beta = \left(\frac{\partial x^0}{\partial v}\right)^2 - \sum_{a=1}^{2} \left(\frac{\partial x^a}{\partial v}\right)^2 , \qquad (8.3.14)$$

so that at least one of the quantities from the right hand side of (8.3.14) is non-zero. Then by direct calculations we find that $Rad\,TN$ is spanned by

$$\xi = \xi^A \frac{\partial}{\partial x^A} ; \quad \xi^A = \sum_{B=0}^{2} \epsilon_B D^{AB} \frac{\partial x^B}{\partial v} , \qquad (8.3.15)$$

where $\{\epsilon_B\}$ is the signature of the basis $\{\frac{\partial}{\partial x^B}\}$ with respect to the Minkowski metric $\bar{g}$. The corresponding 1-dimensional screen distribution $S(TN) = TH$ (see equation (8.3.11)) is spanned by a spacelike vector field

$$U = \frac{\partial x^0}{\partial v} \frac{\partial}{\partial u} - \frac{\partial x^0}{\partial u} \frac{\partial}{\partial v} = D^{10} \frac{\partial}{\partial x^1} + D^{20} \frac{\partial}{\partial x^2} . \qquad (8.3.16)$$

Finally, by using the decomposition (8.3.9) we obtain the spacelike vector bundle $TH^{\perp} = TV$ spanned by

$$V = D^{20} \frac{\partial}{\partial x^1} - D^{10} \frac{\partial}{\partial x^2} . \qquad (8.3.17)$$

Summing up, we have a quasi-orthonormal Frenet frames field, adapted to the composition (8.3.9), as follows:

$$F = \left\{ \xi, \ W_1 = \frac{1}{(\Delta)^{1/2}} U, \ W_2 = \frac{1}{(\Delta)^{1/2}} V \right\}, \quad \Delta = \sum_{a=1}^{2} (D^{0a})^2 . \qquad (8.3.18)$$

Using the terminology of differential geometry, we say that $\mathbf{x} = \mathbf{x}(u, v)$ is a class C^m $(m \geq 1)$ regular parametric representation of N, defined on a coordinates neighborhood $\mathcal{U}$, if
(1) $\mathbf{x}$ is of class $C^m \in \mathcal{U}$.
(2) At least one 2×2 determinant D^{AB} of (8.3.12) is non-zero.

A curve $\mathbf{x}(u, v = v_0)$ is called a u-parameter curve on N. Similarly, a curve $\mathbf{x}(u = u_0, v)$ is called a v-parameter curve on N. To cover any possible singular points, we take necessary overlapping allowable coordinate patches and use elementary topology so that N is at least smooth. If this is not possible then we restrict the parameters to regular points. Thus, subject to above topological constraints, this parametric representation can cover a regular N with a null family and a spacelike family of curves. Then $\mathbf{x}_u(u_0, v_0)$ and $\mathbf{x}_v(u_0, v_0)$ are vectors tangent to the u-parameter and the v-parameter curves respectively at their intersecting point $P(u_0, v_0)$ of N. At this point, we assume that u-parameter curve is null.

Null Tangent Plane. Let $P \in M$ and ℓ be a null vector of $T_P M$. A plane $T_P(N)$ of $T_P M$ is called a null plane directed by ℓ if it contains ℓ, $g_P(\ell, W) = 0$ for any $W \in T_P(N)$ and there exists $W_0 \in T_P(N)$ such that $g(W_0, W_0) \neq 0$. In particular, given a regular parametric representation $\mathbf{x}(u, v)$ of a null surface N and any point $P \in N$, a null plane $T_P(N)$ through P parallel to $\mathbf{x}_u$ and $\mathbf{x}_v$ at P is called the null tangent plane to N at P. One can verify that it is independent of the patch containing P and that a nonzero vector of M is tangent to N at P if and only if it is parallel to $T_P(N)$. Thus, $T_{\mathbf{x}}(N)$ at any $\mathbf{x}$ on N is given by

$$\mathbf{y} = \mathbf{x} + A\,\mathbf{x}_u + B\,\mathbf{x}_v, \quad -\infty < A, B < \infty.$$

Example 4. Construct a 4-dimensional Minkowski spacetime $(\mathbf{R}_1^4, \bar{g})$, with local coordinates $(x^0, x^1, x^2, y = a)$, where (x^0, x^1, x^2) are coordinates on a 3-dimensional globally null hypersurface (M, g) of $\mathbf{R}_1^4$ and a is a constant. Let N be a surface of M given by

$$\mathbf{x} = \mathbf{x}(u, v) = \left(x^0 = f(u, v), \ x^1 = -f(u, v), \ x^2 = -v^2\right),$$

where f is an arbitrary smooth function of two variables u and v and of class $C^m(m \geq 1)$. This parameterization is admissible since from (8.3.14) $\alpha = 0$ but $\beta = -4\,v^2 \neq 0$. Then from (8.3.12) we obtain

$$D^{10} = 0, \quad D^{20} = v\,f_u \neq 0, \quad D^{12} = -v\,f_u \neq 0, \quad f_u \equiv \frac{\partial f}{\partial u}.$$

Using above values in (8.3.13) we verify that N is a null surface of M. Moreover, since not all D^{AB}'s vanish, N is a regular null surface of class $C^m(m \geq 1)$. From (8.2.12) and (8.3.15)-(8.3.18) we obtain

$$\xi = -v^2 f_u \left(\frac{\partial}{\partial x^0} + \frac{\partial}{\partial x^1}\right), \quad U = f_u \frac{\partial}{\partial x^2}, \quad V = f_u \frac{\partial}{\partial x^1}.$$

Thus, we have the following quasi-orthonormal Frenet frames field on M along the null surface N:

$$F = \left\{\xi, \ W_1 = \frac{\partial}{\partial x^2}, \ W_2 = \frac{\partial}{\partial x^1}\right\}.$$

Also,

$$\mathbf{x}_u = (f_u, \ -f_u, \ 0); \quad \mathbf{x}_v = (f_v, \ -f_v, \ -2v)$$

are null and spacelike vectors respectively. Thus, u-parameter and v-parameter curves are null and spacelike respectively. N is regular, of class $C^m(m \geq 1)$, since not all D^{AB}'s vanish. $\{\mathbf{x}_u, \mathbf{x}_v\}$ is a linearly independent set. In particular, let $f(u, v) = u^2 - v^2$ and $P(1, 1)$ a point on N. Then,

$$\mathbf{x}(1, 1) = (0, 0, -1); \quad \mathbf{x}_u(1, 1) = (2, -2, 0); \quad \mathbf{x}_v(2, 1) = (-2, 2, -2).$$

Thus,

$$\mathbf{y} = \mathbf{x}(1, 1) + A\,\mathbf{x}_u(1, 1) + B\,\mathbf{x}_v(1, 1) = \left(2A - 2B, \ 2B - 2A, \ -(1 + B)\right)$$

is the equation of tangent plane at $P(1, 1) \in N$.

Contrary to the Riemannian or semi-Riemannian case, unfortunately, the normal vector field $\mathbf{x}_u \times \mathbf{x}_v$ falls back in $T_P(N)$. Thus, one fails to use, in the usual way, the theory of non-degenerate surfaces to study the geometry of N. To overcome this difficulty, we proceed as follows:

Let $\mathbf{x}(u, v)$ be a coordinate patch on N of class ≥ 1. Then, the differential $d\mathbf{x} = \mathbf{x}_u \, du + \mathbf{x}_v \, dv$ is parallel to $T_{\mathbf{x}}(N)$ at $\mathbf{x}(u, v)$ and the quantity

$$\begin{aligned} \mathbf{I} = d\mathbf{x} \cdot d\mathbf{x} &= (\mathbf{x}_u \, du + \mathbf{x}_v \, dv) \cdot (\mathbf{x}_u \, du + \mathbf{x}_v \, dv) \\ &= h_{11} \, dv^2, \end{aligned}$$

since $\mathbf{x}_u \cdot \mathbf{x}_u = \mathbf{x}_v \cdot \mathbf{x}_v = 0$; $(\mathbf{x}_v \cdot \mathbf{x}_v) \equiv h_{11} \neq 0$.

As in the Riemannian case, we call the function $\mathbf{I} = h_{11} \, dv^2$ the first fundamental form of the null surface N, whose only one surviving component is always in the direction of the family, say $\phi(u, v) = c$(constant), of spacelike curves on N. It is known that $\mathbf{I}$ is independent of any coordinates transformation, however, the coefficient h_{11} varies from point to point on N. Since any 2-dimensional manifold is Einstein, the Ricci tensor of N will also have one surviving component in the direction of $\phi(u, v) = c$. Thus, we state

Theorem 3.1. *Let (N, h) be a null 2-surface of a 3-dimensional globally null manifold (M, g). Then, the geometry of N is closely related to the geometry of its 1-dimensional spacelike integral manifold, say N', generated by the family of its spacelike curves $\phi(u, v) = c$.*

Using above theorem and a well-known procedure to compute curvature quantities of a family of spacelike curves, one can find curvature properties of a null 2-surface N.

Ruled Null Surfaces. Let $\alpha(u)$ be a null curve in a 4-dimensional Lorentz manifold $(\bar{M}, \bar{g})$, where $u \in I \subset \mathbf{R}$. Consider a null vector field $\ell(u)$ of $\alpha(u)$. Then,

$$\mathbf{x}(u, v) = \alpha(u) + v \, \ell(u); \quad v \in I \subset \mathbf{R},$$

is called a ruled null 2-surface, say N, of $\bar{M}$, which is generated by $\ell(u)$ and α is called the base curve of N. Its u-parameter and v-parameter curves both are families of null curves. Therefore, it is also called a totally null surface with $Rad \, TN = TN$. In particular, N is ruled by null geodesics if α is a geodesic curve. The notion of a null geodesic ruled surface was first introduced by Schild [104] in the form of a geodesic null string of a null hypersurface of a 4-dimensional Minkowski or curved spacetime. By null strings we mean 2-dimensional ruled null surfaces on the null cone (with one dimension suppressed) of $\bar{M}$. Since Schild's paper, there has been considerable work done on geodesic and non-geodesic null strings (see, for example, Ilyenko [60] and many others cited therein). Also, in section 8.5 we discuss physical use of ruled surfaces as photon surfaces.

Example 5. Consider a Minkowski spacetime $(\mathbf{R}_1^4, \bar{g})$ with metric

$$ds^2 = -(dx^0)^2 + (dx^1)^2 + (dx^2)^2 + (dx^3)^2.$$

Here we set $x^0 = t$ the time coordinate and x^1, x^2 and x^3 are the three space coordinates. It is well-known that, for this metric, the spacelike hypersurfaces ($t =$ constant) are a family of Cauchy hypersurfaces which cover the whole of $\mathbf{R}_1^4$. Thus, $\mathbf{R}_1^4$ is a product space

$$\left(\mathbf{R}_1^4 = \mathbf{R} \times B, \ \bar{g} = -dt^2 \oplus G\right),$$

where (B, G) is a 3-dimensional Euclidean space. Note that not every spacelike hypersurface of $\mathbf{R}_1^4$ is a Cauchy hypersurface (for details see Hawking-Ellis [56, page 119]). Choose a spherical coordinate system (t, r, θ, ϕ) with $x^1 = r \sin \theta \sin \phi$, $x^2 = r \sin \theta \cos \phi$, $x^3 = r \cos \theta$. Then, above metric transforms into

$$ds^2 = -dt^2 + dr^2 + r^2 \left(d\theta^2 + \sin^2 \theta d\phi^2\right),$$

which is singular at $r = 0$ and $\sin \theta = 0$. We, therefore, choose the ranges $0 < r < \infty$, $0 < \theta < \pi$ and $0 < \phi < 2\pi$ for which it is a regular metric. Actually two such charts are needed to cover the full $\mathbf{R}_1^4$. Now we take two null coordinates u and v, with respect to a pseudo-orthonormal basis, such that $u = t + r$ and $w = t - r$ ($u \geq w$). Thus, above metric transforms as

$$ds^2 = -dudw + \frac{1}{4}(u - w)^2 \left(d\theta^2 + \sin^2 \theta d\phi^2\right),$$

where $-\infty < u, w < \infty$. The absence of du^2 and dw^2 in above transformed metric imply that the hypersurfaces $\{u = $ constant$\}$ and $\{w = $ constant$\}$ are null hypersurfaces since $v_{;a}v_{;b}\eta^{ab} = 0 = u_{;a}u_{;b}\eta^{ab}$. Thus, there exists a pair of null hypersurfaces of $\mathbf{R}_1^4$. We say that a leaf of the 2-dimensional screen distribution S is topologically a 2-sphere S^2, with coordinate system $\{\theta, \phi\}$, and is the intersection of the two hypersurfaces $\{u = $ constant $\}$ and $\{v = $ constant $\}$. In relativity, the null coordinates $u(w)$ are called advanced (retarded) time coordinates and are physically related to incoming (outgoing) spherical waves traveling at the speed of light. Suppose $\alpha(u)$ is the null curve representing incoming spherical waves and $\ell(u)$ any of its null tangent vector field. Then, by definition $\mathbf{x}(u, v) = \alpha(u) + v\ell(u)$; $v \in I \subset \mathbf{R}$ is a ruled surface (also called null string where one dimension suppressed) on the null cone of $\mathbf{R}_1^4$. Similarly, one can construct another null string using retarded coordinate w.

Since a ruled surface N has $\dim(Rad\,TN) = 2$ and any globally null manifold M has exactly 1-dimensional null distribution, M can not carry any ruled null surface. However, in the following we show that there is a direct interplay between ruled null surfaces of $\bar{M}$ and 4-dimensional globally null manifolds. Let (M, g) be a class of 4-dimensional globally null manifolds with integrable screen distribution $S(TM)$. Then, it follows from theorem 2.2 that $M = L \times M'$ is a global product manifold, where (M', g') is a 3-dimensional integral manifold of $S(TM)$. We first deal with the geometry of 3-dimensional M' which, by Definition 2.1, is a complete Riemannian manifold. Here we follow Yamabe [111] for the existence of constant curvature metrics on (M', g'). Denote by $\mathcal{M}$ the space of all smooth Riemannian metrics on M' and $\mathcal{M}_1 \subset \mathcal{M}$ the space of metrics satisfying $vol_{g'} = 1$. Define the total scalar curvature or Einstein-Hilbert action $\mathcal{S} : M' \to \mathbf{R}$ by

$$\mathcal{S}(g') = v^{-1/3} \int_{M'} S^{M'} \, dV_{g'},$$

where $S^{M'}$ is the scalar curvature of M', $dV_{g'}$ is the volume element and v is the volume of M'. The critical points of $\mathcal{S}$ are *Einstein metrics*. Moreover, only in dimension 3 these Einstein metrics are of constant scalar curvature. There is a well-known procedure to obtain Einstein manifolds. Following Yamabe [111], suppose $[g']$ is a conformal class of any metric $g' \in \mathcal{M}_1$. Then there exists a metric $g' \in \mathcal{M}_1$ which achieves its infimum $\mu[g'] \equiv \mathcal{S}|_{[g'] \cap \mathcal{M}_1}$. Such metrics are called *Yamabe metrics*. However, there are restrictions on the existence of Yamabe metrics. Denote by $\sigma(M') = sup(\mu[g'])_{\mathcal{C}_1}$, where $\mathcal{C}_1$ is the subset of unit volume Yamabe metrics. If $\sigma(M') \leq 0$, it has been proved (see Besse in [111]) that any Yamabe metric $g_0 \in \mathcal{C}_1$ such that $S^{M'}_{g_0} = \sigma(M')$ is Einstein. Otherwise, this problem still remains open. Under these restrictions, it is reasonable to say that there exists a 4-dimensional globally null manifold (M, g) whose 3-dimensional Riemannian hypersurface (M', g') is an Einstein manifold with a constant curvature, say k, and g' is a Yamabe metric. Indeed, one can construct null manifold (M, g) by gluing the Riemannian metric g' with the degenerate metric g as follows:

$$g = \begin{pmatrix} O_{1,1} & O_{1,3} \\ O_{3,1} & g' \end{pmatrix}$$

Now consider a ruled null surface N of a 4-dimensional spacetime manifolds $\bar{M}$ of constant curvature, such that its Cauchy hypersurface, say $(\Sigma, \Sigma_{\bar{g}})$, is conformal to M', that is, its induced metric $\Sigma_{\bar{g}} \in [g']$. With this construction, it follows that $(M = L \times M', g)$ is tangent to the null cone $\Lambda_{\bar{M}}$ of $\bar{M}$, that is,

$$T_x(M) \cap T_x(\Lambda_{\bar{M}}) = L_x - \{0\}, \tag{8.3.19}$$

for any common point of contact x of the pair $(M, \bar{M})$.

Definition 3.1. *Let $(M, \bar{M})$ be a pair of 4-dimensional globally null and spacetime manifolds, satisfying (8.3.19) and $\alpha : [a, b] \to M$ be a null curve segment. A piecewise smooth variation f, defined by a two parameter function*

$$f : [a, b] \times (-\epsilon, \epsilon) \to \bar{M}$$

is said to be admissible if all the neighboring curves $f_v : [a, b] \to \bar{M}$, given by $f_v(u) = f(u, v)$ are null for each $v \neq 0$ in $(-\epsilon, \epsilon)$ and $f_0(u) = f(u, 0) = \alpha(u)$ is the null curve common to M and $\bar{M}$ for all $a \leq u \leq b$.

We call u-parameter null curves $f(u, v = \text{constant})$, v-parameter null curves $f(u = \text{constant}, v)$ and $\alpha(v)$ the *longitudinal*, the *transversal* and the *base* curves respectively. Thus, given a point $x \in \alpha$, we have an admissible net of neighboring longitudinal and transversal curves, all of them belonging to the null cone $\Lambda_{\bar{M}}$ with the single base null curve $\alpha(u)$ common to M. Let $\{\mathcal{N}_{\alpha(x)}\}$ denote the set of all nets for all points $x \in \alpha$. Now, consider a maximum sequence of globally null manifolds $\{M_i, g_i\}$ such that each M_i satisfies the equation (8.3.19) and $\{g'_i\}$ is a maximum sequence of unit volume Yamabe metrics on each M'_i. In this way we cover the surface of the cone $\Lambda_{\bar{M}}$ with a global net of all its longitudinal and transversal null curves. Let $\ell(u)$ be a null tangent vector field of $\alpha(u)$. Then, as $\alpha(u)$ moves on the null cone $\Lambda_{\bar{M}}$, it will generate a ruled null 2-surface N, which establishes a link between the pair $(M, \bar{M})$ with its common u-parameter curve $\alpha(u)$. Observe that, based on proposition 3.3, it is possible to consider a maximal sequence of special parameters $\{p^i\}$ such that each longitudinal curve is a null geodesic. With this possibility, all longitudinal curves of the global net are geodesics which means that f is a 1-parameter family of null geodesics. Consequently, N is a null geodesic string in the sense of Schild [104].

8.4 Scalar curvature and null warped products

Consider a globally null manifold (N, g_N) and a complete Riemannian manifold (F, g_F) of dimensions n and m respectively. Using theorem 2.3, construct an $(n + m)$-dimensional globally null warped product manifold $(M = N \times_f F, g)$, where f is a smooth function on N. Then, we have the following result (proof is similar to the proof of theorem 2.2):

Theorem 4.1. *Let $(M = N \times_f F, g, S(TM))$ be an $(n + m)$-dimensional globally null warped product manifold, where f is a smooth function on N and $S(TM)$ a chosen screen distribution. Then, the following assertions are equivalent:*

(a) *The screen distribution $S(TM)$ is integrable.*

(b) *$M = L \times M'$ is a global null product manifold, where L is 1-dimensional integral manifold of the global null curve C in M and (M', g') is a complete Riemannian hypersurface of M which is a triple warped product*

$$(M = L \times B \times_f F, g), \quad M' = (B \times_f F, g') \qquad (8.4.1)$$

where (B, g_B) is a complete Riemannian hypersurface of $N = L \times B$.

We assume that there exists a metric (Levi-Civita) connection ∇ on the above triple warped product $(M = L \times B \times_f F, g)$, with respect to its degenerate metric tensor g. As in the Riemannian case, the Ricci tensor of M is given by

$$Ric(X, Y) = trace\{Z \to R(X, Z)Y\}, \quad \forall X, Y, Z \in \Gamma(TM).$$

where $R(X, Z)Y$ is the curvature tensor of M. In terms of a quasi-orthonormal Frenet frame $\{\xi, W_1, \ldots, W_{n-1}\}$ along a null curve C (see equation (8.3.2)), the degenerate metric g and the Ricci tensor are expressed by

$$g(X, Y) = \sum_{a=1}^{n-1} g(X, W_a) g(Y, W_a)$$

$$Ric(X, Y) = \sum_{a=1}^{n-1} g(R(W_a, X) W_a, Y) = Ric'(X', Y'). \qquad (8.4.2)$$

Using (8.4.2) and theorem 4.1, we have the following important result:

Proposition 4.1. Let $(M = L \times M' = B \times_f F, g)$ be a triple warped product globally null manifold, with M' its complete Riemannian hypersurface satisfying Theorem 4.1. Then, the metric g and the Ricci tensor restricted to M' can be determined from a set of data specified entirely on M'. Also, M and M' have same Ricci tensors. Moreover, the geometry of M, related to the Ricci tensor of M reduces to the Riemannian geometry of its warped product hypersurface $M' = (B \times_f F, g')$, as defined by (8.4.1).

Consequently, (M', g') is an invariant hypersurface of (M, g). Thus, one can essentially do all the analysis, related to the metric tensor g and the Ricci tensor of M, on the complete Riemannian hypersurface M' of M. Moreover, we have

(1) The null leaves $N \times q$, $q \in F$, of warped product M, can be induced to the spacelike leaves $B \times q$, and are totally geodesic in M.

(2) The fibers $(p_o, p) \times F$, $p_o \in L$ and $p \in B$ can be induced to the spacelike fibers $p \times F$, and are totally umbilical in M.

(3) For each $(p, q) \in M'$, the induced leaf $B \times q$ and the induced fiber $p \times F$ are orthogonal at (p, q).

(4) The gradient of the lift $h \circ \pi$ of a smooth function $h \in N$ is the lift to M of the gradient of h on its induced Riemannian manifold B.

In general, for a covariant tensor $T \in N$, its lift $\bar{T} \in M$ is the pullback $\pi^*(T)$ under the map $\pi : M \to B \subset N$. This is why, even if the metric g_N of N is degenerate, all tensors and geometric objects and their pullback are things with respect to the induced Riemannian metric g_B of B. The vectors tangent to leaves and fibers are called *horizontal* and *vertical* respectively. The lift to M of the Hessian of a smooth function f on N, denoted by H^f, agrees with the Hessian of the lift $f \circ \pi$ on the horizontal vectors of B. We denote Ric^B for the pullback by π of Ric' and similarly for Ric^F.

Proposition 4.2. Let $M = L \times B \times_f F$ be an $(n+m)$-dimensional triple warped product manifold with $\dim(F) = m > 1$. Then

(1) $Ric(\xi, \xi) = Ric(\xi, X) = Ric(X, U) = 0$, $\xi \in L$

(2) $Ric(X, Y) = Ric'(X, Y) = Ric^B(X, Y) - \frac{m}{f} H^f(X, Y)$

(3) $Ric(U, V) = Ric^F(U, V) - <U, V> \left\{ \frac{\triangle f}{f} + (m-1) \frac{<\nabla f, \nabla f>}{f^2} \right\}$,

where $\triangle f = trace(H^f)$ is the Laplacian of f, $\nabla f = grad(f)$, X, Y horizontal and U, V vertical vector fields.

Proof. Use proposition 4.1 and follow corollary 7.43 in [82].

We use the following identifications of $T_x(M)$ for any $x = (p_0, p, q) \in M$.

$$T_x(L \times B \times_f F) \cong T_x(L \times B \times F) \cong T_{p_0}(L) \times T_p(B) \times T_q(F)$$

$$T_x(L \times B \times_f F) \stackrel{projected}{\longrightarrow} T_{(p, q)}(B \times_f F) \cong T_{(p, q)}(B \times F)$$

A Frenet frame $\{\xi, W_1, \ldots, W_{n-1}\}$ on $T_{(p_0, p)} N$ (see equation (8.3.2)) is identified to an orthonormal basis $\{W_a\}, (a = 1, \ldots, n-1)$ on $T_p B$. Any horizontal vector $X_{(p_0, p, q)} \in M$ is identified to a horizontal vector $\bar{X}_{(p, q)} = (X_p, 0_q) \in B$. Similar notations follow for vertical vectors and tensors. We denote S^B the pullback by π of the scalar curvature of B and similar for S^F. For the degenerate metric g, at a point $x = (p_0, p, q) \in M$, we have

$$g_x \stackrel{projected}{\longrightarrow} g'_{(p, q)} = (g_p, g_q)$$

$$\bar{g}_{B(p, q)} = (g_p, 0_q), \quad g_{F(p, q)} = (0_p, g_q)$$

where g', g_B and g_F are Riemannian metrics on M', B and F respectively.

Proposition 4.3. *Suppose S is the scalar curvature of a triple warped product manifold $M = L \times B \times_f F$, with $\dim(F) = m > 1$. Then*

$$S = S' = S^B + \frac{S^F}{f^2} - 2m \frac{\triangle f}{f} - m(m-1) \frac{< \nabla f, \nabla f >}{f^2}, \qquad (8.4.3)$$

where S' is the induced scalar curvature of $M' = (B \times_f F, g')$.

Proof. Let $\{\xi; W_a\}$ be a pseudo-orthonormal basis for $T_{(p_0, q)}(L \times B)$ so that $\{W_a\}$ is an orthonormal basis for $T_p B$. Then, by isomorphism, $\{\bar{W}_a = (W_a, 0_q)\}$ is an orthonormal set in $T_{(p, q)}(B \times_f F)$. Choose a set $\{W_i\}$ of m vectors on $T_q F$ such that $\{\bar{\xi}; \bar{W}_a; \bar{W}_i\}$ forms a pseudo-orthonormal basis for $T_{(p_0, p, q)} M$. Thus, $\{\bar{W}_a; \bar{W}_i\}$ is an orthonormal basis for $T_{(p, q)}$. Since

$$g_F(\bar{W}_i, \bar{W}_i) = f^2(W_i, W_i) = g_F(f W_i, f W_i) = 1$$

we conclude that $\{f W_i\}$ is an orthonormal basis for $T_q F$. Using Proposition 3, for each a and each i, we get

$$Ric(\bar{W}_a, \bar{W}_a) = Ric'(\bar{W}_a, \bar{W}_a) = Ric^B(\bar{W}_a, \bar{W}_a) - \frac{m}{f} H^f(\bar{W}_a, \bar{W}_a)$$

$$Ric(\bar{W}_i, \bar{W}_i) = Ric^F(\bar{W}_i, \bar{W}_i) - f \left[\triangle f + (m-1) \frac{< \nabla f, \nabla f >}{f} \right]$$

Hence, using $g(\xi, \xi) = 0$, $g(W_a, W_a) = g(W_i, W_i) = 1$, we get

$$S(p_0, p, q) \xrightarrow{projected} S'(p, q) = R_{\alpha\alpha} \quad (2 \le \alpha \le n+m)$$

$$= Ric(\bar{W}_a, \bar{W}_a) + Ric(\bar{W}_i, \bar{W}_i)$$

$$= S^B + \frac{S^F}{f^2} - 2m\frac{\triangle f}{f} - m(m-1)\frac{<\nabla f, \nabla f>}{f^2}.$$

For physical reason, we restrict to $\dim(M) = n = 4$ as this case has an interplay with some known exact solutions of the static vacuum Einstein equations and the event horizon in general relativity. We set $\dim(B) = 1$ and $\dim(F) = 2$.

Theorem 4.2 ([24]). *Let $M = (L \times B \times_f F, g)$ be a 4-dimensional globally null warped product manifold, $B = (a, b)$ an open connected subset of real line with positive definite metric dr^2 and $-\infty \le a < b \le +\infty$ and the fiber space F be of constant scalar curvature $c \ne 0$. Then, g admits the following warping functions $f(r)$ for which M has a constant scalar curvature k.*

(i) $k > 0$, $f(r) = \sqrt{\frac{3c}{k}} \left[\tan^2(\pm(\frac{k}{6})^{\frac{1}{2}} r + c_1) + 1 \right]^{-\frac{1}{2}}$, $c > 0$,

(ii) $k = 0$, $f(r) = \pm \left(\sqrt{\frac{c}{2}} \right) r + c_1$, $c > 0$,

(iii) $k < 0$, $f(r) = c_1 \exp\left(\sqrt{\frac{-k}{6}} r \right) - \frac{3c}{4c_1 k} \exp\left(-\sqrt{\frac{-k}{6}} r \right)$,

where c_1 is a constant such that $f(r)$ is real and positive.

Proof. Let $f(r) - u^{\frac{2}{3}}$. Then, $\triangle f = f''$ and $< \nabla f, \nabla f >= (f')^2$. Using this with $S^B = 0$, $S^F = c$ and $S = k$ in (8.4.3), we obtain

$$u'' + \frac{3}{8} k u - \frac{3}{8} c u^{-\frac{1}{3}} = 0. \tag{8.4.4}$$

Let $u' = y$ so that $\frac{dy}{dr} = u''$. Using this in (8.4.4), separating variables and then integrating both sides, we obtain

$$y = \pm (3/8)^{\frac{1}{2}} u \sqrt{3c\,u^{-\frac{4}{3}} - k} = \frac{du}{dr}$$

Therefore,

$$\frac{du}{u \sqrt{3c\,u^{-\frac{4}{3}} - k}} = \pm (3/8)^{\frac{1}{2}} dr.$$

Following are three cases of the integral of above equation:

$$k > 0, \quad v^2 = k \tan^2\left(\pm \left(\frac{k}{6}\right)^{\frac{1}{2}} r + c_1 \right),$$

$$k = 0, \quad f(r) = \pm \left(\sqrt{\frac{c}{2}} \right) r + c_1,$$

$$k < 0, \quad \ln \left| \frac{v - \sqrt{-k}}{v + \sqrt{-k}} \right| = \pm \left(-\frac{k}{6} \right)^{\frac{1}{2}} r,$$

where we set $v^2 = 3c\,u^{-\frac{4}{3}} - k$. From above three equations the results of this theorem follow easily for the case of Riemannian warped product manifold (M', g'). To complete the proof, we now show how to glue g' with the degenerate metric g of M. It follows from the proposition 4.3 that the scalar curvatures of M' and M are same. The warping function $f_p \in B$ can be glued with the warping function $f_{(p_0, p)} = (0_{p_0}, f_p) \in L \times B$. Then, the Riemannian metric g' can be glued with the degenerate metric g, of M, as follows:

$$ g = \begin{pmatrix} O_{1,1} & O_{1,3} \\ O_{3,1} & g' \end{pmatrix} $$

where $g'(X, Y) = g_B(\pi_\star X, \pi_\star Y) + (f \circ \pi)^2 g_F(\eta_\star X, \eta_\star Y)$.

Corollary. *If $c = 0$, then following are the warping functions $f(r)$ for which M has a constant scalar curvature k.*

(i) $k > 0$, $\quad f(r) = \left[c_1 \cos\left(\sqrt{\frac{3k}{8}}\, r \right) + c_2 \sin\left(\sqrt{\frac{3k}{8}}\, r \right) \right]^{\frac{2}{3}}$,

(ii) $k = 0$, $\quad f(r) = (c_1 r + c_2)^{\frac{2}{3}}$,

(iii) $k < 0$, $\quad f(r) = \left[c_1 \exp\left(\sqrt{\frac{-3k}{8}}\, r \right) + c_2 \exp\left(-\sqrt{\frac{-3k}{8}}\, r \right) \right]^{\frac{2}{3}}$,

where c_1 and c_2 are constants such that $f(r)$ is real and positive.

Physical Model. Let $M = (L \times M', g)$ be a 4-dimensional globally null manifold, with (M', g') its complete spacelike hypersurface. Also, let $(\tilde{M}, \tilde{g})$ be a 4-dimensional globally hyperbolic spacetime manifold of general relativity. By definition, $\tilde{M}$ has a complete spacelike hypersurface H (called *Cauchy surface*) such that $\tilde{M} = R \times H$. In the following we show, by means of a physical example, that H is a warped product manifold of case 1, and the set $\{M, M', \tilde{M}\}$ of these three manifolds has the following interplay.

$$ (M, g) \supset (M' = B \times_f F, g') \subset (\tilde{M} = R \times H, \tilde{g}), \tag{8.4.5} $$

where $(H = B \times_{\tilde{f}} F, g_H)$ and $\tilde{f}$ is a warping function on $R \times B$.

Example 6. Let $(\tilde{M}, \tilde{g})$ be the Schwarzschild spacetime with the metric

$$ \tilde{g} = -A(r)\, dt^2 + A^{-1}(r)\, dr^2 + r^2\, d\Omega_{s^2}^2, \quad A(r) = 1 - 2mr^{-1} > 0, $$

where S^2 is totally geodesic 2-sphere of radius $2m$ and m is positive mass. This metric represents the most important non-trivial solution of the static vacuum Einstein field equations. It is well-known that $\tilde{M}$ is a globally hyperbolic manifold [10]. To relate this with the equation (8.4.5), we consider the following conformal deformation metric $\bar{g}_H$ defined by

$$ \bar{g}_H = A(r)\, g_H = dr^2 + A(r)\, r^2\, d\Omega_{s^2}^2 $$

If we set B a 1-dimensional space with metric dr^2 and S^2 a 2-dimensional fiber space F of $\tilde{M}$, then using (8.4.5) we conclude that the Schwartzchild spacetime $\tilde{M}$ has a complete Cauchy hypersurface

$$(\bar{H} = B \times_{\bar{f}} F, \ \bar{g}_H), \quad \bar{f} = \tilde{f} \sqrt{A(r)},$$

and it has an interplay with a 4-dimensional globally null manifold M. It is well-known that static solutions of spacetimes are closely connected with an open Riemannian 3-manifold containing a 2-sphere, occuring at the event horizon or the boundary of a black hole in general relativity. This physical relation is apparent in above example and, more generally, in many solutions of the static vacuum equations of asymptotically flat spacetimes, which have 2-spheres near infinity. Moreover, we have further demonstrated, through the equation (8.4.5), that 2-sphere can act as a common link between the three manifolds M, M' and $\tilde{M}$, relating the geometries of globally null and the globally hyperbolic manifolds. Finally, to relate this example with theorem 4.2, consider the case $k = 0$ (others are similar) so that

$$f(r) = \pm \left(\sqrt{\frac{c}{2}} \right) r + c_1.$$

In this example, $F = S^2$ whose scalar curvature is $\frac{1}{4m^2}$ and

$$\tilde{f} = r \sqrt{1 - 2mr^{-1}}.$$

Matching $f = \tilde{f}$, and $c = \frac{1}{4m^2}$, we obtain

$$(1 - 8m^2) r^2 + (16m^3 \pm 4\sqrt{2}mc_1) r + 8mc_1^2 = 0$$

where c_1 is such that r has real solutions from above equation.

8.5   Null geodesics and photon surfaces

Let $C(p)$ be a null curve in a 4-dimensional Lorentzian manifold (M_1^4, g), where $p \in I \subset \mathbf{R}$ is a special parameter, $\{\xi, N, W_1, W_2\}$ is a pseudo-orthonormal Frenet frame along $C(p)$, ξ and N are null vectors such that $g(\xi, N) = 1$, $C = Span\{\xi\}$, and W_1, W_2 are unit spacelike vectors. Recall that if N moves along C, then, it generates a ruled surface given by the parameterization $((I \times \mathbf{R}), f)$ where $f : I \times \mathbf{R} \to M_1^4$ is defined by

$$(p, u) \to f(p, u) = C(p) + uN(p), \quad u \in I_u \subset \mathbf{R}.$$

Above ruled surface is called a *null scroll* which we denote by $\mathcal{S}_c$. It is clear by the above defining equation that the null scroll $\mathcal{S}_c$ is a timelike ruled surface in M_1^4 (see a paper by Tugut and Hacisalihoğlu [108] on timelike ruled surfaces). In particular, if $\mathcal{S}_c$ is ruled by null geodesics, then, there is an important link of the mathematical concept of null scrolls with the literature of physics as follows:

In general relativity, null geodesics are interpreted as the world lines of photons and a timelike null scroll is called a *photon surface* if each null geodesic tangent to $\mathcal{S}_c$ remains within $\mathcal{S}_c$ (for some parameter interval). Here we assume special parameter for each null geodesic. For the notion of 2-dimensional photon surfaces and its physical use, we refer [42] and some more referred therein. The concept of null scrolls and photon surfaces can be generalized for the cases of timelike submanifolds of dimension $k \geq 2$ of M_1^n (see [21] for the case $k = \dim M_1^n - 1$). The objective of this section is to present some physical examples of null scrolls and photon surfaces. First we need following preliminaries to understand examples:

We shall state and prove a main result (needed for constructing examples) in any finite dimensional Lorentzian manifold (M, g). For physical reasons, M_1^4 should be 4-dimensional, time-oriented, and connected; mathematically , however, this assumption is not needed. The results are non-trivial only if M is at least 3-dimensional. Each null scroll will be fully immersed in M as opposed to an embedded submanifold, i.e., we allow for self-intersections. This is necessary as our construction methods for photon surfaces may yield ruled surfaces with self-intersections. A vector field on $\mathcal{S}_c$ assigns to each point $x \in \mathcal{S}_c$ a vector in the tangent space $T_x \mathcal{S}_c$ whereas a vector field along $\mathcal{S}_c$ assigns to each $x \in \mathcal{S}_c$ a vector field in the tangent space $T_{i(x)} M$ where i denotes the immersion $\mathcal{S}_c \to M$. Given a timelike vector field $\mathbf{n}$ on $\mathcal{S}_c$, i.e., $g(\mathbf{n}, \mathbf{n}) = -1$, the condition

$$g(\mathbf{n}, \mathbf{v}) = 0 \quad \text{and} \quad g(\mathbf{v}, \mathbf{v}) = 1 \tag{8.5.1}$$

defines a spacelike vector field $\mathbf{v}$ on $\mathcal{S}_c$ uniquely up to sign. Since $\mathbf{n}$ and $\mathbf{v}$ are orthonormal, the relation

$$L^{\pm} = \frac{\mathbf{n} \pm \mathbf{v}}{\sqrt{2}} \tag{8.5.2}$$

defines two null vector fields L^+ and L^- on $\mathcal{S}_c$ such that at each point x of $\mathcal{S}_c$, the vectors L_x^+ and L_x^- span the two different null lines tangent to $\mathcal{S}_c$ and

$$g(L^+, L^-) = -1. \tag{8.5.3}$$

Since for any choice of $\mathbf{n}$, $\mathbf{v}$ is unique up to sign, so L^+ and L^- are unique up to interchanging. Suppose we replace $\mathbf{n}$ by another vector field $\bar{\mathbf{n}}$ on $\mathcal{S}_c$ such that $g(\bar{\mathbf{n}}, \bar{\mathbf{n}}) = -1$ (and take corresponding vector field $\bar{\mathbf{v}}$), then, following transformation equations will hold:

$$L^+ \mapsto A L^+ \quad \text{and} \quad L^- \mapsto A^{-1} L^-, \tag{8.5.4}$$

where A is a now where vanishing scalar function on $\mathcal{S}_c$.

Proposition 5.1 ([42]). *A point x in a Lorentzian manifold (M, g) admits a neighborhood that can be foliated into timelike photon 2-surfaces if and only if on some neighborhood of x there are two linearly independent null geodesic vector fields L^+ and L^- such that the Lie bracket $[L^+, L^-]$ is a linear combination of L^+ and L^-. In this case, the photon 2-surfaces are the integral manifolds, say $\mathcal{S}_c$, of the 2-surfaces spanned by L^+ and L^-.*

Proof. L^+ and L^- being linearly independent they generate a timelike 2-surface S_c at each point x of M. Then, by the well-known Frobenius theorem these 2-surfaces admit local integral manifolds if and only if the Lie bracket $[L^+, L^-]$ is a linear combination of L^+ and L^-. Since, by hypothesis, L^+ and L^- are geodesic, it makes sure (by definition) that these integral manifolds, say S_c, are timelike photon 2-surfaces, which proves the if part. To prove the only if part, one just has to verify that the null geodesic vector fields L^+ and L^-, given on each leave of the respective foliations up to transformation (8.5.4), can be chosen such that they make up to two smooth vector fields L^+ and L^- on some neighborhood of x, which completes the proof.

To relate above proposition with the subject of this section, suppose C is a null curve of a Lorentzian manifold (M_1^{m+1}, g). Consider an orthonormal basis $\{\partial_t, \partial_{x^1}, \ldots, \partial_{x^{m+1}}\}$, with respect to a local coordinates system $(t, x^1, \ldots, x^{m+1})$ of $T_x M$ for any point $x \in M$. Then, as per case 1 in section 5 of chapter 1, there exists a quasi-orthonormal basis

$$B = \{\xi, N, \partial_{x^2}, \ldots, \partial_{x^{m+1}}\}$$

such that ξ and N are real null vectors satisfying

$$\xi = \frac{1}{\sqrt{2}}\{\partial_t + \partial_{x^1}\}, \quad N = \frac{1}{\sqrt{2}}\{\partial_t - \partial_{x^1}\}$$

$$g(\xi, \xi) = g(N, N) = 0, \quad g(\xi, N) = -1.$$

We adjust such that $C = Span\{\xi\}$ and N is the unique null transversal vector field of C with respect to ξ. Now we set $L^+ = \xi$ and $L^- = N$ and consider a distinguished parameter for which C is a null geodesic curve of M, with ξ and N null geodesic vector fields, which are obviously linearly independent. Thus, the first part of the proposition 5.1 is satisfied.

On the existence of timelike photon surfaces in a Lorentzian manifold (M, g), one can characterize such surfaces in terms of their second fundamental form **B** as follows: Let Σ be a 2-dimensional timelike submanifold of (M, g). Then, the orthogonal projections

$$P^\perp : T_x M \to (T_x \Sigma)^\perp,$$

define a tensor field $P^\perp$ along Σ that maps vector fields along Σ to itself. Then, using (8.5.1) one can show that $P^\perp$ takes the form

$$P^\perp(Y) = Y - g(v, Y)v + g(\mathbf{n}, Y)\mathbf{n}, \quad \forall v \in \Gamma(\Sigma). \tag{8.5.5}$$

Using (8.5.1), (8.5.5) takes the form

$$P^\perp(Y) = Y + g(L^-, Y)L^+ + g(L^+, Y)L^-.$$

Since $L^\pm$ are null, above implies

$$P^\perp(\nabla_{L^\pm} L^\pm) = \nabla_{L^\pm} L^\pm + g(L^\mp, \nabla_{L^\pm} L^\pm)L_\pm. \tag{8.5.6}$$

Based on above, the second fundamental form **B** can be written as (see [82])

$$\mathbf{B}(u, w) = P^\perp(\nabla_u w), \quad \forall u, w \in \Gamma(\Sigma), \tag{8.5.7}$$

Definition 5.1 ([82, page 106]). *A non-null submanifold Σ of a semi-Riemannian manifold (M, g) is called totally umbilical if there is a normal vector field Z along Σ such that*

$$\mathbf{B}(u, w) = g(u, w)Z, \quad \forall u, w \in \Gamma(\Sigma). \tag{8.5.8}$$

If $Z = 0$, then, Σ is totally geodesic. It is easy to see that the property of being totally umbilical is conformally invariant whereas the property of being totally geodesic is not. Following is a characterization result for timelike photon 2-surfaces:

Proposition 5.2 ([42]). *A 2-dimensional timelike submanifold Σ of a Lorentzian manifold (M, g) is a photon 2-surface if and only if Σ is totally umbilical.*

Proof. Let $L^\pm$ be the two null vector fields on Σ, unique up to transformations of the form (8.5.4), which are normalized according to (8.5.2). Suppose Σ is totally umbilical. Then, (8.5.8) requires that $\mathbf{B}(L^\pm, L^\pm) = 0$. Using (8.5.6) and (8.5.7) we conclude that L^+ and L^- are geodesics, so Σ is a photon 2-surface.

Conversely, assume Σ is a photon 2-surface. Let $u = aL^+ + bL^-$ and $w = cL^+ + dL^-$ be any two vector fields on Σ, where a, b, c, d are scalar functions on Σ. Since $\nabla_{L^+}L^+$ and $\nabla_{L^-}L^-$ are tangents to Σ, the second fundamental form (8.5.7) reduces to

$$\mathbf{B}(u, w) = ad\, P^\perp(\nabla_{L^+}L^-) + bc\, P^\perp(\nabla_{L^-}L^+).$$

The fact that $L^\pm$ and $[L^+, L^-]$ must be tangent to Σ implies

$$P^\perp(\nabla_{L^+}L^-) = P^\perp(\nabla_{L^-}L^+) =: -Z.$$

On the other hand, $g(u, w) = -ad - bc$. Thus, we conclude that (8.5.8) holds, which completes the proof.

Now we show that there are specific physical examples satisfying the second part, that is, $[L^+, L^-]$ is a linear combination of L^+ and L^- which justifies the existence of timelike photon 2-surfaces.

Physical examples. Consider a product manifold (M_1^{m+2}, g) given by

$$(M_1^{m+2} = B \times F, \quad g = B_g \oplus F_g).$$

Let the distance element for a coordinates system $(t, x^1, \cdots, x^{m+2})$ be

$$ds^2 = -e^\lambda\, dt^2 + e^\mu(dx^1)^2 \oplus d\Sigma^2,$$

where λ and μ are functions of t and x^1 alone, (B, B_g) is a 2-dimensional Lorentzian submanifold and (F, F_g) its Riemannian submanifold of codimension 2.

Let $\mathbf{E} = \{e_0, e_1, \cdots, e_{m+1}\}$ be an orthonormal basis of $T_x M$, such that e_0 is timelike and all others are spacelike unit vectors. Let us call $\mathcal{G} = \{\lambda, \mu, d\Sigma^2\}$ the generating set for a family of Lorentzian manifolds with prescribed values of its element in above distance element ds^2 of M.

Example 7. Suppose $\mathcal{G} = \{0, 0, d\Sigma^2\}$ such that F is a Euclidean space. Then, $(M = \mathbf{R}_1^{m+2}, \delta_{ij})$ is a Minkowski spacetime. A foliation of the Minkowski spacetime $\mathbf{R}_1^{m+2}$ into timelike photon 2-surfaces can be easily found using proposition 5.1 and setting $\xi = L^+$ and $N = L^-$, the two null vector fields associated to a null curve C of $\mathbf{R}_1^{m+2}$, such that

$$\xi = \frac{1}{\sqrt{2}} \{\partial_t + \partial_{x^1}\}, \quad N = \frac{1}{\sqrt{2}} \{\partial_t - \partial_{x^1}\}$$

and have vanishing Lie bracket, i.e., $[L^+, L^-] = 0$. Moreover, for all basis unit vector fields ∂_i with $(0 \leq i \leq m+1)$ we obtain from the Minkowski metric

$$g(\nabla_{L^\pm} L^\pm, \partial_i) = \frac{1}{2} \partial_i \left(g(L^\pm, L^\pm)\right) = 0,$$

so $\nabla_{L^\pm} L^\pm = 0$. Thus, ξ and N are null geodesic vector fields. Here ∇ is the Levi-Civita connection satisfying the identity [82, page 157]:

$$\begin{aligned} 2g(\nabla_X Y, Z) &= X(g(Y, Z)) + Y(g(X, Z)) - Z(g(X, Y)) \\ &+ g([X, Y], Z) + g([Z, X], Y) - g([Y, Z], X) \end{aligned}$$

for any vector fields X, Y and Z on M. Then, by the proposition 5.1, the surfaces $\{x^i = \text{constant}\}$ for every $i \in \{2, \ldots, m+1\}$ are timelike photon 2-surfaces.

Note that a photon 2-surface in $(m+1)$-dimensional Minkowski spacetime is, in particular, a surface in $(m+1)$-dimensional affine space that admits two different rulings by straight lines. It is known that in 3-dimensional affine space the only such surfaces are planes, rotational hyperboloids and hyperbolic paraboloids.

Conformally flat spacetimes. Since a photon 2-surface is invariant if the given spacetime metric is conformally transformed, the photon 2-surfaces in a conformally flat spacetime are locally the same as in the Minkowski spacetime.

To determine all timelike photon 2-surfaces in a Minkowski space $\mathbf{R}_1^{m+2}$, we first refer O'Neill [82, page 117] who proved that a connected, complete non-null hypersurface in a vector space with non-degenerate scalar product of arbitrary signature is totally umbilic if and only if it is either a hyperplane or a hyperquadratic. By proposition 5.1, this result implies that the connected and complete timelike photon surfaces in $\mathbf{R}_1^3$ Minkowski spacetime are the timelike planes and the timelike quadratics.

By adding more spatial dimensions, one can extend this construction for all timelike photon 2-surfaces in $(m+2)$-dimensional Minkowski spacetime, under the required condition that its totally umbilical timelike 2-surface must be completely contained in a 3-dimensional affine space. Using above, all timelike photon

2-surfaces in conformally flat spacetimes can be constructed.

Example 8. A spacetime is called *Gödel Spacetime* whose metric is

$$ds^2 = -(dt)^2 - 2\, e^{Ax}\, dt dx^2 + (dx)^2 + \frac{1}{2}\, e^{2Ax}(dy)^2 + (dz)^2,$$

where A is a non-zero constant and (t, x, y, z) are local coordinates. It is a rotating dust solution of Einstein's field equations with cosmological constant [56]. A foliation of this spacetime into timelike photon 2-surfaces can be found with the help of Proposition 5.1 Indeed, It is obvious from above metric that the vector fields

$$L^\pm = \frac{\partial_z \pm \partial_t}{\sqrt{2}}$$

are null and have vanishing Lie brackets, $[L^+, L^-] = 0$. Moreover, for all basis vector fields ∂_i with $i = t, x, y, z$ we read from above metric that

$$g(\nabla_{L^\pm} L^\pm,\ \partial_i) = L^\pm g(L^\pm,\ \partial_i - \frac{1}{2}\partial_i\, (g(L^\pm,\ L^\pm)) = 0,$$

so $\nabla_{L^\pm} L^\pm = 0$. By proposition 5.1, the surfaces $\{x = \text{constant}\}$, $\{y = \text{constant}\}$ are timelike photon surfaces.

Example 9. Consider a class of 3-dimensional spacetimes with metric (see details in the introduction)

$$ds^2 = -e^{2\lambda(r)}dt^2 + e^{-2\lambda(r)}dr^2 + r^r d\theta^2.$$

which includes, as particular examples, the restriction to the equatorial place of the Schwarzschild spacetime,

$$e^{2\lambda(r)} = 1 - \frac{2m}{r}, \tag{8.5.9}$$

and of the Reissner-Nordström spacetime,

$$e^{2\lambda(r)} = 1 - \frac{2m}{r} + \frac{e^2}{r^2}.$$

Since both ∂_t and ∂_θ are hypersurface orthogonal Killing vector fields, we can consider two vector fields

$$L^\pm = A(r)(\partial_t + X(r)\partial_r) \pm \frac{1}{r}\partial_\theta,$$

where $A(r)$ and $X(r)$ are functions of r to be specified later in this example. Then,

$$[L^+, L^-] = \frac{A(r)X(r)}{r}(L^+ - L^-).$$

Therefore, L^+ and L^- are surface-forming for any choice of A and X. Choose A and X such that

$$X^2 = e^{2\lambda}(e^{2\lambda} - A^2) \tag{8.5.10}$$

so that L^+ and L^- are null. With this choice, the 2-dimensional integral surfaces generated by $\{L^+, L^-\}$ are timelike. Since each of these integral surfaces is obviously invariant under the flow of ∂_θ, each can be interpreted as the world sheet of a circular path whose radius changes with time. To complete our example, we add the condition that $L^\pm$ are geodesics, i.e., $\nabla_{L^\pm} L^\pm = f_\pm L^\pm$, for some functions $f_\pm$. This holds if and only if

$$g(\nabla_{L^\pm} L^\pm, \partial_i) = f_\pm g(L^\pm, \partial_i) \qquad (8.5.11)$$

for $i = t, r, \theta$, which we assume. Thus, we have

$$g(\nabla_{L^\pm} L^\pm, \partial_i) = L^\pm g(L^\pm, \partial_i) - g(L^\pm, [L^\pm, \partial_i]). \qquad (8.5.12)$$

With (8.5.12) it is easy to evaluate (8.5.11). For $i = \theta$ and $i = t$ we obtain

$$A(r)X(r) = f_\pm r \qquad (8.5.13)$$

$$X(r)A(r)A'(r) - 2X(r)A(r)^2\lambda'(r) = f_\pm A(r), \qquad (8.5.14)$$

respectively. By eliminating $f_\pm$ from above two equations we obtain an ordinary first-order differential equation for A whose solution yields

$$A(r) = \frac{r}{c} e^{2\lambda(r)}, \qquad (8.5.15)$$

where, without any loss of generality, we assume that the constant of integration c is positive. Then, for any choice of $c(> 0)$ substituting (8.5.15) in (8.5.10) we obtain a positive and a negative solution B on respective part of the spacetime where

$$r^2 e^{-2\lambda(r)} > c^2. \qquad (8.5.16)$$

If (8.5.16) holds on the entire spacetime under consideration, then, this construction provides two families of photon 2-surfaces, one corresponding to the positive and the negative solutions for X, such that the members of either family foliate the spacetime. On the other hand, if (8.5.16) holds on a proper part U of the spacetime, then, it is not difficult to verify that all members of both families either meet at the boundary ∂U tangentially or asymptotically approach the boundary ∂U. In the first case each member of the first family can be glued together with a member of the second family at the boundary. Consequently, there is actually only one family of photon 2-surfaces which covers U twice.

Consider the exterior Schwarzschild spacetime (Reissner-Nordström case is left as exercise), by specifying λ as given in (8.5.9) with the region $r > 2m$. Following are three cases:

(i) $c < \sqrt{27}\, m$: (8.5.16) is satisfied for all $r > 2m$. Here c is associated with two different families of photon 2-surfaces, each one is generated by null geodesics that extend from the horizon at $r = 2m$ to infinity.

(ii) $c = \sqrt{27}\, m$: (8.5.16) is satisfied for all $r > 2m$, except at $r = 3m$ where the left-hand side of (8.5.16) is equal to the right-hand side. Thus, although c is associated with two different families of photon 2-surfaces, but, the surface $r = 3m$ belongs to both families. Therefore, all members of both families asymptotically approach this particular photon 2-surface at $r = 3m$ either for $t \to \infty$ or for $t \to \infty$.

(iii) $c > \sqrt{27}\, m$: (8.5.16) is satisfied for all $r > 2m$, except some interval around $r = 3m$. There is only one family of photon 2-surfaces associated with c which covers the allowed region twice, whose each member is ruled by null geodesics which either come from infinity, reach a minimum radius and go to infinity, or come from the horizon, reach a maximal radius and go down to the horizon.

Note. A graph of each case (discussed above) can be seen in [42]. Readers are invited to construct timelike photon 2-surfaces of *Robertson-Walker spacetimes* whose metric is given by the equations (8.1.1) and (8.1.2).

8.6 Brief notes on research papers

This last section of the book is intended for advanced level readers who have good knowledge of *Global Lorentzian Geometry* as presented in standard books such as [10, 56, 82]. Our objective is to provide sufficient background information for understanding a given research problem under discussion, state the main results and cite the references for details on their proofs and other side results. The choice of the material, in this section, is focused on the applications of null geodesics. We review four papers of Gutiérrez-Palomo-Romero [49, 50, 51, 52] on *conjugate points along null geodesics* and suggest some other related papers for reading, including a paper of Minguzzi-Sánches [81], two papers of Sánchez [102, 103], four papers and a book of Perlick [90, 91, 92, 93, 94] on *relativistic Fermat's principle and light rays* and a paper of Garcia-Rio et al. [44]. All these papers have several other papers (cited therein) which may be of interest to the readers.

Conjugate points along null geodesics. Let $C(s)$ be a curve in a Lorentzian manifold (M, g), parameterized with respect to an arc-length or a special parameter s according as C is non-null or null respectively. A variation of a curve segment $\alpha : [a, b] \to M$, of C, is a 2-parameter mapping

$$f : [a, b] \times (-\delta, \delta) \to M,$$

such that $\alpha(u) = f(u, 0)$ for all $a \leq u \leq a$. The u-parameter and the v-parameter curves of this variation are called the *longitudinal* and the *transversal* curves respectively, whereas its base curve is α. The vector field V on α given by $V(u) = f_v(u, 0)$ is called the *variation vector field* of f, which is the initial velocity of the transversal curve $v \to f(u, v)$. If every longitudinal curve of f is geodesic, then, f is called a *geodesic variation* or *1-parameter family of geodesics*. Suppose C and C' denote a geodesic and its tangent vector field on M respectively. A vector field V on C is called a *Jacobi vector field* if it satisfies the following *Jacobi differential equation*:

$$\nabla_{C'} \nabla_{C'} V = R(C', V)C',$$

where ∇ is a metric connection on M. It easy to prove (see [82, page 216]) that the variation vector field of a geodesic variation is a Jacobi field.

Definition 6.1. *Let $C(s)$ be a geodesic in M. We say that a point p on C is conjugate to a point q along $C(s)$ if there is a Jacobi field along $C(s)$, not identically zero, which vanishes at q and p.*

From a geometric point of view, a conjugate point $C(a)$ of $p = C(0)$ along a geodesic C can be interpreted as an "almost-meeting point" of a geodesic starting from p with initial velocity $C'(0)$. In general relativity, since the relative position of neighboring events of a free falling particle C is given by the Jacobi field of C, the attraction of gravity causes conjugate points, while the non attraction of gravity will prevent them. Although a physical spacetime is generally assumed to be causal (free of closed causal curves), all compact Lorentzian manifolds are acausal, i.e., they admit closed timelike curves [82, lemma 14.10]. Beem et al. [10, chapters 10 and 11, Second Edition] have done extensive work on conjugate points along null geodesics of a general Lorentzian manifold which may be causal or acausal.

Recently, considerable new work has been done by Gutiérrez-Palomo-Romero [49, 50, 51, 52] on *conjugate points along null geodesics* of compact Lorentzian manifolds (not covered in [10]), which we briefly present in this sub-section. It is well-known that a compact manifold M admits a Lorentzian metric if and only if the Euler number of M vanishes [82, proposition 5.57], which we assume. Moreover, since for Lorentzian metrics, compactness does not imply geodesic completeness (a desirable requirement), Romero-Sánchez [97] has proved that a compact Lorentzian manifold which admits a timelike *conformal Killing vector* (CKV) field yields to its geodesic completeness. Recall that (M, g) admits a conformal Killing vector (CKV) field K with conformal function σ if

$$\pounds_K g = 2\sigma g$$

which reduces to homothetic or Killing vector field whenever σ is non-zero constant or zero respectively. For this particular reason, besides geometric and physical uses of conformal Killing vector fields (see [32]) and two papers by Sánchez [102, 103] on Killing vector fields), the above mentioned four authors assumed that M admits a timelike CKV field. The motivation of the choice of compact Lorentzian manifold comes from the following problems in Riemannian geometry:

In 1948, Hopf [59] proved that *the total scalar curvature of a closed surface without conjugate points is nonpositive and vanishes only if the surface is flat.*

This result of Hopf together with the Gauss-Bonnet theorem implies that a *Riemannian torus with no conjugate point must be flat.*

Later on, Hopf's result was extended by Green [46] to any dimension and proved that *if a compact Riemannian manifolds (M, g) has no conjugate points then*

$$\int_M S d\mu_g \leq 0,$$

and it vanishes only if the metric is flat, where S and $d\mu_g$ denote the scalar curvature of M and the canonical measure associated to g. Also, Green [47] solved the famous

2-dimensional Blaschke conjecture, using the following Berger equality

$$Area(M, g) \geq \frac{2a^2}{\pi} \chi(M),$$

where $\chi(M)$ is the Euler-Poincaré characteristic of M, for a 2-dimensional compact Riemannian manifold (M, g) without conjugate points before a fixed distance a in the parameter of any (unit) geodesic. The equality holds only if M has constant sectional curvature $\frac{\pi^2}{a^2}$. Later on, Berger (and independently Green [16, proposition 5.64]) generalized this inequality as follows:

$$Vol(M, g) \geq \frac{a^2}{\pi^2 n(n-1)} \int_M S d\mu_g, \quad \dim(M) = n.$$

Above inequality can be equivalently written as follows:

$$Vol\,(U(M), \hat{g}) \geq \frac{a^2}{\pi^2 n(n-1)} \int_{U(M)} \bar{Ric}\, d\mu_{\hat{g}},$$

where $U(M)$ denotes the unit tangent bundle of M, $\hat{g}$ is the restriction to $U(M)$ of the metric on the tangent bundle TM. $\bar{Ric}$ denotes the quadratic form associated with the Ricci tensor of M and $d\mu_{\hat{g}}$ is the canonical measure associated with $\hat{g}$. Following results on conjugate points along null geodesics are well-known:

- An essential problem of causality theory is to determine if a pair of points can be joined by a timelike curve. This problem is related to conjugate points along null geodesics as following: If $C : [0, M] \to M$ is a null geodesic with $C(0)$ and $C(a)$ conjugate points along C. Then, $\forall \epsilon > 0$ there is a timelike curve from $C(0)$ to $C(a + \epsilon)$, arbitrarily near to C [82, page 296].

- No null geodesic of any 2-dimensional Lorentzian manifold has conjugate points [10, Lemma 10.45].

- There is no conjugate point along null geodesics in any Lorentzian manifold of constant sectional curvature [82, Ex. 10-11].

Recall the following notion of *null sectional curvature* [53, 10]. Let $x \in (M, g)$ and ξ be a null vector of $T_x M$. A plane H of $T_x M$ is called a *null plane* directed by ξ if it contains ξ, $g_x(\xi, W) = 0$ for any $W \in H$ and there exists $W_o \in H$ such that $g_x(W_o, W_o) \neq 0$. Then, the null sectional curvature of H, with respect to ξ and ∇, is defined as a real number

$$\mathcal{K}_\xi\,(H) = \frac{g_x(R(W, \xi)\xi, W)}{g_x(W, W)},$$

where $W \neq 0$ is any vector in H independent with ξ (and therfore spacelike). It is easy to see that $\mathcal{K}_\xi\,(H)$ is independent of W but depends in a quadratic fashion on ξ. An $n(\geq 3)$-dimensional Lorentzian manifold is of constant curvature if and only if its null sectional curvatures are everywhere zero [82, Prop. 8.28].

Unless otherwise indicated, we assume that (M, g) is a time orientable Lorentzian manifold with dimension $n \geq 3$ and K is a timelike vector field on M. Recall [53] that the *null congruence associated with K* is defined by

$$C_K M = \{\xi \in TM : g(\xi, \xi) = 0, \ g(\xi, K_{\pi(\xi)}) = 1\},$$

where $\pi : TM \to M$ is the natural projection. $C_K M$ is an oriented embedded submanifold of TM with dimension $2(n-1)$ and $(C_K M, \pi, M)$ is a fiber bundle with fibre type S^{n-2}. Therefore, for a compact M, $C_K M$ will be compact.

If a null congruence $C_K M$ is fixed with respect a timelike vector field K, then one can choose, for every null plane H, the unique null vector $\xi \in C_K M \cap H$, so that the null sectional curvature can be thought as a function on null planes. This function is called the *K-normalized null sectional curvature*.

Consider a map, associated with ∇, defined by

$$f : TTM \to TM, \quad X \mapsto \left. \frac{\nabla \alpha}{ds} \right|_0,$$

where α is a curve with $\alpha'(0) = X$ and $\frac{\nabla \alpha}{ds}$ is the covariant derivative of the vector field α along the curve $\pi \circ \alpha$ on M. This map f is well defined and the pair (f, π) is a vector bundle morphism from $\pi_{TM} : TTM \to TM$ to $\pi : TM \to M$, π_{TM} being the natural projection. Consider the Sasaki metric $\hat{g}$ on TM defined from g as follows (see [16, Section 1.K]):

$$\hat{g}(X, Y) = g(d\pi(X), d\pi(Y)) + g(f(X), f(Y)), \quad \forall X, Y \in \Gamma(TTM)$$

It is clear from above that $(TM, \hat{g})$ is semi-Riemannian with index 2. We also represent $\hat{g}$ the induced metric on $C_K M$, so $(C_K M, \hat{g})$ is a Lorentzian manifold and the restriction of π to $C_K M$ is a semi-Riemannian submersion with spacelike fibres. Denote by $d\mu_{\hat{g}}$ the canonical measure on $C_K M$ induced from $\hat{g}$. It is known that the cotangent bundle T^*M of M carries a natural symplectic structure given by $d\alpha$ of a 1-form α defined by $\alpha(X) = -q(X)[p_*(X)]$ for all $X \in TT^*M$ where $q : TT^*M \to T^*M$ and $p : T^*M \to M$ are the natural projections. Consider a vector bundle isomorphism $\flat : TM \to T^*M$, by putting $v \mapsto g(v, \cdot)$. Call α_g the pullback by $\flat$ of α. Then $\alpha_g(X) = -g(v, \pi_*(X))$, where X belongs to $T_v(TM)$. A geodesic vector field, denoted by Z_g, on TM is defined by $i_{Z_g} d\alpha_g = dE$, where E is given by $E(v) = \frac{1}{2} g(v, v)$. The flow, denoted by $\{\Phi_t\}$ of Z_g, given by $\Phi_t(v) = C_v'(t)$, is called the *geodesic flow* of (M, g). Now we state the following two results on null conjugate points, related to the conformal Killing symmetry (for proofs see [49]).

- $C_K M$ is invariant by the geodesic flow if and only if the vector field K is a CKV. Moreover, $div(Z_g|_{C_K M}) = 0$, where div denotes the divergence operator of $(C_K M, \hat{g})$. So, If (M, g) is compact and K is a CKV (therefore, (M, g) is geodesically complete [97]) then, the following holds:

$$\int_{C_K M} (\phi \circ \Phi_t) \, d\mu_{\hat{g}} = \int_{C_K M} \phi \, d\mu_{\hat{g}}, \quad \forall \, \phi \in C^0(C_K M) \quad \text{and} \quad t \in \mathbf{R}. \quad (8.6.1)$$

- If K is a CKV, then every null geodesic C_ξ of M, with $\xi \in C_K M$, gives rise to the null geodesic C'_ξ of $(C_K M, \hat{g})$. Furthermore, each null geodesic β of (M, g) may be re parameterized to obtain a null geodesic α satisfying $\alpha' \in C_K M$.

Finally, we recall the following integral inequality (see in [82, pages 290-1]):

Let $C_\xi : [0, a] \to M$ be a null geodesic such that there are no conjugate points of $C_\xi(0)$ in $[0, a)$. Then the *Hessian form* $H^\perp_{C_\xi}$ is positive semidefinite, i.e.

$$H^\perp_{C_\xi}(V, V) := \int_0^a \left[g\left(\frac{\partial V}{ds}, \frac{\partial V}{ds} \right) - g(R(V, C'_\xi)C'_\xi, V) \right] ds \geq 0, \qquad (8.6.2)$$

for every vector field V along C_ξ such that $V(0) = 0$ and $g(C'_\xi, V) = 0$.

Based on above, we state the following main results on null conjugate points:

Theorem 6.1 ([49]). *Let (M, g) be an $n(\geq 3)$-dimensional compact Lorentzian manifold that admits a timelike conformal vector field K. If there exists a real number $a \in (0, +\infty)$ such that every null geodesic $C_\xi : [0, a] \to M$, with $\xi \in C_K M$, has no conjugate points of $C_\xi(0)$ in $[0, a)$, then*

$$Vol(C_K M, \hat{g}) \geq \frac{a^2}{\pi^2 n(n-1)} \int_{C_K M} \bar{Ric}\, d\mu_{\hat{g}}. \qquad (8.6.3)$$

Equality holds if and only if M has K-normalized null sectional curvature $\frac{\pi^2}{a^2}$.

Note 1. In [49, theorem 3.5], those authors have given another version of above theorem, involving U-normalized null sectional curvature and the scalar curvature of M. Moreover, they have proved several side results followed by an application to Lorentzian odd-dimensional spheres (also, see [52] for some more results on Lorentzian odd-dimensional spheres).

Note 2. In [50], the authors used theorem 6.1 and proved several inequalities relating conjugate points along geodesics to global geometric properties. Consequently, they have shown some classification results on certain compact Lorentzian manifolds without conjugate points along its null geodesics.

In [51], the authors asked the following question: *When a Lorentzian manifold, with no conjugate points along the null geodesics, has constant sectional curvature.*

To answer above question, they proved the following two main results:

Theorem 6.2 ([51]). *Let (M, g) be an $n(\geq 3)$-dimensional compact Lorentzian manifold admitting a timelike conformal Killing vector field. If (M, g) has no conjugate points along its null geodesic, then*

$$\int_M [\bar{Ric}(U) + S]\, h^n\, d\mu_g \leq 0,$$

where $h = [-g(K, K)]^{-1/2}$ so that $g(U, U) = -1$ with $U = hK$. Moreover, equality holds if and only if (M, g) has constant sectional curvature $k \leq 0$.

Theorem 6.3 ([51]). *Under the hypothesis of theorem 6.2, if K is a timelike Killing vector field, then*

$$\int_M S\, h^n\, d\mu_g.$$

Equality holds if and only if M is isomorphic to a flat Lorentzian n-torus up to a (finite) covering. In particular, U is parallel, the first Betti number of M is non-zero and the Levi-Civita connection of g is Riemannian.

Example 10. Let $(M = B \times_f F,\ g)$ be a Lorentzian warped product manifold where (B, g_B) is an $n(\geq 2)$-dimensional compact Riemannian manifold and $(F,\ g_F) = (S^1,\ -g_{can})$, g_{can} denotes the canonical metric of S^1. Assume that M has no conjugate points along its null geodesics and $B \times S^1$ admits a timelike vector field K given by the lift of the vector field $z \to iz$ on S^1. Then, $h = 1/\sqrt{-g_f(K, K)} = 1/(f \circ \pi_B)$. It follows from [82, lemma 12.37] that K is Killing and using [82, corollary 4.3] we obtain

$$\int_B \left[\frac{(n-1)\triangle f}{f^{n+1}} + \frac{S^B}{f^n} \right] d\mu_{g_B} \leq 0,$$

where $\triangle$ and S^B are the Laplacian and the scalar curvature of (B, g_B), respectively. Using the relation

$$\triangle \left(\frac{-1}{nf^n} \right) = \frac{\triangle f}{f^{n+1}} - \frac{n+1}{f^{n+2}} \| \, grad\, f \, \|^2$$

and the classical Green divergence theorem, we write

$$\int_B \frac{\triangle f}{f^{n+1}}\, d\mu_{g_B} = (n+1) \int_B \frac{\| \, grad\, f \, \|^2}{f^{n+2}}\, d\mu_{g_B} \geq 0.$$

Therefore, since M has no conjugate points along its null geodesics, using above in theorem 6.3 we obtain

$$\int_B \frac{S^B}{f^n}\, d\mu_{g_B} \leq 0,$$

and equality holds if and only if f is constant and (B, g_B) is flat.

Bibliography

[1] Akivis, M. A. and Goldberg, V. V. The Geometry of Lightlike Hypersurfaces of the de Sitter Space, Acta Appli. Math., 53, 1998, 297-328.

[2] Akivis, M. A. and Goldberg, V. V. On some methods of construction of invariant normalizations of lightlike hypersurfaces, Differential Geometry and its Applications, 12(2), 2000, 121-143.

[3] Artin, E. *Geometric Algebra*, Interscience Publishers, New York, 1975.

[4] Atindogbe, C. and Duggal, K. L. Conformal screen on lightlike hypersurfaces, International J. of Pure and Applied Math., 11(4), 2004, 421-442.

[5] Balgetir, H., Bektas, M. and Ergüt, M. On a characterization of null helix, Bull. Inst. Acad. sinica, 29, 2001, 71-8.

[6] Barros, M. Generai helices and a theorem of Lancret. Proc. Am. Math. Soc., 125, 1997, 1503-9.

[7] Barros, M. Geometry and Dynamics of relativistic particles with regidity, Gen. Rel. Grav., 34, 2002, 837-852.

[8] Barros, M., Ferrández, A., Lucas, P. and Meroño, M. A. Solutions of the Betchov-Da Rios soliton equation: a Lorentzian approach, J. Geom. Phys., 31, 1999, 217-228.

[9] Barros, M., Ferrández, A., Javaloyes, M. A. and Lucas, P. Relativistic particles with regidity and torsion in $D = 3$ spacetimes, Classical Quantum Gravity, 22(3), 2005, 489-513.

[10] Beem, J. K. and Ehrlich, P. E. *Global Lorentzian Geometry*, Marcel Dekker, Inc. New York, First Edition, 1981, Second Edition (with Easley, K. L.), 1996.

[11] Bejancu, A. A canonical screen distribution on a degenerate hypersurface. Scientific Bulletin, Series A, Applied Math. and Physics, 55, 1993, 55-61.

[12] Bejancu, A. Lightlike curves in Lorentz manifolds, Publ. Math. Debrecen, 44, no. f.1-2, 1994, 145-155.

[13] Bejancu, A. and Duggal, K. L. Degenerate hypersurfaces of semi-Riemannian manifolds, Bull. Inst. Politehnie Iasi, (S.1), 37, 1991, 13-22.

[14] Bejancu, A., Ferrández, A. and Lucas P. A new viewpoint on geometry of a lightlike hypersurface in a semi-Euclidean space, Saitama Math. J., 1998, 31-38.

[15] Berger, M. *Riemannian Geometry During the Second Half of the Twentieth Century*, Lecture Series, 17, Amer. Math. Soc., 2000.

[16] Besse, A. *Manifolds all of whose geodesics are closed*, Springer-Verlag, Ergeb. Math. Grenzgeb, 93, Berlin, 1978.

[17] Bishop, R. L. and O'Neill, B. Manifolds of negative curvature, Trans. Amer. Math. Soc., 145, 1969, 1-49.

[18] Bonnor, W. B. Null curves in a Minkowski spacetime, Tensor N. S., 20, 1996, 229-242.

[19] Bonnor, W. B. Null hypersurfaces in Minkowski spacetime, Tensor N. S., 24, 1972, 329-245.

[20] Cartan, E. *La Théorie Des Groupes Finis et Continus et la Géométrie Différentielle*, Gauthier-Villars, Paris, 1937.

[21] Claudel, C. M., Virbhadra, K. S. and Ellis, G. F. S. The geometry of photon surfaces, J. Math. Phys., 42, 2001, 818.

[22] Cöken, A. C. and Ciftci, Ü. On the Cartan curvatures of a null curve in Minkowski spacetime, Geometry Dedicata, 114, 2005, 71-78.

[23] Duggal, K. L. Warped product of lightlike manifolds, Nonlinear Anal., 47, 2001, 3061-3072.

[24] Duggal, K. L. Constant scalar curvature and warped product globally null manifolds, J. Geom. Phys., 43, 2002, 327-340.

[25] Duggal, K. L. Null curves and 2-surfaces of globally null manifolds, International J. of Pure and Applied Math., 1(4), 2002, 389-415.

[26] Duggal, K. L. Harmonic maps, morphisms and globally null manifolds, International J. of Pure and Applied Math., 6(4), 2003, 421-436.

[27] Duggal, K. L. On scalar curvature in lightlike geometry, J. Geom. Phys., 57, 2007, 471-481.

[28] Duggal, K. L. and Bejancu, A. *Lightlike submanifolds of semi-Riemannian manifolds and applications*, Kluwer Academic Publishers, 364, 1996.

[29] Duggal, K. L. and Giménez, A. Lightlike hypersurfaces of Lorentzian manifolds with distinguished screen, J. Geom. Phys., 55, 2005, 107-122.

[30] Duggal, K. L. and Jin, D. H. Geometry of null curves, Math. J. Toyama Univ., 22, 1999, 95-120.

[31] Duggal, K. L. and Jin, D. H. Totally umbilical lightlike submanifolds, Kodai Math. J., 26, 2003, 49-68.

[32] Duggal, K. L. and Sharma, R. *Symmetries of spacetimes and Riemannian manifolds*, Kluwer Academic Publishers, 487, 1999.

[33] Ehrlich, P., Jung, Y. T. and Kim, S. B. Volume comparison for Lorentzian manifolds, Geo. Dedicata, 73, 1998, 39-56.

[34] L. P. Eisenhart, *A treatise on Differential Geometry of Curves and Surfaces*, Ginn and Company, 1909.

[35] Ferrández, A., Giménez, A. and Lucas, P. Null helices in Lorentzian space forms, Int. J. Mod. Phys., A16, 2001, 4845-4863.

[36] Ferrández, A., Giménez, A. and Lucas, P. Degenerate curves in pseudo-Euclidean spaces of index two, Third International Conference on Geometry, Integrability and Quantization, Coral Press, Sofia 2001, 209-223.

[37] Ferrández, A., Giménez, A. and Lucas, P. Null generalized helices in Lorentz-Minkowski spaces, J. Phys. A: Math. Gen., 35, 2002, 8243-8251.

[38] Ferrández, A., Giménez, A. and Lucas, P. Geometrical particle models on 3D null curves, Phys. Lett. B 543, no. 3-4, 2002, 311-317.

[39] Ferrández, A., Giménez, A. and Lucas, P. s-degenerate curves in Lorentzian space forms, J. Geom. Phys. 45, no. 1-2, 2003, 116-129.

[40] Ferrández, A., Giménez, A. and Lucas, P. Null generalized helices and the Betchov-Da Rios equation in Lorentz-Minkowski spaces, Proceedings of the XI Fall Workshop on Geometry and Physics, Madrid, 2004, 215-221.

[41] Ferrández, A., Giménez, A. and Lucas, P. Relativistic particles with rigidity along lightlike curves, In: Horizons in world physics, Vol. 245, Tori V. Lynch(editor), Nova science Publications, 2004, ISBN: 1-5945-063-2.

[42] Foertsch, T., Hasse, W. and Perlick, V. Inertial forces and photon surfaces in arbitrary spcetimes, Class. Quantum Grav., 20, 2003, 4635-4651.

[43] Frank, H. and Giering, O. Generalized ruled surfaces in the large, Weralgmeinerte Regalflächen im Grassen I. Arch. Math., 38, 1982, 106-115.

[44] García-Río, E. Kupeli, D. N. and Vázquez-Abal, M. E. On a problem of Osserman in Lorentzian geometry, Diff. Geom. Appl. 7, 1997, no. 1, 85-100.

[45] Graves, L. K. Codimension one isometric immersions between Lorentzian spaces, Trans. Amer. Math. Soc., 252, 1979, 367-390.

[46] Green, L. W. A theorem of E. Hopf, Mich. Math. J., 5, 1958, 31-34.

[47] Green, L. W. Auf Wiedersehensflöchen, Ann. of Math., 78, 1963, 289-299.

[48] Guggenheimer, H. W. *Differential Geometry*, General Publishing Com., 1997.

[49] Gutiérrez, Manuel; Palomo, Franisco, J. and Romero, Alfonso, A Berger-Green type inequality for compact Lorentzian manifolds, Trans. Amer. Math. Soc., 354(11), 2002, 4505-4523.

[50] Gutiérrez, Manuel; Palomo, Franisco, J. and Romero, Alfonso, Conjugate points along null geodesics on Lorentzian manifolds with symmetry, Proceedings of workshop on geometry and physics, Madrid, 2001, 169-182.

[51] Gutiérrez, Manuel; Palomo, Franisco, J. and Romero, Alfonso, Lorentzian manifolds with no null conjugate points, Math. Proc. Cambridge Philos. Soc., 137(2), 2004, 363-375.

[52] Gutiérrez, Manuel; Palomo, Franisco, J. and Romero, Alfonso, A new approach to the study of conjugate points along null geodesics on certain compact Lorentzian manifolds, Preprint, 2005.

[53] Harris, S. G. A characterization of Robertson-Walker Spaces by null sectional curvature, Gen, Relativity Gravitation, 17, 1985, 493-498.

[54] Hayden, H. A. Deformations of a curve in a Riemannian n-space which displace vectors parallelly at each point, Proc. Lond. Math. Soc., 32, 1931, 321-336.

[55] Hayden, H. A. On a generalized helix in a Riemannian n-space, Proc. Lond. Math. Soc., 32, 1931, 337-345.

[56] Hawking, S. W. and Ellis, G. F. R. *The large scale structure of spacetime*, Cambridge University Press, Cambridge, 1973.

[57] Honda, K. and Inoguchi, J. Deformation of Cartan framed null curves preserving the torsion, Differ. Geom. Dyn. Syst., 5, 2003, 31-37.

[58] Hopf, H. Closed surfaces without conjugate points, Proc. Nat. Acad. Sci. U.S.A., 34, 1948, 47-51.

[59] Hopf, H. and Rinow, W. Über den Begriff des vollständigen differentialgeometrischen Fläche, Comment. Math. Helv. 3, 1931, 209-225.

[60] IIyenko, K. Twistor representation of null 2-surfaces, J. Math. Phys., 10, 2002, 4770-4789.

[61] Ikawa, T. On curves and submanifolds in an indefinite Riemannian manifold, Tsukuba J. Math., 9, 1965, 353-371.

[62] Ikawa, T and Honda, K. Some graph type hypersurfaces in a semi-Euclidean space, Turkish J. Math., 24, 2000, 197-208.

[63] Inoguchi, J. Biharmonic curves in Minkowski 3-space, Int. J. Math. Math. Sci., 21(11), 2003, 1365-1368.

[64] Inoguchi, J. and Lee, S. Null curves in Minkowski 3-space, 2006, preprint.

[65] Israel, W. Event horizons in static vacuum spacetimes, Physl Revl, 164, 1967, 1776-1779.

[66] Israel, W. Event horizons in static electrovac spacetimes, Comm. Math. Phys., 8, 1968, 245-260.

[67] Jin, D. H. Null curves in Lorentz manifolds, J. of Dongguk Univ. vol. 18, 1999, 203-212.

[68] Jin, D. H. Fundamental theorem of null curves, J. Korea Soc. Math. Educ. Ser. B: Pure Appl. Math. vol. 7, No. 2, 2000, 115-127.

[69] Jin, D. H. Fundamental theorem for lightlike curves, J. Korea Soc. Math. Educ. Ser. B: Pure Appl. Math. vol. 10, No. 1, 2001, 13-23.

[70] Jin, D. H. Frenet equations of null curves, J. Korea Soc. Math. Educ. Ser. B: Pure Appl. Math. vol.10, No. 2, 2003, 71-102.

[71] Jin, D. H. Natural Frenet equations of Null curves, J. Korea Soc. Math. Educ. Ser. B: Pure Appl. Math. vol.12, No. 3, 2005, 71-102.

[72] Katsuno, K. Null hypersurfaces in Lorentzian manifolds, Math. Proc. Cab. · Phil. Soc., 88, 1980, 175-182.

[73] Kim, Y. H. and Yoon, D. W. Classification of rules surfaces in Minkowski 3-space, J. Geom. Phys., 49, 2004, 89-100.

[74] Klishevich, S. and Plyushchay, M. S. Zitterbewegung and reduction: 4d spinning particles and 3d anyons on lightlike curves. Phys., Lett. B, 459, 1999, 201-207. hep-th/0111014.

[75] Kobayashi, S. and Nomizu, K. *Foundations of Differential Geometry*, Vol. 1., Inter science Publish., New York, 1965.

[76] Krüger, H. Differential geometry and dynamics of a lightlike point in Lorentzian spacetime, Annales de La Foundation Louis de Broglie, Volume 24, 1999, 39-66.

[77] Kupeli, D. N. Null cut loci of spacelike surfaces, Gen. Rel. Grav., 20, 1988, 415-425.

[78] Kupeli, D. N. On conjugate and focal points in semi-Riemannian geometry, Math. Z., 198(4), 1988, 569-589.

[79] Langer, J. and Perline, R. Poisson geometry of the filament equations, J. Nonlinear Sci., 1991, 71-93.

[80] Langer, J. and Perline, R. Local geometric invariants of integrable evolution equations, J. Math. Phys., 35, 1994, 1732-1737.

[81] Minguzzi, E. and Sánchez, M. Connecting solutions of the Lorentz force equations do exist, ArXiv:math-ph/050514 v1, 5 May 2005 (preprint).

[82] O'Neill, B. *Semi-Riemannian geometry with applications to relativity*, Academic Press, New York, 1983.

[83] Nersessian, A. and Ramos, E. Massive spinning particles and the geometry of null curves, Physics Letters B 445, 1998, 123-128.

[84] Nersessian, A. and Ramos, E. A geometrical particle model for anyons, Modern Phys., Lett. A, 14(29), 1999, 2033-2037. hep-th/9807143.

[85] Nersessian, A., Manvelyan, R. and Müller-Kirsten, H. J. W. Particle with torsion on 3D null curves, Nuclear Phys. B Proc. Suppl., 88, 2002, 381-384.

[86] Nomizu, K. and Ozeki, H. The existence of complete Riemannian metrics, Proc. Amer. Math. Soc., 12, 1961, 889-891.

[87] Nomizu, K. and Sasaki, T. *Affine Differential Geometry*, Cambridge Tracts in Math. **111**, Cambridge Univ. Press, 1994.

[88] Nurowski, P. and Robinsom, D. Intrinsic geometry of a null hypersurface, Class. Quantum Grav., 17, 2000, 4065-4084.

[89] Penrose, R. The twistor geometry of light rays. Geometry and physics, Classical Quantum Gravity, 14(1A), 1997, A299-A323.

[90] Perlick, V. On Fermat's principle in general relativity: 1. The general case, Class. Quantum Grav., 7, 1990, 1319-1331.

[91] Perlick, V. On Fermat's principle in general relativity: 11. The conformally stationary case, Class. Quantum Grav., 7, 1990, 1849-1867.

[92] Perlick, V. Infinite dimensional Morse theory and Fermat's principle in general relativity: 1, J. Math. Phys., 36(12), 1995, 6915-6928.

[93] Perlick, V. *Ray Optics, Fermat's principle and applications to general relativity.* Lecture Notes in Physics. Monographs, 61. Springer-Verlag, Berlin, 2000. x+220 pp. ISBN: 3-540-66898-5.

[94] Perlick, V. A variational principle for light rays in a general-relativistic plasma. Nonlinear Anal., 47(5), 2001, 3019-3030.

[95] Petrovic-Torgrasev, M and Sucurovic, E. Some characterizations of the Lorentzian spherical timelike and null curves, Mat. Vesn. 53(1-20), 2001, 21-27.

[96] Ricca, R. L. The contributions of Da Rios and Lwvi-Civita to asymptotic potential theory and vortex filament dynamics, Fluid Dynamics Research, 18, 1996, 245-268.

[97] Romero, M. and Sánchez, M. Completeness of compact Lorentzian manifolds admitting a timelike conformal Killing vector field. Proc. Amer. Math. Soc., 123, 1995, 2831-2833.

[98] Rosca, R. Sur les hypersurfaces isotropes de défaut l'incluses dans une variété Lorentzienne, C.R. Acad. Sci. Paris, 292, 1971, 393-396.

[99] Rosca, R. On null hypersurfaces of a Lorentzian manifold, Tensor N.S., 23, 1972, 66-74.

[100] Sahin, B., Kilic, E. and Günes, R. Null helices in $\mathbf{R}_1^3$, Differential Geometry-Dynamical Systems, 3(2), 2001, 31-36.

[101] Samuel, J. and Nityananda, R. Transport along null curves, J. Phys. A, Math. Gen., 33(14), 2000, 2895-2905.

[102] Sánchez, M. Structure of Lorentzian tori with a Killing vector field, Trans. Amer. Math. Soc., 349, 1997. 1063-1080.

[103] Sánchez, M. Lorentzian manifolds admitting a Killing vector field, Nonlinear Ana., 30, 1997, 643-654.

[104] Schild, A. Classical null strings, Phys. Rev. D, 16(2), 1977, 1722-1726.

[105] Spivak, M. *A Comprehensive Introduction to Differential Geometry*, vol. III, IV, Publish or Perish, Boston, 1975.

[106] Struik, D. J. *Lectures on Classical Differential Geometry*, New York, 1932.

[107] Trautman, A., Pirani, F. A. E. and Bondi, H. *Lectures on General Relativity*, Vol. 1, Prentice-Hall, New Jersey, 1965.

[108] Turgut, A. and Hacisalihoğlu, H. H. Timelike ruled surfaces in Minkowski 3-space II, Tr. J. of Math., 22, 1998, 33-46.

[109] Urbantke, H. Local differential geometry of null curves in conformally flat spacetime, J. Math. Phys., 30(10) 1989, 2238-2245.

[110] Walker, A. G. Completely symmetric spaces, J. London Math. Soc., 19, 1944, 219-226.

[111] Yamabe, H. On a deformation of Riemannian structures on compact manifolds, Osaka Math. J., 12, 1960, 21-37.

Index

A

admissible variation, 262
anti-De Sitter spacetime, 145
anyonic, 179
arc length parameter, 8
associated matrix of metric, 1
associated Cartan null curve, 148
asymptotically flat (spacetime), 244
atlas, 3
axis of helix, 159

B

base curve, 263
basis, 2
Bertrand curve, 147
Bertrand mate, 147, 151
Betchov-Da Rios equations, 166
Bianchi's identities, 184
binormal vector, 147
B-scroll, 166

C

canonical lightlike transversal vector
    bundle, 210
canonical representation for null
    curve, 31
canonical screen distribution, 212
canonical transformation, 68
Cartan curvature, 142
Cartan equations, 14
Cartan frame, 14
Cartan Frenet equation, 31
Cartan Frenet frame, 29
Cauchy surface, 241
causal character, 5
causal structure, 5
chart, 2

Christoffel symbols of first and second
    type, 185
C^k-compatible, 2
complete, 225
complete vector field, 235
complexified vector space, 15
compound general Frenet frame (and
    equations), 105
conformal Killing vector field, 276
conformally flat, 187
conformally flat manifold, 187
conformally flat spacetime, 272
congruence of curve, 235
conjugate point, 275
conserved linear momentum law, 177
coordinate system, 2
cotangent space, 4
covariant derivative, 4
covariant derivative operator, 4
critical point of action, 177
critical curve, 173
curvature functions, 20

D

degenerate curve, 95, 161
degeneration degree, 95
De-Sitter spacetime, 144
diagonal transformation, 36
differentiable manifold, 2
differential 1-form, 4
distinguished parameter, 27, 74
distribution, 192
dual basis, 4
dual vector space, 4

E

Einstein manifold, 187